KU-517-173

Leeds Metropolitan University

17 0336644 4

AFTER POPPER, KUHN AND FEYERABEND

AUSTRALASIAN STUDIES
IN HISTORY AND PHILOSOPHY OF SCIENCE

VOLUME 15

General Editor:

R. W. HOME, *University of Melbourne*

Editorial Advisory Board:

W. R. ALBURY, *University of New South Wales*
D. W. CHAMBERS, *Deakin University*
S. GAUKROGER, *University of Sydney*
H. E. LE GRAND, *University of Melbourne*
A. MUSGRAVE, *University of Otago*
G. C. NERLICH, *University of Adelaide*
D. R. OLDROYD, *University of New South Wales*
E. RICHARDS, *University of Wollongong*
J. SCHUSTER, *University of Wollongong*
R. YEO, *Griffith University*

AFTER POPPER, KUHN AND FEYERABEND

Recent Issues in Theories of Scientific Method

Edited by

ROBERT NOLA

University of Auckland, New Zealand

and

HOWARD SANKEY

University of Melbourne, Australia

KLUWER ACADEMIC PUBLISHERS

DORDRECHT/BOSTON/LONDON

Library of Congress Cataloging-in-Publication Data is available.

ISBN 0-7923-6032-X

Published by Kluwer Academic Publishers,
PO Box 17, 3300 AA Dordrecht, The Netherlands

Sold and distributed in North, Central and South America
by Kluwer Academic Publishers,
101 Philip Drive, Norwell, MA 02061, USA

In all other countries, sold and distributed
by Kluwer Academic Publishers,
PO Box 322, 3300 AH Dordrecht, The Netherlands

Printed on acid-free paper

All Rights Reserved
© 2000 Kluwer Academic Publishers
No part of the material protected by this copyright notice may be reproduced or
utilized in any form or by any means, electronic or mechanical,
including photocopying, recording or by any information storage and
retrieval system, without written permission from the copyright owner.

Printed and bound in Great Britain by
Antony Rowe Limited

TABLE OF CONTENTS

FOREWORD

Australia and New Zealand boast an active community of scholars working in the field of history, philosophy and social studies of science. *Australasian Studies in History and Philosophy of Science* aims to provide a distinctive publication outlet for their work. Each volume comprises a group of essays on a connected theme, edited by an Australian or a New Zealander with special expertise in that particular area. In each volume, a majority of the contributors are from Australia or New Zealand. Contributions from elsewhere are by no means ruled out, however, and are indeed actively encouraged wherever appropriate to the balance of the volume in question. Earlier volumes in the series have been welcomed for significantly advancing the discussion of the topics they have dealt with. I believe that the present volume will be greeted equally enthusiastically by readers in many parts of the world.

R.W. Home
General Editor
Australasian Studies in History and Philosophy of Science

ACKNOWLEDGEMENTS

We would like to acknowledge the assistance of Rod Home, the General Editor of the series *Australasian Studies in History and Philosophy of Science*, in helping us to get this book under way. Also Alan Musgrave, a member of the Editorial Advisory Board of ASHPS, took his advisory role conscientiously; we owe him a debt of gratitude, especially for his critical comments on the 'Selective Survey'. We would also like to thank Paul Hoyningen-Huene for his comments on the Survey's section on Kuhn and Feyerabend.

Both editors have been, on two occasions, appointed to Visiting Fellowships at the Center for Philosophy of Science, University of Pittsburgh. The contacts we were able to make with other philosophers in the Pittsburgh region are reflected in this book. Part of Howard Sankey's editing task was carried out while at the Center during the Fall of 1998 while some of the ideas in the 'Selective Survey' were explored by Robert Nola during an earlier visit at the Center. We would like to thank Nicholas Rescher for comment on his work on methodology and hope that we have it largely right despite the short space we could give to it. Also part of Howard Sankey's editorial work was carried out while a visitor at Zentrale Einrichtung für Wissenschaftstheorie und Wissenschaftsethik, Universität Hannover. We would like to thank both centres for the support they have given us while preparing this book.

We dedicate this book to Jan and Noëlle.

ROBERT NOLA AND HOWARD SANKEY

INTRODUCTION

Some of the papers in this volume were initially given at a combined one-day meeting of the *Australasian Association for Philosophy* and the *Australasian Association for the History, Philosophy and Social Studies of Science* held in Auckland in July 1997 as part of 'Philfest 97'. The organisers thought it would be appropriate to commemorate the three well known philosophers of science who had made a major contribution to theories of scientific method and whose recent departure from the world scene in effect marked the end of an era in the subject. The theme title, which is the first half of the title of this book, reflected the influence that Karl Popper (1902–94), Thomas Kuhn (1922–96) and Paul Feyerabend (1924–94) have had on our conception of scientific method. Also both Popper and Feyerabend had had personal associations with New Zealand owing to the time they had spent in the country as teachers of philosophy. The papers presented at the joint meeting appear in a revised form in this book. Since other papers at the conference were promised elsewhere we also contacted other philosophers who were willing to make a contribution to more recent developments in theories of method in line with the announced conference theme.

The basis upon which this collection was put together reflects three guidelines, two positive and one negative. The first positive guideline was to explore some of the unfinished business of the debates over methodology that arose out of the work of Popper, Kuhn and Feyerabend. The papers by Fox, Forster, Pyle and Worrall address some of the issues raised in those debates, and debates with other conceptions of method. The second positive guideline was to investigate some of the new ways in which theories of method have developed since that time; this the papers by Norton, Sankey, Laudan, and Kelly do. The third was a constraint concerning what should be excluded from the book.

Despite their many differences the one thing that Popper, Kuhn and Feyerabend had in common was that they rejected inductivist methods in science. To those working in the field of methodology, the most prominent development of the last 20 years or so has been the emergence of Bayesian accounts of scientific method. Though Bayesianism is different in many ways from classical inductivism, it is today's heir of the inductivism against which Popper, Kuhn and Feyerabend developed (in part) their views on method. Thus the negative guideline which constrained the choice of contributions to the book: it would not cover areas in recent approaches to methodology which were either inductivist,

Robert Nola and Howard Sankey (eds.), After Popper, Kuhn and Feyerabend, xi–xix.
© 2000 *Kluwer Academic Publishers. Printed in Great Britain.*

or probabilist or Bayesian. There is in any case a readily available burgeoning literature on Bayesian methods in science. From this exclusion it should not be concluded that the contributors to the volume are necessarily anti-Bayesian; as will be seen some are while others are not. Rather there is ongoing research in methodology which does not necessarily take its cue from the probabilistic approach that informs Bayesianism. Since this alternative area of research is quite broad and has many facets, not all of it can be represented in a collection of this sort; what appears here is merely indicative of work in the area.

Some have commented that the title does not reflect adequately the work of others who were also important contributors to the Popper, Kuhn and Feyerabend debates. One was Lakatos whose contribution was considerable but was cut short by his untimely death in 1974. In the light of this it has been suggested that we should have entitled the book 'After Pop-Lak-Kuhn-Abend', using David Stove's irreverent name for the quartet of 'irrationalists' he liked to lampoon. As tempting as this suggestion was, we have resisted it. Again some have commented that we fail to acknowledge the considerable contribution made by Carl Hempel (who died in November 1997) to philosophy of science, particularly in the areas of confirmation, induction and scientific rationality which are relevant to the book's theme. But since our project had already been launched earlier in 1997, the first half of the book's title has remained as originally planned with its imperfect connotations.

In reviewing the original proposal for a book on the announced theme, some on the Editorial Board of the ASHPS series, and others, wondered whether our proposal was more flogging of the well dead horse of scientific method. It is a common view that the heydays of theories of scientific method are truly over and that current conceptions of science leave little, or no, room for a role for methodology. Kuhn and Feyerabend are said, rightly or wrongly, to have played a significant role in methodology's demise. Methodology, it is commonly thought, has been superseded by sociological studies, or by a post-modernist approach (whatever that might be). Since this is a widespread view, it gave extra urgency to our project which is to show that there is still life to be found in research into methodology despite the sociologists of science and despite a sense of *fin de méthode* engendered by the Pop-Lak-Kuhn-Abend debates. There are issues still to be addressed in their debates, and there is still work to be done in bringing methodological theories into accord with history of science. But more than this, new paths are being struck in research into methodology that have an independence and vitality of their own.

Again other members of the same Board suggested that the two editors write a survey of theories of scientific method. Initially we did not take that suggestion seriously since it appeared to be too daunting a task. But we became aware that there are very few overviews of the field of scientific method currently available – and so decided to 'give it a go'. In setting out to write a survey we found why few have been attempted; the field is too broad to be captured in the span of a review article even as long as ours became. So once we decided to embark upon writing an overview, we agreed that its title would definitely have to have the word 'selective' in it. And selective it is, in ignoring many important things that do appropriately fall under the broad rubric of 'recent issues in theories of scientific method'. However, what the 'Selective Survey' and the other collected papers show is that methodology is a live and active field of investigation and that reports of its demise have been greatly exaggerated.

The paper 'A Selective Survey of Theories of Scientific Method' by the editors sets out a general schema (sections 3 and 4) which distinguishes between three levels of the scientific enterprise. The bottom level is the actual history of each of the various sciences. The second level is that of methodological principles, or theories of scientific method, that have allegedly been applied during the historical growth of the sciences. Just as each of the sciences has a history, so there is a history of our theories of method, beginning with Aristotle's *Organon*. Finally there is the level of meta-methodological principles which have been used to adjudicate between theories of method, or to adjudicate between second level theories of method and their fit with the first level sciences and their history. Such a tripartite schema can be found embryonically in the work of Popper; but it comes into its own in the writings of Lakatos, Laudan and Hempel, as we indicate.

Theories of scientific method, we argue, can be set out as a number of principles; and these in turn comprise two elements – methodological rules, and values (or goals or ends) that the rules are supposed to realise. An example of such a methodological principle would be Popper's anti-*ad hoc* rule which bids us to introduce only those saving hypotheses which increase, and do not decrease, the testable content of a theory modified by the introduction of the saving hypothesis as compared with the original theory; the goal of such a rule, or the value it realises, is extra testability of the modified theory plus saving hypothesis over the original unmodified theory.

There are a number of ways of assessing such a principle of method. One might ask how desirable the goal is, or how well, or how reliably, the rule realises the goal, and whether or not some other rule might do as well or better in realising the goal. One might ask whether or not the measures of testable content can be made unproblematically. Again one might ask, as Feyerabend does, whether the principle has been improperly expressed in that there needs to be a time limit on the testing of any modification by saving hypotheses. Thus the failure to observe the angular parallax of stars, a consequence noted by Copernicus when he proposed that the Earth orbits the Sun, was rescued by making the stars very distant; though the rescue is correct the requisite observations were not made until three hundred years later. Without a reasonable time limit a rule purporting to ban saving hypotheses would have little effect. Though these are important issues to consider, the approach adopted in the 'Selective Survey' is slightly different. In terms of the tripartite schema, theories of method can be broadly classified as having either an *a priori* or an empirical meta-methodological justification.

There are a range of *a priori* approaches to the justification of methodological principles, with some being treated in the 'Selective Survey' in greater detail than others. The first *a priori* approach (section 5) might be dubbed 'transcendental' in that it attempts to show, in a Kantian or a related manner, that the bare possibility of science makes some methodological principles a necessary presupposition. Again we might call 'logicism' (section 5) the *a priori* approach to methodology that tries to show that the principles of method can be viewed on a par with the principles of deductive logic. There may be few theories of method that are logicist in this sense; though Carnap's logic of probability can be viewed as *a priori*, the methodological principles which tell us how to apply the probability logic in the context of science are not *a priori*. Some might view the principles of method as analytic because they are connected by means of meaning relations to the very notion of science itself; surprisingly the later Kuhn seems to have held a view akin to this in which methodological principles are justified analytically by the relation of meaning holism they bear to the word 'science' (section 8). Finally there are conventionalists who view methodological principles as conventions to be adopted, or decided upon, relative to some purpose; this is the view of the early Popper of the *Logic of Scientific Discovery* in which he argues that methodological rules are neither empirical nor logicist but are conventions adopted for the purpose of making theories testable and revisable (section 6).

Empirical approaches to the justification of methodological principles are currently more widely supported. They also attempt to link methodologies to the history of the sciences to which they apply, which in turn provides a basis for their justification. Thus the later Popper justified his methodology on the grounds of whether or not it captured what he called the 'great' science of Newton, Einstein, Bohr and others, but did not capture what he called 'unacceptable' science, examples of which were allegedly proposed by Marx and Freud (section 7). Popper used his theory of falsification as a meta-methodology in order to test his principles of method against his intuitions as to what was great and unacceptable science. Lakatos extended this procedure by adopting as a test basis the judgements of the scientific élite; various methodologies could then be tested against this basis using themselves as their own meta-methodology (section 7). Then the best methodology is the one which, on the basis of his own criterion of a progressive scientific research programme, renders more of the history of science rationally explicable by its own lights than any other. In this way meta-methodological justification proceeds by invoking quite strong meta-methodological principles in order to test both themselves and other methodologies against the test basis.

The method of reflective equilibrium also uses a similar procedure (section 7). It requires both a set of methodological principles and a set of particular judgements (perhaps made by some élite) about what is and what is not an acceptable move in the game of science. Then the principles and the particular judgements are brought into relation with one another through a process of either deleting principles or dropping particular cases until a 'best fit' is found. Since there are different methodologies to be considered, there is a further meta-methodological process of deciding which of the various 'best fits' is the best overall. While reflective equilibrium has some common features with the quasi-empirical approach of Lakatos, it does invoke a quite different meta-methodological principle in deciding between rival methodologies.

The most recent empiricist approach is that of Laudan's normative naturalism (section 11). It rejects Lakatos' appeal to the judgements of a scientific élite as a test basis and instead turns to the actual strategies that have been employed in the various sciences to realise some goal, or value. The task is then to test any proposed principle of methodology, understood as containing a rule and a value, against the historical record of strategies actually employed. Originally this procedure was employed to test a range of methodological principles advocated by members of the 'historical school' (such as Lakatos, Kuhn, etc.) as opposed to members of the earlier 'positivist school' against which they reacted. But the method of normative naturalism can be extended to any principles of method of any methodology whatever. Laudan proposes a meta-inductive rule, a version of Reichenbach's Straight Rule, as the preferred meta-methodological principle of test. Once some alleged principle has passed the meta-inductive test, it can then be employed both as a methodological principle and as a supplement to the meta-inductive rule thereby increasing the means available for testing further methodological principles. In this way methodology becomes a means–ends science which, like any science, has its conceptual and empirical aspects, which are open to test, and which grows like any other science through the revision or replacement of its hypotheses (which, in effect, reflect methodological principles as rules for achieving ends).

There are also pragmatic approaches to the meta-methodological justification of methods that are definitely not *a priori* in character but which do not differ greatly from empirical approaches. The 'Selective Survey' looks at both Quine (section 11) and Rescher (section 12) as advocates of methodological pragmatism. Rescher in particular adopts a complex schema in which the success of science is used to justify, and to refine, principles of method. We also form a metaphysical picture of the world which makes possible inquiry of the sort carried out in science using our best methodologies to date. In turn, meta-methodological principles are invoked to explain not only the success of science using our current methods, but also those methods themselves and the metaphysical picture we have of the world that makes such inquiry possible. The procedures here invoke only the methods of empirical inquiry, along with a little help from evolutionary epistemology and a Darwinian view of methodological progress.

Kuhn and Feyerabend are commonly thought to have undermined the prospects of any theory of method. In the case of Kuhn (section 8) there is evidence for this in the first 1962 edition of his *The Structure of Scientific Revolutions*. But the 1970 second edition 'Postscript' has the beginnings of a theory of weighted values that Kuhn developed further subsequently. Finally Kuhn thought that, in the light of the doctrine of meaning holism that he adopted, it was possible to give a quasi-analytic account of his methodology. In the case of Feyerabend (section 9), it is clear that his dictum 'anything goes' does not apply to his own views on method; rather, he argues, it is a principle that his more 'Rationalist' opponents should adopt. However the quip has backfired against him, since many think it is an expression of his own views. Feyerabend's position is that there are a host of contextual and defeasible rules of method, each of which governs some move in the game of science; but there is no universally applicable rule(s) of method which governs every move in the game of science. Insofar as our methods must be closely applicable to actual scientific practice, and that our practice be criticisable by appeal to rules, Feyerabend adopts a meta-methodological position in which there is a close 'dialectical' interaction between principles and practice akin to that found in reflective equilibrium. Feyerabend also adopts a 'relativist' stance; whatever else this might mean, it at least indicates that there can be no *a priori*, empirical or even pragmatic overarching justification of methodological principles.

The Strong Programme in the sociology of scientific knowledge, as well as other sociologically inspired accounts of science, look to the social causes of a person's scientific beliefs, thereby hoping to undercut any role for an appeal to the application of methodological principles as a ground for belief in science. Such theories have widespread advocacy, and the studies they inspire into episodes in the history of science are often said to expose the poverty of philosophically based approaches to science. While the tenets of the Strong Programme can be quickly expressed it is not clear what they mean. Section 10 attempts some clarification, and distinguishes its main tenet about the social causes of belief from a quite different 'interests' thesis in which cognitive items such as interests are said to cause scientific belief. Thus it is not always clear whether, in the case of some particular scientist, it is their social circumstance, or their interests in their social milieu, that is the cause of their beliefs. One case study is briefly explored, that of Forman's claim that the cultural milieu of Weimar Germany caused some German physicists to believe in acausality in physics. It is argued that this case study is badly flawed by an improperly applied method of test for the causes of belief in acausality. Given the sociologists' rejection of any internalist explanation based on methodological principles, Forman's failure to properly apply causal methods of test in his case study helps undermine the claim that there is a new approach to science studies that signals the *fin de méthode*.

The 'Selective Survey' ends with a brief account of orthodox subjective Bayesianism, mentioning two philosophers who have deviated from orthodoxy. The first is Abner Shimony who wishes to supplement subjective Bayesianism with a 'tempering condition' that bids us to be restrained in our adoption of values for our prior probabilities. The second is van Fraassen who suggests that we adopt a 'new epistemology' freed from some of the constraints of orthodox Bayesianism. Thus even within the Bayesian camp there are variations to explore. There are many other current theories of method that we do not explore in the 'Selective Survey', such as the role of learning theory in methodology as explored by Kevin Kelly; however readers can glean much about this approach from Kelly's own contribution to this volume.

John Norton's 'How We Know About Electrons' investigates the methodological principles that are employed to justify the existence (and not the mere construction) of entities in science. In particular he reviews two strategies used to establish the existence and properties of electrons. The first, discussed recently by Salmon in the context of the reality of atoms, requires that a diversity of evidence massively overdetermine the numerical value of properties of the electron.

In the second, discussed independently by many scholars including Dorling, Gunn, Harper and Norton himself in the context of other examples, regularities in evidence are translated directly into law-like properties of electrons, so that theory is inferred from phenomena. The method of inference from phenomena goes by a number of names in the literature, Norton's preferred designation being 'demonstrative induction'. This mode of inference is explored particularly with reference to Bohr's 1913 theory of the atom. Norton also briefly surveys the grounds for scepticism about micro-entities such as electrons.

Andrew Pyle's paper 'The Rationality of the Chemical Revolution' takes up the issue of relating an historical episode in science to a particular theory of method, that found in Kuhn's *Structure of Scientific Revolutions*. Kuhn's book has attracted a lot of criticism from philosophers of science, but as Pyle argues, much of this criticism has failed to hit its proper target. Much criticism that has been directed towards Kuhn has alleged philosophical sins, from relativism and irrationalism to subjective idealism; in contrast Kuhn's account of the history of science has not been subject to the careful and detailed criticism it deserves. Pyle also argues that the case made in the *Structure* rests firmly on a few key examples of scientific revolutions; so it is important to get these historical case studies right. One of Kuhn's central examples is the chemical revolution of the late eighteenth century. But was this scientific revolution a Kuhnian paradigm-shift? Moreover what role is there for Kuhnian notions such as 'Gestalt switch' around which so much anti-rationalist rhetoric has turned? Pyle's study of the history leads to the conclusion that even though there was a revolution there were no paradigm shifts, nor any Gestalt switches, or the like. The episode of the chemical revolution can be given a rational account, but not a Whiggish rational account of the sort often prescribed by many current theories of method. Pyle concludes that his account of the chemical revolution brings him closer to the later Kuhn to be found in his post-*Structure* writings.

The title of John Worrall's paper raises a query: 'Kuhn, Bayes and "Theory Choice": How Revolutionary is Kuhn's Account of Theory Change?'. It begins with an account of Brewster's not so elderly hold-out against the wave theory of light and in favour of an emissionist/corpuscularian theory, and some of his reasons for this. Using this example Worrall examines two theories of method: Kuhn's model of theory change which combines objective values with subjective weightings of these values, and subjective Bayesianism. For both of these share a common feature, subjectivism, that has been of concern to more objectively minded critics of both models of theory change. Worrall provides an account of how some theorists, such as Salmon and Earman, have tried to incorporate Kuhn's theory within subjective Bayesianism and some difficulties with this project. For Worrall Kuhn's model taken along subjective Bayesian lines does not provide a sufficiently radical account of theory change, nor a satisfactory account of the rationality of Brewster's hold-out. He concludes with the important suggestion that what objectivists need to examine is a complex entity 'the state of the argument' which can exist between competing views in science.

John Fox in his paper 'With Friends Like These . . ., or What is Inductivism and Why is it Off the Agenda?' describes how Popper attacked what were at the time a number of orthodox claims about science that went under the name 'inducti-vism'. This is characterised in terms of twelve theses most of which we have come to reject under the influence of Popper – even the 'Carnapians', Fox argues, had rejected most of them. He then sets out ten of Popper's anti-inductivist theses, and shows that in attacking Popper in the name of 'inductivism' the subjective Bayesians actually rejected all of Popper's twelve theses of inductivism and were in broad agreement with many of Popper's theses. The thrust of the paper is that traditional inductivism is dead. Full adjudication between the rival claims of Popper and the Bayesians is, however, a more nuanced matter. Fox ends by discussing a respect in which inductivism had merit, but concerning which if the Popperians are badly off then the Bayesians are even worse off.

Larry Laudan's paper 'Is Epistemology Adequate to the Task of Rational Theory Evaluation?' ends provocatively with a negative answer. The paper explores the following three questions. Is it reasonable for scientists, in evalua-ting theories, to expect that a good or acceptable theory will save the known phenomena? Does such a requirement have any epistemic rationale? Finally, if the answer to the first question is positive and the second is negative, does that imply anything important about the pervasive twentieth-century project for reducing scientific methodology to epistemology? In answering these questions Laudan investigates the role that non-refuting anomalies can play in theory evaluation, these being phenomena that the theory ought to explain but does not, or puzzles it ought to solve but does not. This raises a sharp problem for the appraisal of theories if they are to be assessed only by their consequences, for non-refuting anomalies are not part of the consequences of theories. Since episte-mology is characteristically concerned with truth, but the issue of non-refuting anomalies is concerned more with the incompleteness of a theory and not its truth, then non-refuting anomalies raise non-epistemic considerations about theories. Thus epistemology, as standardly understood, cannot do justice to theory appraisal in science and needs to be superseded by methodology.

Kevin Kelly's paper 'Naturalism Logicized' is an application of formal learning theory which leads to an account of reliable inquiry, in particular an investigation of those methods which guarantee that eventually one arrives at the truth. The focus in this paper is on normative naturalism, the view that meth-odological directives are justified insofar as they promote scientific goals. Two different styles of normative naturalism are contrasted: Larry Laudan's histor-iographical approach, according to which history is gleaned for evidence about the problem-solving effectiveness of different scientific strategies, and formal learning theory, which provides a logical framework within which to determine the solvability of empirical problems and the effectiveness of solutions to them. The main contribution of the paper is a learning theoretic analysis of both finite and infinite epistemic regresses of the sort that arise when scientific methods are employed to determine whether the presuppositions of other scientific methods are met. According to this analysis, a regress is *vicious* if it cannot be turned into a single method that succeeds in the best possible sense. For various sorts of methodological regresses including potentially infinite (unfounded) ones, it is then determined what the best kind of single method the regress can be turned into.

In his paper, 'Methodological Pluralism, Normative Naturalism and the Realist Aim of Science' Howard Sankey argues that Laudan's normative naturalist account of epistemic warrant may be combined with a scientific realist conception of the aim of science as advance on truth. Such an approach shows how a pluralist account of scientific methodology of the kind proposed by Kuhn and Feyerabend may be integrated into a scientific realist framework while avoiding epistemological relativism. According to normative naturalism, methodological rules are to be analysed instrumentally as hypothetical imperatives which recommend appropriate means for the realisation of desired epistemic ends. To incorporate normative naturalism within scientific realism, it suffices to treat the ultimate epistemic aim served by methodological rules as the realist aim of truth about the world. However, Laudan has argued that theoretical truth is a trancendent aim incapable of serving as the legitimate object of rational scientific pursuit. The main task of the paper consists in responding to Laudan's objections to the realist aim of truth. In particular, it is argued that (a) theoretical knowledge is possible, (b) even if it were not, it may be rational to pursue an unattainable ideal, and (c) it may be rational to pursue theoretical truth even in the absence of infallible criteria which permit the recognition of such truth.

In 'The Hard Problems in the Philosophy of Science' Malcolm Forster argues the 1960s' Kuhn maintained that there is no higher standard of rationality than the assent of the relevant community. Realists have sought to evaluate the rationality of science relative to the highest standard possible – namely the truth, or approximate truth, of our best theories. Given that the realist view of rationality is controversial, it seems that a more secure reply to Kuhn should be based on a less controversial objective of science – namely, the goal of predictive accuracy. Not only does this yield a more secure reply to Kuhn, but it also provides the foundation on which any realist arguments should be based. In order to make this case, it is necessary to introduce a three-part distinction between theories, models, and predictive hypotheses, and then ask some hard questions about how the methods of science can actually achieve their goals. As one example of the success of such a programme Forster explains how the truth of models can sometimes lower their predictive accuracy. As a second example, he describes how one can define progress across paradigms in terms of predictive accuracy. These are examples of hard problems in the philosophy of science, which fall outside the scope of any social psychology of science.

ROBERT NOLA AND HOWARD SANKEY

A SELECTIVE SURVEY OF THEORIES OF SCIENTIFIC METHOD

1. WHAT IS THIS THING CALLED SCIENTIFIC METHOD?

For some, the whole idea of a theory of scientific method is yester-year's debate, the continuation of which can be summed up as yet more of the proverbial deceased equine castigation. We beg to differ. There are reasons for the negative view, however, some of which will be canvassed in this selective survey of the territory the debate about theories of method has traversed in the second half of the twentieth century. The territory is very wide-ranging. It is hard to find a perspective from which one can get an overall view. If one focuses on one part of the philosophical debate about method then others go out of focus or do not come into view at all. What will be attempted here is a number of snapshots of the philosophical landscape which hopefully convey, if not the whole picture, something of the debate over method that has taken place.

One focus concerns the debate surrounding the three figures whose names form the title of this book, Popper, Kuhn and Feyerabend – and to which one can add Lakatos. Both Popper and Lakatos were advocates of what one can call the 'grand' approach to theories of scientific method; grand in the sense that not only did they wish to propose some substantive universal and binding principles of scientific method but also to arrive at the ultimate goal of methodology – a demarcation criterion which draws a sharp line between science and non-science or pseudo-science. Such a goal is evident in Aristotle's attempt to characterise science as that which has apodictic certainty, or in Newton's attempt in Rule VI of his *Rules of Reasoning in Philosophy* to employ rules of inductive inference as criteria of demarcation. Popper and Lakatos reject such proposed demarcation criteria but still insist that there is some demarcation criterion to be found. On certain interpretations of their work, Kuhn and Feyerabend have been taken to undermine the pretensions of such grand theories of method – along with the principles of method upon which such demarcation criterion relied. That such an understanding of Kuhn's and Feyerabend's position has won wide acceptance is one reason for the common view that discussions of theories of method, coupled to criteria of demarcation, are part of yester-year's debate. However such an interpretation of their work needs qualification, as will be seen in sections 8 and 9. Overall, what will be maintained here is that science can be demarcated by the

1

Robert Nola and Howard Sankey (eds.), After Popper, Kuhn and Feyerabend, 1–65.
© 2000 *Kluwer Academic Publishers. Printed in Great Britain.*

methodological principles it employs, notwithstanding other attempts at the grand project of demarcation which might be deemed to have failed.[1]

Whatever differences there are between Popper, Kuhn and Feyerabend, they are united by a common opposition to an inductivist and/or probabilistic approach to methodology. This suggests that one can shift the focus of the debate about method from matters to do with demarcation to matters to do with the nature of scientific inference. And one can refocus again so that the above-mentioned philosophers of science blur into the background while the sharp foreground is occupied by those who have developed inductivist, probabilistic and Bayesian accounts of a scientific method.[2] Refocusing once more, one can bring into view the approaches that are currently being taken by the heirs to both schools of thought. Contemporary methodologists may be openly inductivist and/or probabilistic and/or Bayesian[3]; or they may explicitly develop rival theories of method; or they might remain silent on this rivalry and be content to investigate piecemeal particular principles of method. A glimpse of the latter two positions can be found in the papers collected in this book.

Refocusing yet again, all of the above sorts of philosophers of science can disappear into the background while the foreground is occupied by those who reject not only the demarcationist pretensions of methodology but the whole enterprise of scientific method itself. Some take their cue from a particular understanding of Kuhn or Feyerabend. Surprisingly others take their cue from a radical Bayesianism which says that there is nothing to scientific method except the accommodation of one's beliefs to the probability calculus and a rule of conditionalisation.[4] But by far the majority are influenced by sociological and postmodern theories of science which declare the end of methodology as we have known it. There is a conflict between the view of many philosophers that there is something to be said on behalf of the rationality of science, which theories of method try to capture, and the view of many sociologists of science (and those influenced by them) that there is very little to be said for it and much against it. From the sociological standpoint, the very content of our science is (and has been) determined by personal, professional, social and cultural and other such factors or interests. Sociologists also claim that the sciences we adopt are nothing but the result of negotiation between contending parties.

For many it is the sociological turn in science studies that has done most to turn debates about scientific method into yester-year's issues. Two positions are most commonly adopted. The weaker is that yester-year's debate has shown that there are principles of scientific method, but they are not universally binding and they do not lead to substantive demarcation criteria. The stronger is that yester-year's debate has shown that there are no substantive principles of method at all, and no demarcation to be drawn. Either 'anything goes', or the debate was misguided since it did not look to the social and cultural influences that determine the very content of scientific belief. Both positions will be explored more fully later. But the strong position can readily be shown to be far too strong and cannot strictly be the case, given what the word 'method' might mean.

The English word derives from the Ancient Greek 'methodos' (μέθοδος) 'the pursuit of knowledge' or 'a way of inquiry' (literally 'way of pursuit'). The Oxford

English Dictionary tells us that a method is a way of doing something in accordance with a plan, or a special procedure; or it is a regular systematic way of either treating a person or attaining a goal. It is also a systematic or orderly arrangement of topics, discourses or ideas. Thus one can speak of a method for gutting fish, baking a cake, wiring a fuse box, or surveying a given piece of landscape. There are also methods for teaching languages, e.g., the immersion method; there are methods for teaching the violin to the very young, e.g., the Suzuki method; there are methods for learning, and playing, tennis or golf; and so on. In general there are the methods whereby one can best teach (or learn) a given subject, present an effective case in a law court, write up a report, and so on.

It would be very surprising if one did not find methods of the above sorts in the sciences. Thus for astronomers who use optical telescopes there are methods for making observations of the positions of heavenly bodies and the times at which they are observed,[5] and methods for recording the information. For biologists there are methods for staining tissue for viewing under a microscope, or preparing cellular matter for DNA analysis. For chemists there are methods for preparing solutions with a specific pH value. For sociologists there are methods for preparing questionnaires and there are statistical methods for analysing their results. Mathematicians give us methods for solving various kinds of differential equations, or for finding the curve which best fits some given data. There are also methods for presenting data, methods for presenting the outcome of an experiment and methods for setting out papers for publication. The sciences are full of special methods, techniques and procedures for conducting experiments, analysing data, preparing results, etc. Not to follow these methods, techniques or procedures is, in some sense, to be unscientific – or at the very least to be sloppy and unsystematic in a way which undermines the goal of the activity in which one is engaging.

Most of the above examples fall either into the category of what one might call 'the material practice of science' (e.g., making a solution of a given pH value, or conducting a particular experiment), or into the category of mathematical methods (techniques for finding solutions to differential equations, or methods such as the method of least squares in curve fitting, etc.). But philosophers and scientists are also interested in a range of methods which transcend the material practices of any particular science, or transcend the mathematical methods used to solve particular problems in each of the sciences. More generally they might want to know about the methods for making inferences in science, say from given data to either the truth or falsity, or the probability or improbability, of certain hypotheses. It is here that the topic of scientific inference, in part falling within the domain of statistical inference, comes into contact with what philosophers regard as part of the domain of scientific method. In turn such an idea of a scientific method becomes part of the province of epistemology, viz., there are methods of justifying, accepting and rejecting beliefs not only in science but elsewhere. It is method as a province of, or a close relative to, epistemology, understood as the 'knowledge-getting enterprise', that is usually under discussion by philosophers and not so much the material practices of science or mathematical methods which are also abundant in science.

Philosophers distinguish at least two broad goals, aims or ends[6] of the scientific enterprise: the non-cognitive and the cognitive. Non-cognitive goals are various. They might concern the personal goals any scientist might have in engaging in their science, from getting a good salary to job satisfaction. Or they might concern the professional goals scientists have, from becoming a member of the Royal Society, getting a Nobel prize or being able to attract large amounts of funding. Or the goals may be humanitarian, social or political, as in using science to improve the health of people, increase productivity, clean up pollution or enhance defence capability. Again political and social goals are involved in the funding of research into one kind of science rather than another (research related to women's health has often received a lower priority in research funding than many other medical research projects). In some cases there may well be methods for making such choices in science in order to arrive at the chosen personal, professional, humanitarian or political goals, or in satisfying if not maximising these goals. Methods for making such choices commonly fall within the province of decision-making under varying degrees of risk. There is a well established body of literature on such decision-making which has grown up in fields such as engineering, economics and philosophy and which can be applied in this area.[7]

From the philosophers' point of view (but perhaps not that of the sociologist of science), such non-cognitive goals appear to be extrinsic to the scientific enterprise. Though they might be external engines driving the enterprise, they do not constitute its intrinsic features. Cognitive goals in science are of a different character and are commonly held to be intrinsic to the scientific enterprise. Such goals have already arisen in talk of the mathematical techniques to be employed in assessing rival hypotheses. Here the cognitive goal is to find the hypothesis best supported by the data, the notion of *best support* being one of the cognitive aims of science. Again they have arisen in talk of methods for *finding* or *discovering* hypotheses in the first place. Here an old distinction comes to the fore between the context of justification and the context of discovery. The crucial question here is: are there methods for inventing or discovering hypotheses as well as for justifying hypotheses (whether the same or different)?

The arch methodologist Popper surprises his readers when he says: 'As a rule I begin my lectures on Scientific Method by telling my students that scientific method does not exist'. But it transpires that what he means by this is the following three claims:

(1) There is no method of discovering a scientific theory;
(2) There is no method for ascertaining the truth of a scientific hypothesis, i.e., no method of verification;
(3) There is no method for ascertaining whether a hypothesis is 'probable', or probably true (Popper 1983, pp. 5, 6).

Popper's first point reiterates his long-held view about the context of discovery, viz., there can be no methods for discovering hypotheses or theories. Popper's position is highly contested. His opponents point to the existence of computer-based procedures for the scientific discovery of hypotheses to fit data, for example the BACON programmes which have even been used on Kepler's data to arrive at his Third Law of planetary motion (Langley *et al.* 1987). There are also non-algorithmic procedures for finding curves which best fit given data by using methods such as that of the least mean squares. And there are methods (suggested in the papers by Norton and Forster in this book) of 'deducing hypotheses from phenomena' (first adumbrated by Newton), or making 'deductions' from phenomena along with other heuristic principles.

Popper's second point takes us away from the context of discovery and back to the context of justification. It has wide acceptance, if by 'verification' is meant 'can be shown to be true by human powers'. It is readily granted that since theories are unrestrictedly general then we cannot show them to be true, even though they may be true. But could we not confirm, or probabilify, them? Popper's third point introduces a further substantive disagreement between his own anti-probabilistic stance in which only deductive reasoning is needed in the sciences,[8] and those who adopt either an inductivist or probabilistic or Bayesian view of how hypotheses are to be assessed. Despite his view that there is no method in any of these three senses Popper goes on to say that 'the so-called method of science consists in ... criticism' (*op. cit.*, p. 7), and then spells out what his critical method is – on which more later.

The Popperian aim of criticisability of our theories, and the specification of canons of criticism, gives us one conception of what a theory of scientific method would look like. But criticisability is not the only cognitive aim of science, even for Popper. Philosophers look to a number of aims for science of which the most general are descriptive and explanatory aims. The descriptive goal of science is to find out the truths about the world. These truths might be restricted to those about observable phenomena only, in which case we have a refinement of this aim adopted by constructive empiricists such as van Fraassen, viz., to find theories which are empirically adequate in the sense that they fit all observable phenomena (van Fraassen 1980, chapter 2). Realists have a broader aim; not only do they aim for truths about the observable but they also aim for truths about an unobservable realm of objects, properties and processes and for truths about laws governing these. The dispute between realists and non-realists (such as constructive empiricists) about the realisability of the realist aim has often turned on the viability of a methodological principle, that of inference to the best explanation.[9] Like many others, Popper also says that 'the aim of science is to find *satisfactory explanations*' (Popper 1972, p. 191). Whatever account of explanation is adopted, the goal of increasing our understanding of how the world works, or of increasing our knowledge of how and why the world is the way it is, is a further central cognitive aim of science. The broad cognitive aims of adequate description and explanation give substance to the philosophical conception of scientific method, supplemented by the means whereby these aims can be reliably realised.

2. THE IDEA OF PRINCIPLES OF METHOD

One way of focusing on much of the broad terrain covered by the last fifty years of debate concerning scientific method is by using the following schema.

The scientific enterprise can be analysed into three different levels, as illustrated in the accompanying table.

Level 3 Meta-methodologies	*Nihilist*: Sociologists of Science, Postmodernists, etc. *A Priori*: Transcendentalism, Logicism, etc. *Conventionalist*: early Popper. *Empiricist*: Historical Intuitionism (later Popper and Lakatos); Reflective Equilibrium; Varieties of Naturalism; etc. *Pragmatist*: Rescher.
Level 2 Scientific Methodologies (SMs)	Aristotle's Organon, . . . , Bacon's Methods, Descartes' *Regulae*, Newton's *Rules*, . . . , Popper's Critical Rationalism, Lakatos' Scientific Research Programmes, Kuhn's Weighted Values, Feyerabend's Methodological Pluralism, Bayesianism, Decision-Theoretic Methods, etc.
Level 1 Historical Sequence of Scientific Theories (of some domain)	*Dreams*: Homer, Bible, . . . , Aristotle, . . . , Freud, Jung, . . . , Crick, Hobson, . . . *Motion*: Aristotle, . . . , Kepler, Galileo, Descartes, Newton, . . . , Laplace, Lagrange, Hamilton, . . . , Einstein, . . . , etc.

The bottom level concerns the actual historical sequence of our choices of scientific theories concerning some domain of phenomena. Amongst all the possible theories that might have been given individual or communal assent by scientists over some period of time, the actual historical sequence is a path through the 'tree' of alternative possible histories of science. Thus consider the phenomena of dreaming. Our actual historical sequence of theories about the nature and causes of dreams begins with the works of Homer, the Bible and Plato. Aristotle seems to have been the first to have proposed a theory about the causes of dreams which did not appeal to spirit-like entities and which has affinities with modern scientific theories.[10] Skipping the intervening centuries and focusing on our own, at the beginning of this century there are the psychoanalytic theories of Freud and Jung while in our own time Crick, Hobson and others have proposed theories of dreaming based on the neurophysiological functioning of the brain. If one considers a different domain of phenomena, such as motion, then there has been an equally long sequence of theories amongst the most prominent of which are the theories of Aristotle, Kepler, Galileo, Newton, Laplace, Lagrange and Einstein. Similarly for other domains of phenomena. The idea that there is an historical sequence of scientific theories is fairly uncontroversial; it yields the raw material on which the two higher levels are based.

Amongst all the possible theory choices scientists might have made at any time, what determines the historical sequence of scientific theories that were actually chosen by the community at large? (This does not preclude two or more theories being chosen by members of a given community at any one time.) Or in other words, what has led to the historical sequence of changes in theory? The word 'change' is used deliberately since talk of growth, development or improvement in our choices involves appeal to methodology, and in particular values, to provide criteria for such growth, etc. So, what values, if any, do the changes exemplify? Let us agree that the historical sequence of theories at least displays, over time, the value of increase in instrumental success, where such success is either our increased ability to predict, and/or our improved ability to manipulate, the natural and social worlds to our own ends. (This allows that not all our attempts at prediction and manipulation have been successful.) Rephrasing the above questions we can ask: what explains our choice of a sequence of theories which yield such instrumental success (given that there are many pathways through all possible theories that would not yield such success)?

Is the successful sequence of theories due to luck, or to the tacit incommunicable 'feel for the right thing' that some such as Polanyi (1958, 1966) alleged is indicative of scientific genius? Methodologists suggest that it is neither of these but rather the use we have made of principles of scientific method in choosing theories which are successful. Sociologists of science not only reject appeal to luck or tacit 'feel'; they also reject the appeal to principles of method, offering instead a rival explanation. Our non-cognitive interests, or our social circumstances have largely produced the sequence of historical theories, not principles of method. Which of these last two rivals explains such success better?

Our non-cognitive personal, professional and political interests are often such that we want theories with high instrumental success. But it is not very probable that merely having such interests would, on the whole, lead us to choose those very theories which, when we examine and apply them, turn out to be successful, thereby realising our interests. Personal, professional and political interests seem, on the face of it, to be quite randomly linked to whether the theories chosen on the basis of such interests are also those which are successful. Methodologists would, in contrast, argue that something else intervenes producing the sequence of theories that satisfy our desire for instrumental success. These are principles of method, some of which do link some of a theory's epistemic and cognitive features to the theory being successful in the above sense. Our use of principles of method provides a more plausible explanation of why we have chosen a highly successful sequence of theories rather than some other possible sequence which might have been less successful. In sum, employing non-cognitive interests in choosing theories makes it improbable that the theories are successful; in contrast employing principles of method in choosing theories makes it more probable that the theories are successful. This is tantamount to saying that the methods we use explain success better than our non-cognitive interests. Using the principle of Inference to the Best Explanation, we can say that there is something true, right or valid about those principles of method we use which explain, much better than our non-cognitive interests do, why we have chosen a successful historical sequence of scientific theories.

It remains, of course, to show *how* our use of principles of method has produced our historical sequence of successful theories, what the link is between the methods and such success, and what it is about the principles that is right, true or valid (see section 12). But the argument above helps undermine the claims of sociologists of science that talk of principles of method can be relegated to yester-year's debates – a claim which will be discussed more fully later. Thus principles of method have an important role in explaining much of the historical growth of science. But what are these principles like? And why is their use so efficacious in producing successful theories? The first question will be addressed now while the latter is addressed in different ways in subsequent sections.

Just as there is a historical sequence of scientific theories at the bottom level, so at the next level up, the level of scientific method (SM), there has been proposed a number of theories of SM which also form a historical sequence. (The various SMs are at a higher level only in the sense that principles of method at the second level have as their object of application scientific theories at the first level.) Though the historical sequence of SMs begins with Plato's methodological remarks in his Socratic dialogues, the first fully set out theory of method for science can be found in Aristotle's *Organon*. Methodological precepts can be found in the heirs to the Aristotelian tradition up until the late Renaissance. But with the advent of the sixteenth- and seventeenth-century 'scientific revolution' methodological matters come to the fore in the work of Bacon, Galileo, Descartes and Newton, just to mention a few.[11]

In the second half of the twentieth century there has been a proliferation of theories of method. In a 1986 publication, a team of eight researchers did us the service of working through some of the writings on scientific method by five philosophers of science, Popper, Lakatos, Kuhn, Feyerabend and the early Laudan (Laudan *et al.* 1986). They collected together well over 250 theses about what allegedly does, or what ought to, happen when one theory (paradigm, research programme or whatever) is followed by another. The goal of the team was to compare the 'positivist models' of scientific change with the models of the post-positivists, also dubbed the 'historical school' (founders of which were Hanson, Feyerabend, Toulmin and Kuhn), by testing some of their theses against the historical sequence of scientific changes. Such testing is an important part of what Laudan has called 'normative naturalism', a meta-theory about SMs discussed in section 11.

One reason for the need to test was the degree to which the methodological claims uncovered by the research team rivalled one another. Of the 39 sets of theses the team identify, consider the 20th which lists six claims which according to the methodologist named in brackets, tell us how scientists have behaved with respect to particular theories and their ability to solve problems:

Scientists prefer a theory which

(20.1) can solve some of the empirical difficulties confronting its rivals (Laudan, Kuhn);

(20.2) can turn apparent counter-examples into solved problems (Laudan);
(20.3) can solve problems it was not invented to solve (Laudan, Lakatos);
(20.4) can solve problems not solved by its predecessors (Kuhn, Lakatos, Laudan);
(20.5) can solve all the problems solved by its predecessors plus some new problems (Lakatos);
(20.6) can solve the largest number of important empirical problems while generating the fewest important anomalies and conceptual difficulties (Laudan).

(Laudan *et al.* pp. 171, 172; references to work cited by the bracketed methodologists has been omitted.)

While having some features in common the claims differ in important ways. So which are correct, if any, or more probable, given the choices scientists have actually made? Moreover, the above methodological remarks are expressed as factual claims about how scientists are alleged to behave with respect to the problem-solving abilities of a pair of rival theories; as such these claims are of a historical or sociological character and are open to test against the record of the actual behaviour of scientists in some particular context[12] (a task carried out by normative naturalism in examining many of the 250 theses uncovered by the research team). Though they are not expressed this way, the above can also be interpreted as imperatives about what scientists ought to do when choosing between a pair of theories with respect to their problem-solving abilities. Thus (20.5) can be re-expressed: scientists *ought* to prefer that theory which can solve all the problems solved by its predecessors plus some new problems. This brings us to the issue of what are the methodological principles that comprise SMs, and how they are to be formulated.

3. RULES, VALUES AND METHODOLOGIES

In what follows SMs will have the following characteristics. Each SM has associated with it the pair $\langle R, V \rangle$ where R $(= \{r_1, r_2, ..., r_m\})$ is a set of methodological rules and V $(= \{v_1, v_2, ..., v_n\})$ a set of epistemic *values* (or *goals*, or *aims* – these will be treated as equivalent).[13] A *principle* of a methodology will then be a hypothetical imperative of the form: if one wishes to realise value v_i then one ought to follow rule r_j; SMs can then be characterised as a set of principles $\{P_1, ..., P_r\}$.[14]

As noted the theses (20.1)–(20.6) above are all declarative rather than imperative; and they contain no explicit reference to a value. On the second point, there are good grounds to suppose that each thesis has a suppressed reference to some unspecified value. There are various reasons why they might not contain an explicit reference to a value, or values. Perhaps the value is not explicit because it is a general presupposition, or the thesis has been expressed elliptically. However values (ends, aims) must be understood in the context; otherwise the declarative sentence merely contains a rule 'do x' without telling us what the purpose of doing x is. On the first point, we can regard the declarative claim (with reference to a value made explicit) as having the following form: following r_j will realise value v_i.

What is the connection between the methodological principle, which is a hypothetical imperative, 'you ought to follow r_j if you want to realise value v_i' and the declarative which is an empirical claim 'following r_j will realise value v_i'? This issue is important for normative naturalism and is discussed in section 11.

Examples of values adopted by SMs include: truth, or increased verisimilitude; empirical adequacy; generality; testability; falsifiability; coherence; explanatory breadth; high probability on available evidence; the capacity to withstand severe tests; openness to revisability; and so on. Could an SM lack any values? As just indicated, though SMs may lack explicitly stated epistemic values, they are generally presupposed. Principles of method which lack any value whatever would become mere rules with no point to their application.

Rules of method tell us what we ought to do to realise some value. Examples of methodological rules are: avoid *ad hoc* modifications to theories; prefer theories which make surprising novel predictions over those which predict what we already know; prefer double-blind over single- or zero-blind experiments; accept a new theory only if it explains all the successes of its predecessors; reject unfalsifiable theories; for the same kind of effect postulate, as far as possible, the same kind of cause; and so on.

Principles of method are then hypothetical imperatives concerning the link between rules and values. They are instrumental in character telling us about the means we ought to adopt in order to realise some end. How reliable are they? This is a matter for further investigation. In our actual world they could be 100% reliable, or less than 100%. Thus in their declarative form they could be viewed as akin to statistical generalisations with varying degrees of reliability such as: following r will realise v n% of the time (where $50 < n \leq 100$). The principles might also be statistical comparative in form: following r is more likely to realise value v than following some rival r^*.

Are methodological principles *a priori* or empirical in character? This is a large question to be addressed in the next and subsequent sections since in the history of methodology they have been understood in both ways. Are they necessary, or contingent? That is, for any enquirer in any possible world are some principles always available since they hold in all possible worlds? Or are the principles contingent in that they hold for only some worlds and that enquirers in sufficiently different possible worlds will have to adopt different principles of method in order to make discoveries about their world? Deductive rules, which aim at the preservation of the truth of the conclusion providing the premises are true, will hold for all enquirers in all possible worlds. But clearly not all principles of method are necessary in this sense. This points to another sense in which principles of method can be said to be reliable. They may be reliable not only because they hold 100% in this world but because they also hold 100% (or nearly so) in a range of possible worlds other than our actual world. Deductive rules with true premises are clearly reliable with respect to their conclusions in both senses.

The sciences propose laws, one feature of laws being their counterfactual robustness, that is, the laws are not merely generalisations about this world because they tell us what will happen in a sub-class of possible worlds varying in distance from our actual world, viz., the physically (or naturally) possible worlds.

If principles of method are akin to scientific laws then they would be contingent and hold only in some sub-class of possible worlds; but they would also exhibit a similar counterfactual robustness. For all enquirers equipped with some principles of method in each of the sub-class of possible worlds, an investigation into their world would be widely reliable; the principles hold for that sub-class of possible worlds. But suppose the principles of method we adopt are not very counterfactually robust in that they apply only in our world, or perhaps only in a small range of worlds close to this one; outside this world, or the small range of possible worlds, they are not reliable for any enquirer. It would then be a matter of epistemic luck that our principles of method do yield theories which do exhibit some success. Our principles would be highly contingent generalisations in that in only slightly different circumstances they would be quite unreliable for use in inquiry.

The matter of reliability becomes important because we are concerned with reliability across all those worlds which are possible relative to the information enquirers have, including all the information provided by our scientific theories. Up to a point an enquirer can ignore consideration of those worlds which are inconsistent with their best scientific information, or even those which are highly improbable given their best scientific information. But this still leaves a range of worlds which are consistent, or are probable, with respect to that information. Any enquirer would want principles which are reliable in all these worlds. So principles of method cannot be just a matter of contingent luck; they must also exhibit some degree of counterfactual robustness and apply reliably in a sufficiently broad range of possible worlds.

How are rules and values to be distinguished? In 'Objectivity, Value Judgement and Theory Choice' (Kuhn 1977), Kuhn canvasses the idea that there might be rules governing the choice between theories on grounds of accuracy, consistency, scope, simplicity and fruitfulness; but then he abandons the idea that these can be expressed as rules, preferring to understand each of these as values (Kuhn 1977, p. 331). In surveying the literature it is common to find that what one writer regards as a value another treats as a rule, and conversely. Thus while Kuhn ultimately treats simplicity as a value Lycan treats it as a rule saying: 'Other things being equal, prefer T_1 to T_2 if T_1 is simpler than T_2' (Lycan 1988, p. 130). Perhaps we can admit a great deal of interchange between what counts as a rule and what counts as a value from one SM to another. Values can get expressed as rules; and rules can be expressed as values. What we must not admit within the same SM are redundant values and rules which are trivially linked with one another, e.g., within the same SM a Kuhnian value of simplicity associated with a Lycan-type rule of simplicity.

Popper's account of his methodology yields an example of a redundant rule and value. In one place he tells us that falsifiability is one of his aims for science: 'I propose to adopt such rules as will ensure the testability of scientific statements; which is to say, their falsifiability' (Popper 1959, p. 49). However later Popper proposes a supreme rule which serves as a norm for other rules of method: '. . . the other rules of scientific procedure must be designed in such a way that they do not protect any statement in science against falsification' (*ibid.*, p. 54). A rule which says 'do not protect against falsification' will, if successfully applied, trivially realise the goal of adopting only falsifiable statements. Popper's method might have other values which are not redundant with respect to this rule (e.g., high explanatory power or increased verisimilitude); or his method might retain the value of falsifiability but realise it with rules different from the supreme rule just cited. However a value of falsifiability linked to a rule which bids us to seek falsifiability is vacuous – or we might say that it involves redundancy. Thus for an SM with its associated rules and values $\langle R, V \rangle$, if there is a principle P_1 which contains a rule r_1 which trivially realises value v_1 but another principle P_2 which non-trivially realises value v_2 then we are to discount P_1, but not P_2, on the grounds of redundancy.

In eliminating redundancy appeal is made to a higher third level rule, or meta-methodological rule, which bans such trivial rule-value linkages. The idea of a meta-methodological rule has yet to be introduced; but such a ban on redundancy is an example of a meta-rule, though admittedly not an exciting substantive meta-rule.

Presumably there is also a meta-methodological rule of consistency which bids us to formulate second level SMs which comprise a consistent sets of principles, i.e., there cannot be pairs of principles one of which bids us to follow r while another bids us not to follow r in order to realise some value v. We will also want, for any SM, sets of values, and sets of rules, each of which taken as a whole are consistent.

There has been much debate amongst methodologists concerning consistency as a second level rule (or value, or principle) which applies to scientific theories. Some rules would prohibit us from adopting theories which are internally inconsistent while others would prohibit us from adopting theories which are inconsistent with other prevailing theories. Concerning the latter, Feyerabend, in one of his methodological moods, has advocated what he calls 'methodological pluralism' which positively invites the proliferation of theories which are inconsistent with prevailing theories as a condition for the empirical advance of science – this advance being a value allegedly realised by the rule (Feyerabend 1975, chapter 3). That is, for Feyerabend the value of empirical advance in science is (more likely) to be advanced by adopting a rule which bids us to proliferate inconsistent theories rather than the more conservative rule which bids us to entertain only theories which are consistent with one another.

Whether Feyerabend's principle, or its more conservative opposite, is to be adopted is not of immediate concern. What is of interest is whether there should be a *meta-rule* of consistency that is to be imposed on all our values, rules and principles of SMs at the lower level. If we grant that there should be,[15] then it is still an open matter as to whether we adopt Feyerabend's lower level principle of SM that we proliferate inconsistent scientific theories, or we adopt the more conservative principle against inconsistent theories. Thus it would appear to be quite possible that there is a rule of consistency at the meta-level which applies to all values, rules and principles of SMs, but that there be no principle of consistency for any adequate SM concerning what scientific theories we are to adopt or countenance, if science is to advance. But this is only possible if higher level meta-rules are distinguished from the principles of SMs to which they apply. This amplifies the suggestion that there is a third level discipline of meta-methodology to be imposed on the rules, values and principles of SMs at a lower level.

Thus meta-methodology is not a completely empty discipline; there appear to be at least two meta-methodological rules banning redundancy and inconsistency for principles of SMs. Whether there are more substantive principles remains to be seen. In fact it will be argued that philosophical theories which attempt to justify particular theories of scientific method are best viewed as providing meta-methodological justifications.[16]

4. META-METHODOLOGY

We have now ascended to the quite rarefied atmosphere of meta-methodology which contains at least the meta-rule of non-redundancy and (perhaps) of consistency. As rarefied as it might be, it is not unfamiliar territory for those who have followed the many discussions about how induction is to be justified. In the schema above, rules of inductive inference, such as the simple rule of enumerative induction, find their place as second level principles of an inductivist SM. Hume's philosophical challenge that induction cannot be justified is itself a meta-methodological claim about the status of second level inductive principles – and so are most of the arguments philosophers have advanced to rebut Hume's claim. For Humeans, even though we may still make inductive inferences (and perforce must do so), there is no rational basis for making such inferences. This meta-methodological claim is the core of Humean scepticism about induction. However we will call Hume's position 'nihilistic' because of the meta-methodological claim it makes, viz., there is no justification at the meta-level for the second level inductive inferences we in fact use and whose propositional contents are the claims of the sciences at the first level.

Attempted philosophical rebuttals of Hume's view also have the character of meta-methodological arguments, only two of which will be mentioned here. One of Hume's main lines of attack is that the Principle of the Uniformity of Nature cannot be justified. In his attempt at a justification Kant tried to argue that this Principle, in the form of a principle of universal causation, could be shown to be a synthetic *a priori* truth. Our task here is not to judge the success or failure of Kant's argument (a good account of its failure can be found in Salmon 1967, pp. 27–40). Rather if we generalise Kant's procedure then we can take any principle of method and look for its justification, from a meta-methodological viewpoint, either on analytic or synthetic grounds, or (as is the case in the schema set out in the table) on *a priori* or *a posteriori* grounds.

Again, attempts have been made to justify induction by investigating a hierarchy of inductive principles. Thus first level inductions about observed white swans or observed green emeralds are justified by appeal to second level inductive principles which are about the success of the arguments at the first level. These second level principles must themselves be inductively strong by their own system of inductive rules at the second level, and have as their conclusion that arguments at the first level will work well the next time they are used. An account of such an inductive justification (at a higher level) of inductive inferences (at a lower level) can be found in Skyrms (1975, pp. 30–41) and Black (1954, pp. 191–208). Generalising this procedure we can view the connection between methodology and meta-methodology in the same light. There is a hierarchy of rules in which those at a higher level apply to, and provide a justification for, those at the next level down; in particular the same type of rule (e.g., an enumerative inductive rule) can appear at more than one level in the hierarchy.

This suggests that third level meta-methodologies have the following features. They are at a different level only in the sense that meta-methodological principles apply to, and are about, principles of SMs at the second level. But they are also intended by some methodologists to bring the principles of SMs into relation with certain historical features of the sciences at the first level; so they apply to more than just principles of SMs. The possibility should also be left open that some methodological rule, or some value, can appear at both Level 2 and Level 3, e.g., the rule, or value, of consistency which applies both to scientific theories and to methodologies. Finally meta-methodologies can embody philosophical theories which in turn yield reasons for adopting some particular set of principles of scientific method. Meta-methodologies thus provide an answer to the 'legitimation problem' for theories of method, viz., the grounds on which SMs can be justified, or legitimated.

If meta-methods are to provide justifications, then there is an apparent dilemma concerning justifications which they should avoid. If we are to adjudicate between truth and falsity in meta-methodology then we must be able to justify any claims we make. Consider some claim M in meta-methodology. This stands in need of justification. Either we appeal to M itself or to some other Principle M' to provide the justification. If we appeal to M the justification is circular. If we appeal to M' to justify M then we need a reason for accepting M'. Thus an infinite regress of justifications threatens. So the acceptance of any meta-methodology is threatened by either circularity or a regress of justifications. This raises difficulties for attempts to provide justifications in terms of methodological principles elevated to meta-methodology; it remains to be seen how each meta-methodological theory might deal with this difficulty.

Are there any substantive theories or principles of meta-methodology, other than the two suggested principles of consistency and non-redundancy? The *nihilist* position is that there is no substantive theory or principle to be found at the meta-methodological level to adjudicate between, or to justify, principles of SM. We will need to investigate the extent to which Feyerabend, Kuhn, sociologists of science and postmodernist accounts of science adopt nihilism with respect to the project of meta-methodology. The other approaches to be considered adopt meta-methodologies of varying degrees of strength; either there is some philosophical position which gives us some purchase on principles of SM, or there are quite strong meta-methodological justifications that are available to support principles of SM. The first of these is that meta-methodology is an entirely *a priori* discipline. This might mean that we can give *a priori* justifications for the rules and values of an SM. Or less strongly it might mean that we can show *a priori* that the rules do realise their associated values. Allied to this position is the view that principles of SM have the status of conventions, the most prominent advocate of this view being the early Popper. Rivalling these *a priori* accounts is the view that meta-methodological matters are to be decided in an empirical manner like other matters in first level science. One such view is that of the later Popper and Lakatos. However there are now a number of varieties of naturalism concerning scientific method which are empirical in character, the clearest example being the position adopted by the later Laudan known as 'normative naturalism'. Finally there is an approach that can be broadly characterised as 'pragmatic'; it admits, contrary to nihilism, that there is a meta-methodological story to be told but it is different from either of the *a priori* or empirical approaches just mentioned. Aspects of the positions of Laudan, Rescher and Quine could be characterised under the heading of pragmatism. However not all methodologists we wish to consider fall neatly into one or other of these divisions, as will be seen. We begin by considering meta-methodologies which are *a priori*.

5. SOME *A PRIORI* APPROACHES TO META-METHODOLOGY

There is a long history of *a priori* attempts to provide an answer to the legitimation problem for SMs, three of which will be mentioned here. The first of these we can dub 'transcendentalism'. Transcendentalists attempt to mount some kind of Kantian argument from the bare possibility of science to the presuppositions of the scientific enterprise, amongst which hopefully will be found some principles of scientific method.[17] However transcendental arguments have a low success rate, including arguments in this context to some principles of method. It is hard to see how such an argument could yield any substantive methodological principle.

A second *a priori* approach might be dubbed 'logicism'. Logicism in the theory of scientific method is the view that just as there are principles of deductive reasoning which (on certain views of the nature of logic) can be given an *a priori* justification, so there are principles of non-deductive scientific inference that can be given an *a priori* justification. One *a priorist* approach might be through the Probability Calculus and some of its theorems such as Bayes' Theorem. Since the theorems are simple consequence of the axioms of the Probability Calculus, and if we assume that the axioms themselves can be given an *a priori* justification, then we already have the makings of an *a priori* justification.

Consider the simplest version of Bayes' Theorem, viz., $p(H, E) = p(E, H) \times p(H)/p(E)$. Are the posterior and prior probabilities which make up the Theorem known *a priori* or *a posteriori*? If we could establish *a priori* the numerical values of '$p(E, H)$', '$p(H)$' and '$p(E)$', then there would be a totally *a priori* argument to the value of $p(H, E)$, and thus, it might seem, the beginnings of an *a priori* probabilistic methodology for science. Such a position arises naturally for those, such as Carnap, who think of expressions like '$p(H, E)$' on a model of partial entailment, with total entailment being the special case of logical deduction. On this account deductive logic will give an account of when it is appropriate to claim that E logically implies H. Similarly an inductive logic will tell us when E *partially* logically entails *H*. That is, it ought to be possible to provide a theory of a confirmation function that will give us a quantitative value for '*r*' for expressions such as '$p(H, E) = r$'.

It is now well known that the programme of finding *a priori* a numerical value for expressions such as '$p(H, E)$', explored by Keynes, Carnap and others for any *H* and *E*, faces many difficulties; so this logicist justification of a probabilistic methodology for science has little hope of success. Most present-day advocates of probability in methodology are Bayesians (a position which Carnap also explored); they take a quite different subjectivist approach in which the key issues of justification turn on matters such as avoiding 'Dutch Books' and the like.

However, even if the programme of finding *a priori* numerical values for the probability expressions were to succeed, it does not follow automatically that all of methodology would be *a priori*. This is particularly the case for Carnap who carefully distinguished between logical and methodological aspects of logic (Carnap 1962, sections 44 and 45). Thus in the case of deductive systems there are two 'fields' to consider. The first concerns the theorems which pertain to the system; the second concerns the methods which might be used to prove theorems under various conditions and for various purposes. The second field Carnap calls the 'methodology' of deductive logic. (Given the way in which we have already introduced the term 'methodology' in sections 2 and 3, it would be better for us to refer to Carnap's second field for deductive logic as the 'heuristics' of theorem-proving.) Similarly there is a methodology accompanying inductive logic. Thus in testing a given hypothesis *h* against new and old evidence Carnap says: 'methodology tells us which kinds of experiments will be useful for this purpose by yielding observational data e_2, which if added to our previous knowledge e_1, will be inductively highly relevant for our hypothesis *h*, that is, $c(h, e_1 \cdot e_2)$ is either considerably higher or considerably lower than $c(h, e_1)$' (Carnap 1962, p. 203). Thus an inductive logic will yield the values for Carnap's confirmation functions and tell us which is greater; in contrast methodology will tell us both what evidence we should look for and what we should do with the hypotheses with the higher and the lower numerical values for the two *c*-functions.

Given the further matters he addresses, it is clear that for Carnap the task of methodology in inductive logic goes beyond the role it plays in deductive logic as a mere heuristics of theorem-proving, thus bringing his conception of methodology into closer contact with the concept we have introduced in previous sections. This becomes evident when one discovers that the methodology of inductive logic contains rules or principles of the following sort. There is a Requirement of Total Evidence which bids us to take into account the total evidence in calculating the degree of confirmation (*ibid.*, p. 211). There is also a requirement concerning the variety of evidence to be used in testing a hypothesis (*ibid.*, p. 230). Carnap's inductive logic also needs supplementing with methodological rules concerning the decisions that are to be made in the light of one's observations and available hypotheses. To this end Carnap investigates a number of methodological rules from The Rule of Maximum Probability (i.e., from a set of possible events, expect the most probable), to rules of Maximising Estimated Utility (*ibid.*, sections 50 and 51).

Such methodological rules, which Carnap must add to his account of inductive logic, go beyond methodology understood merely as the heuristics of theorem-proving as in deductive logic. In fact some methodology is required if Carnap's inductive logic is to have any application at all in either scientific or every-day contexts. Thus Carnap's account of the methodology of inductive logic can be brought into relation with the conception of methodology outlined above with its rules which realise values (such as truth). Given this, it is far from obvious that all the methodological principles Carnap discusses have an *a priori* justification; it is clear that some do not. Thus it would be wrong to regard the Carnapian programme as being committed to an entirely *a priori* account of scientific method, though elements of it will have an *a priori* justification (or would have if, for example, there were a satisfactory account of Carnapian confirmation functions).

Finally, there is a further sense in which methodology might be *a priori*. Most methodologists assume that a rule *r* of an SM will, when correctly followed, realise some value *v*. However there is rarely a proof that this is so. One task of meta-methodology would be to show that rules are reliable realisers of certain goals. This suggests another weaker *a priori* approach in which an *a priori* proof might be given for some particular principle of scientific method. Suppose our aim is to get considerably increased support for our theory. How should we go about this? If we adopt a rule which says 'one ought always to take into account new evidence which is unexpected in the light of known evidence' then there is a proof in the Probability Calculus that this rule will realise the aim. The proof is immediately provided by Bayes' Theorem in the form:

$$p(T, E \,\&\, K) = p(E, T \,\&\, K) \times p(T, K)/p(E, K).$$

On certain conditions concerning the numerator on the right hand side, $p(T, E \,\&\, K)$ is inversely proportional to $p(E, K)$, i.e., the expectedness of new evidence E with respect to old evidence K. If the expectedness is high, i.e., $p(E, K)$ is close to 1, then the goal of increased probability will be realised, but only to a very small extent. However if the expectedness is low, i.e., $p(E, K)$ is close to 0, then the goal of increased probability will be realised in a quite striking way. Using the Probability Calculus in this way, an *a priori* proof can be given of the important principle of many SMs that new expected evidence will realise the goal of increase in probability of our theories; but there is an additional boon in that the proof sets out conditions under which the principle holds.

6. METHODS AS CONVENTIONS: THE EARLY POPPER

Even though Popper remained fairly consistent about what were his preferred principles for a theory of scientific method for critically evaluating scientific theories, commonly called 'Critical Rationalism' (CR), he adopted different meta-methodological accounts of his CR. The early Popper regarded his methodological principles as conventions and justified his SM in much the same way in which one would adjudicate between rival conventions. But the later Popper adjudicated between his own and rival SMs by comparing them with historical judgements about what was, and what was not, great science. In this respect the later Popper and Lakatos have closely similar views about meta-methodology – it is 'quasi-empirical', as they say, and not *a priori* or conventionalist in character. In this section we will examine the views of the early Popper and in the next the later Popper and Lakatos together.

The early Popper was impressed by the way in which science, unlike many other bodies of belief, was open to radical revision (on the whole), such revisions being best exemplified by the overthrow, in the first quarter of the twentieth century, of the highly confirmed Newtonian mechanics by the Special and General Theories of Relativity and by Quantum Mechanics. What makes such radical revision possible? Popper viewed the testing of scientific theories hypo-thetico-deductively; hypotheses and theories are tested in the context of other auxiliary assumptions by drawing out their test consequences for comparison by observation or experiment. (Observation and experimentation, along with the auxiliaries, might also involve more theory; but this would not be currently under test. In addition Popper allowed for the testing of statistical hypotheses and non-deductive, as well as deductive, drawing out of test implications to be compared against observational reports, or what Popper called 'potential falsifiers'.) Because of the unrestrictedly general character of our theories, Popper argues that they cannot be verified; but they could be open to falsification.[18] Thus what makes a scientific hypothesis open to revision is the fact that in principle it has test consequences, each of which has a potential falsifier; these may either remain potential if the test implication is correct or become actual if it is false. Having potential falsifiers which are actual is a necessary, but not a sufficient, condition for falsification; but once the further conditions for falsification are realised[19] then the demand for revision becomes imperative.

Being open to revision is thus linked to being open to tests – the more open to tests, the more opportunities there are for revision if actual falsification occurs. Popper promotes the logico-epistemic property that scientific theories possess of being falsifiable into a central role in forming his methodologically based demarcation criteria for science.[20] Thus the value of radical revisability is cashed out in terms of falsifiability (or testability, of which a theory needs to be given, including an account of degree of testability). This logico-epistemic property in turn becomes one of Popper's conventionally adopted values: 'My criterion of demarcation will accordingly have to be regarded as a *proposal for an agreement or a convention*' (Popper 1959, p. 37). What is important here is that the mere logico-epistemological property of falsifiability is not Popper's demarcation criterion in, say, the same fashion as the related Verification Principle was proposed as a criterion of demarcation. Rather demarcation arises from conformity to a set of methodological principles, in which the demarcation proposal plays a central role; but it is the principles that do the work of demarcation for Popper and not the mere logico-epistemological property of falsifiability.

Though falsifiability is a key Popperian value, other values are also endorsed. One such value is increase in explanatory depth (Popper 1972, 'The Aim of Science'), which Popper alleges can be cashed out in terms of increasing falsifiability.

In the 1950s Popper also came to adopt increased verisimilitude as a value; and this is also linked to increased falsifiability. What rules of method realise these values? Though they play an important role in Popper's theory of Critical Rationalism they get scant attention, and are not carefully formulated. Popper eschews any rules of inductive or probabilistic support, claiming that science can get by with only rules of deduction and his theory of non-inductive, non-probabilistic, corroboration (even though probability relations are employed in more formal attempts at the definition of corroboration). It is over the issues of inductive support and confirmation that much of the debate between Popper and his opponents has taken place.

But there are more positive rules that Popper sets out (especially in Popper 1959, section 11 entitled 'Methodological Rules as Conventions') in which he speaks of 'the rules of the game of science'. Such rules are said to differ from the rules of pure deductive logic and are more akin to the rules of chess: '... an inquiry into the rules of chess could perhaps be entitled 'The Logic of Chess', but hardly 'Logic' pure and simple. (Similarly, the result of an inquiry into the rules of the game of science – that is, of scientific discovery – may be entitled 'The Logic of Scientific Discovery.)' (Popper 1959, section 11) Popper's first example of a rule is not even expressed in rule form: 'the game of science is, in principle, without end' (*ibid.*, p. 53). If we were to stop subjecting our scientific claims to test and regard them as 'verified' then we would give up the critical stance. We can take Popper to be proposing an anti-dogmatism rule which bids us: 'subject all claims to test'. However such a rule, to be at all practicable, must be qualified by considerations of diminishing returns. Popper's second example of a rule (*loc. cit.*) spells out one way in which his rules are a 'logic of *discovery*'. The discovery is not so much the *invention* of hypotheses (Popper has ruled this out), but rather the discovery of either which hypotheses we should provisionally accept (they pass tests) or which we should reject (we reject hypotheses because either they have been falsified through hypothetico-deductive testing, or we have rival hypotheses which are more testable).

When Popper introduces his proposal for a demarcation criterion, he recognises that it is always possible to evade falsification by decreasing the degree of testability of a hypothesis through adopting various 'saving' stratagems (*ibid.*, section 6). To combat these stratagems he adds a necessary methodological supplement to his demarcation criterion in the form of a supreme rule about all other rules of method: 'the other rules of scientific procedure must be designed in such a way that they do not protect any statement in science from falsification' (*ibid.*, p. 54). This supreme rule is rather contentless; and, as has been mentioned in section 3, is a rule which is redundant with respect to the value it realises. However Popper's more specific anti-*ad hoc* rules are neither contentless nor redundant. The first of these concerns the introduction of saving hypotheses to rescue a theory which has been refuted: 'only those [saving hypotheses] are acceptable whose introduction does not diminish the degree of falsifiability or testability of the system in question, but, on the contrary increases it'[21] (*ibid.*, pp. 82, 83). Theories can also be saved by a number of stratagems directed not at the theory under test but the observations or experiments which allegedly refute them (e.g., questioning the competence of the observers or experimenters). To combat this Popper proposes a second anti-*ad hoc* rule: 'intersubjectively testable experiments are either to be accepted, or to be rejected in the light of counter-experiments'. A third anti-*ad hoc* rule is introduced to combat the saving of theories by altering the meanings of their constituent terms. Though no rule is specifically proposed, Popper intends a two-part prescription. The first is that there be a rule requiring semantic stability of terms in hypotheses which are undergoing test. The second is that theories not be presented as sets of implicit definitions which cannot be tested; rather theories must be regarded as a set of (largely) empirical claims open to test (*ibid.*, sections 17 and 20).[22] All three anti-*ad hoc* rules have the same value of increasing testability.

Why the three anti-*ad hoc* rules? Popper recognises that there is a view of scientific method which rivals his own and against which, unlike inductivism, he has no argument. This is a Conventionalist view of method against which Popper can only say: 'underlying it is an idea of science, of its aims and purposes, which is entirely different from mine' (*ibid.*, p. 80), and 'the only way to avoid conventionalism is by taking a *decision*: the decision not to apply its methods' (*ibid.*, p. 82). Popper uses the term 'conventionalist' in at least two ways. The first concerns the status of his rules of method; these are conventions in the sense of proposals for an agreement. In Popper's view SMs could not have the status of empirical claims of science; such a view he criticises as 'naturalistic' (*ibid.*, section 11). And it was evident to him that principles of an SM could not be known to be *a priori* true (or even analytically true). Since the *a priori*/empirical distinction is exhaustive, the only alternative to declaring such principles to be meaningless was to view them as conventions which we could adopt for various purposes. Thus despite disclaimers to the contrary, Popper was still imbued with some of the positivism of his day. The second use of 'Conventionalism' is as the name of an SM. We can glean what Popper takes this to be since a Conventionalist SM adopts rules and values opposed to those of Popper's Critical Rationalism, viz., whatever a Conventionalist SM has as its values, falsifiability is not one of them; and it positively advocates what Popper's anti-*ad hoc* rules prohibit (see *ibid.*, p. 81).

That there are rival theories of SM is a central idea concerning the table drawn up in section 2. What is of interest here is Popper's meta-methodological claim that there is no way of adjudicating between rival SMs such as Conventionalism and Critical Rationalism, except to make a decision to adopt one, and not to adopt, or reject, the others. In Popper's view the values embodied in these SMs are so fundamental that no argument can be given for adopting one value over another without a prior commitment to some value. However Popper does not always adopt such a decisionist conception of meta-method, for elsewhere he suggests ways in which rival SMs with their disparate rules and values might be compared. On what grounds should one adopt the rules and values of Popper's SM of Critical Rationalism? Popper says of his own proposal: 'it is only from the consequences of my definition of empirical science and from the methodological decisions which depend upon this definition, that the scientist will be able to see how far it conforms to his intuitive idea of the goal of his endeavours' (*ibid.*, p. 55). For example, if scientists intuitively favour the exposure of a theory to test, especially where the test concerns novel test consequences, or they like the challenge opened by a falsification, then Popper says that they will favour his methodology of Critical Rationalism which incorporates these intuitions over the Conventionalist SM which plays them down (*ibid.*, p. 80).

Thus it appears that Popper does adopt a meta-methodological stance that is not merely based on decisions which lie beyond evaluation. He adopts a hypo-thetico-deductive meta-methodology in which the consequences of any set of rules and values defining some SM are to be drawn out and compared on a number of grounds with some test bases. The first of these are the intuitions of scientists about values and rules embodied in their own scientific endeavours. Other 'test bases' are more philosophical; they concern the ability of any pro-posed SM, which Popper treats as akin to a theory of knowledge, to uncover inconsistencies and inadequacies in previous theories of knowledge, and to solve problems within epistemology. Thus if a meta-methodology is to be attributed to the Popper of *The Logic of Scientific Discovery* for adjudicating between rival SMs, it is a version of his own theory of SM which he applies to the sciences; but it is elevated to a higher level of meta-methodology and adapted so that it can deal with the assessment of theories which are themselves not empirical in character. On these grounds Popper is able to dismiss Inductivist methodology since his own SM of conventionally adopted rules and values allegedly solves the problems which Inductivism faces (largely by allegedly bypassing them); and he dismisses Conventionalism not because of any problem he can detect in it but because of its alleged inconsistency with the intuitions of scientists about values and the rules to be adopted to preserve those values. It is this last idea which comes to the fore in a slightly different form in the later Popper and Lakatos, and in the context of a more overtly empirical meta-methodology.[23]

The idea that principles of method are expressions of means for some epistemic or cognitive end is implicit in the account Popper gives of methodology in *The Logic of Scientific Discovery* (see especially sections 9–11 and 20). It is more explicit in later works when he says in criticism of some doctrines of essentialism in science: 'I am concerned here with a *problem of method* which is always a problem of the fitness of means to ends' (Popper 1963, p. 105, fn. 17). And later he emphasised that the 'ought' of methodology is a '*hypothetical* imperative' with the growth in scientific knowledge as its single goal and its means the critical method of trial and error (Popper 1974, p. 1036). The idea that principles of method are hypothetical imperatives will become important in the discussion of normative naturalism in section 11.

7. EMPIRICAL APPROACHES I: INTUITIONISM IN POPPER, LAKATOS, AND REFLECTIVE EQUILIBRIUM

Popper's final extended treatment of issues to do with method occurs in his 'The Problem of Demarcation' (Popper 1974). Popper revisits his earlier defi-nition of science in terms of his criterion of demarcation, as well as the definitions of others, and casts aspersions on them all saying: 'A discussion of the merits of such definitions can be pretty pointless' (Popper 1974, p. 981). The reasons for this have to do with his change in meta-methodological justifications for adopting SMs, since he continues: 'This is why I gave here first a description of great or heroic science and then a proposal for any criterion which allows us to demarcate – roughly – this kind of science'. For Popper the paradigms of great or heroic science are the laws of Kepler and the theories of Newton and Einstein; instances of disreputable science are those of Marx, Freud and Adler.

Popper's new meta-methodological criterion for assessing SMs is the following procedure: any acceptable SM must capture those sciences on the great/heroic list and miss none out (otherwise some other SM might do better); and it must capture none of the sciences on the disreputable list (on pain of refutation). Popper does not say how the two lists are drawn up; but given the lists they constitute a (fallible) foundation against which SMs can be tested in much the same way that scientific observations are a (fallible) foundation against which scientific theories can be tested. Thus once more Popper's method of Critical Rationalism has been elevated to meta-methodology, but this time it has been provided with a different test basis for theories of SM in the form of the two lists of heroic and disreputable science. Given such a fallible test basis, Popper's approach to the status of SMs is thereby empirical (or 'quasi-empirical' as Lakatos puts it) rather than *a priori* or conventionalist.

Popper's intuitions about what is, and what is not, great science seem to be plucked out of the air. In contrast Lakatos suggests a different way of drawing up the fallible foundation against which SMs are to be tested. Appeal is made to the general community of scientists working in a given field, the 'scientific élite', and their judgement as to what is, and what is not, an acceptable move with each change within their prevailing web of scientific belief. Such judgements need not, and better not be, informed by some theory of SM if they are to test SMs. Rather the judgements arise out of the day-to-day workings of scientists independently of methodological or philosophical reflection upon their scientific enterprise. While there is considerable dispute, even amongst scientists, over what is an acceptable SM, there is, alleges Lakatos, no comparable dispute over whether some move in 'the game of science' is scientific or unscientific (Lakatos 1978, p. 124). Granted such an admittedly fallible 'foundation' of value-judgements, Lakatos argues that Popper's own theory of SM, when elevated to a meta-method for testing SMs against the 'foundation', is falsified (see Lakatos 1978, pp. 123–131). Lakatos' case against Popperian Critical Rationalism in part turns on case histories. For example, the alleged Popperian requirement that all theories must have falsifiability conditions laid down before any testing can take place is violated by Freudian psychology; so it is declared to be non-scientific. But equally, argues Lakatos, no Newtonian has ever laid down falsifiability conditions for Newtonian mechanics. So it is equally unscientific – thus removing one of the paradigm cases of heroic and great science. So Critical Rationalism is criticised and found wanting by its own criteria.

Using the same Popperian meta-method, Lakatos argues that other theories of SM, in particular Inductivism, Conventionalism and his own SM of Scientific

Research Programmes (SRP) can also be falsified. However Lakatos can see no reason why he should accept Popperian meta-method while rejecting Popperian SM; so he replaces it with his own methodology of SRP elevated to a meta-method, but retains the test basis in the same 'foundation' of judgements made by a scientific élite. SRP as a meta-method is a research programme within the historiography of science. According to Lakatos all histories of science are written with some implicit theory of SM in mind. Each SM will render episodes in the history of science rational by its own lights, thus providing an 'internalist' explanation of why some episode occurred in the history of science. However it will not be able to explain all episodes, even some of those judged to be acceptable moves by the scientific élite. These will be relegated to an 'externalist' approach to history of science in which the 'irrational' leftovers are available for social or psychological explanation. There will always be a residue of irrational leftovers; however the task of a SRP applied to the historiography of science will be to discover which of the various SMs maximise the number of episodes in the history of science its makes 'rational' by its own lights, i.e., maximise internalist explanations of scientific change.

Just as an SRP requires that there be some empirical progress, i.e., that it uncovers novel facts, in order for it to be dubbed 'scientific', so a successful historiographical SRP will uncover some novel historiographical facts. The notion of novelty need not be confined to the discovery of previously unknown facts; it also includes those facts which are known but which get their first explanation within some new progressive SRP while rival SRPs of longer standing failed to explain them. The same applies to judgements made by the scientific élite; these can be novel in the sense that they get their first internalist explanation in terms of some SM while all previous SMs treated them as an irrational leftover for psycho-social explanation. In terms of his own criteria for competing SRPs, Lakatos bids us accept that SRP which is progressive with respect to its rivals. In the case of historiographical SRPs, we are to accept that SRP which is progressive in the sense that it renders rational (by its own lights) judgements of the scientific élite that no other historiographical SRP was able to render rational (by their own lights). As Lakatos puts it: '*progress in the theory of scientific rationality is marked by discoveries of novel historical facts, by the reconstruction of a growing bulk of value-impregnated history as rational*' (Lakatos 1978, p. 133).

What Lakatos has done here is to take his own theory of SM, which he applies to ordinary sciences, and elevate it to a meta-methodology whereby he can assess rival SMs according to whether or not they show empirical progress by rendering rational more of the historical changes in science. Once more an important role is played by the basic value-judgements of the scientific élite about scientific change. The history of science that is 'rationally reconstructed' according to some SM is said to be 'value-impregnated' because it is based in the élite's value-judgements. Lakatos recognises that no SM will capture all such value-judgements, adding that while this would be a problem for Popper's SM elevated to meta-methodology (it gets falsified by its own meta-criterion), it is not a problem for his own SM of SRPs elevated to meta-methodology (since it allows for progress in an ocean of anomalies): '*rational reconstructions remain for ever submerged in an ocean of anomalies. These anomalies will eventually have to be explained either by some better rational reconstruction or by some "external" empirical theory*' (Lakatos 1978, p. 134).

Finally the role played by the basic value-judgements of the scientific élite explains why Lakatos' meta-method is 'quasi-empirical' and not *a priori* or conventionalist. The history of such judgements provides an empirical basis against which SMs are to be tested, using whatever meta-criterion of test. It has been assumed that such judgements are unproblematic and readily available. But are they? Lakatos cannot relegate all sociological considerations in science to external factors; some sociology is needed to survey the scientific élite to discover what are their judgements about particular moves in the game of science. Nor should it be assumed that there would be unanimity amongst the élite; a socio-logical survey might show that the views of scientists ranged from strong con-sensus to equal division for and against. Nor is it clear what the lowest threshold for agreement might be; if there is less than, say, 80% agreement then some of the value-judgements might not be useable to decide important methodological matters. In addition allowance might have to be made if the scientific community were to change its views about some episode over time. Moreover, how is the scientific élite to be determined? We should not admit that all scientists can make value-judgements about all moves in all the sciences, including the many sciences with which they are unacquainted. Nor should we allow the élite to be chosen by the fact that they are good scientists, for what counts as 'good' could well turn on whether they are appropriate users of some SM – the very matter over which the judgements of a scientific élite have been invoked in order to make adjudi-cations. Nor should any methodology be invoked by the élite as the grounds on which their judgements are made on pain of similar circularity. Presumably philosophers are to be excluded from the ranks of the scientific élite since most of them are untutored in the ways of science; if this is the case then Popper's own list of heroic and disreputable science is hardly to be given the significance he gives it as a test basis.

The earlier Laudan of *Progress and Its Problems* also assumed that we have '*our preferred pre-analytic intuitions about scientific rationality*' (Laudan 1977, p. 160) based on episodes in the history of science which are much more firm than any intuitions we have about the theories of SM that embody such rationality. But he later abandoned any such role for pre-analytic intuitions for the above reasons, and others such as the following. SMs are deprived of any substantive critical role; this is to be played by the intuitions which no SM should overturn. Nor is it clear that the intuitions will single out some preferred SM above all others; it might well be the case that given all the widely acceptable intuitions, SMs as disparate as those advocated by Bayesians, Popperians, Lakatosians and Kuhnians might fit them equally well. That is, in Quinean fashion our intuitions, which play a role similar to that of observations in science, might underdetermine SMs in that two or more SMs are tied for best fit with the intuitions (Laudan 1986). In a subsequent section we will look at the different view developed by the later Laudan, normative naturalism, which makes no appeal to intuitions.

The idea that we can pit intuitions about particular cases against rules (or principles, of some SM) is the core idea behind another meta-methodological approach known as 'Reflective Equilibrium' (RE), in either a more narrow or a broader version depending on how much is allowed into the equilibrating process. Nelson Goodman originally expressed the idea of RE as a meta-methodological principle for adjudicating between the particular inferences we make and the deductive rules we adopt, saying: '... rules and particular inferences alike are justified by being brought into agreement with each other. *A rule is amended if it yields an inference we are unwilling to accept; an inference is rejected if it violates a rule we are unwilling to amend*' (Goodman 1965, p. 64). Goodman's meta-methodological principle RE can be extended from the case of deductive to inductive logic and to other areas, its best known extension being to the case of ethics in which RE 'brings into agreement' moral principles and intuitions about particular moral examples.

How might RE adjudicate between rival SMs? It is not clear that it could do a better job than the meta-criteria proposed either by Popper or Lakatos. Suppose we are given a set of intuitions I (about moves in the game of science based, for example, on the judgements of some scientific élite) and a set of principles (rules) P of an SM. First, there is the unclear notion of what might be meant by I and P being 'brought into agreement'. In the case of deductive logic the notion of logical form plays a crucial role; a necessary condition for principles and particular arguments to be 'brought into agreement' is that the same logical form needs to be found in both. In the case of methodology 'bringing into agreement' can only mean that an intuition I_i about a particular move in the game of science is such that a principle P_j of an SM is able to render that move rational in the sense of giving an internalist explanation of that move.

In addition the notion of 'bringing into agreement' is said to have justificatory force. Thus if I_i and P_j were to be brought into agreement with one another then they would both receive justificatory support. This is a significant additional claim since justification does not flow merely from being brought into agreement. In the case of deductive logic, if all our unacceptable intuitions and unacceptable (invalid) rules were to be brought into agreement with one another through systematising all invalid arguments, it could not be said that they would thereby be justified. Further RE might be made more realistic by giving different weightings to intuitions and principles; this would affect whether or not they can be brought into agreement. Thus given the same weightings to I_i and P_j it might not be possible to bring them into agreement and, say, I_i is discarded. But such importance might be attached to preserving I_i in any theory or SM that a weighted I_i is accepted along with P_j which perhaps does not do too good a job of explaining the episode about which such a strong intuition has been expressed.

If such an approach can be made to work, at best it will suit only one SM at a time; it cannot compare SMs. To compare them, one needs to investigate, first, the extent to which each SM is able to bring its P's and the I's 'into agreement' and, second, to be able to compare SMs according to the extent they are able to do this. That is, there needs to be some consideration about principles which pass the RE test and intuitions which are to be dropped (or vice versa) – and the significance, if any, to be attached to which intuitions are dropped or maintained. Thus the meta-methodological criterion of RE needs supplementing with a further principle about how well each SM produces its 'agreements'; that is, we need a meta-criterion which is not merely RE, but *best overall* RE. And it will be the notion of which RE does *best overall*, and not RE *simpliciter*, that will do most of the work in determining which SM we should adopt.

In the critical literature (Stich 1990, chapter 4 and Siegel 1992), other issues are raised about RE pertinent to their application to methodology. The first is whether passing the RE test is *constitutive* of the 'correctness', 'validity' or justification of the principles of method, or merely *good evidence* for their justification or 'validity' or 'correctness'. The second concerns the status of RE (either supplemented or not), viz., whether it is some kind of conceptual truth which can be known *a priori*, or whether it is non-conceptual and knowable only *a posteriori*, or whether it is a non-conceptual truth which is necessary but is knowable only *a posteriori*. Such questions are important given the schema set out in the table in section 2 for classifying meta-theories used to adjudicate between SMs. If we follow the critique provided by Stich and Siegel, we are led to the negative conclusion that, whatever the status of RE, it is not a principle that should be adopted to adjudicate between even principles of logic, as originally proposed by Goodman. Thus the prospects of RE as a meta-methodology are not promising, even though some meta-methods do adopt a way of proceeding that is reminiscent of aspects of RE.

8. KUHN'S THEORY OF WEIGHTED VALUES

Those who view Kuhn as holding either an irrationalist or anti-methodology stance, or endorsing a paradigm-relative account of method, can find passages in the Kuhn of 1962 that support these views. Using a political metaphor to describe scientific revolutions Kuhn says of scientists working in different paradigms that 'because they acknowledge no supra-institutional framework for the adjudication of revolutionary difference, the parties to a revolutionary conflict must finally resort to the techniques of mass persuasion, often including force' (Kuhn 1962, p. 93). Continuing the metaphor, there is also a suggestion that the methods of evaluation in normal science do not carry over to the evaluation of rival paradigms:

> Like the choice between competing political institutions, that between competing paradigms proves to be a choice between incompatible modes of community life. Because it has that character, the choice is not and cannot be determined merely by the evaluative procedures characteristic of normal science, for these depend in part upon a particular paradigm, and that paradigm is at issue. When paradigms enter, as they must, into a debate about paradigm choice, their role is necessarily circular. Each group uses its own paradigm to argue in that paradigm's defense. (Kuhn 1970, p. 94)

Later he speaks of paradigms 'as the source of the methods ... for a scientific community' (*ibid.*, p. 103). These and other passages tell us that methodological principles might hold within a paradigm but that there are no paradigm transcendent principles available. Thus it would appear that Lakatos was right to say of paradigm change: '*in Kuhn's view scientific revolution is irrational, a matter for mob psychology*' (Lakatos 1978, p. 91).

By the time he came to write the 'Postscript' for the 1970 edition of his book, Kuhn effectively abandoned talk of paradigms in favour of talk of exemplars and disciplinary matrices. Values are one of the elements of a disciplinary matrix; and contrary to the impression given above, they are 'widely shared among different communities' (*ibid.*, p. 184). That is, scientists in different communities, and so working in different 'paradigms', value theories because of their following features: they yield predictions (which should be accurate and quantitative rather than qualitative); they permit puzzle-formation and solution; they are simple; they are self-consistent; they are plausible (i.e., are compatible with other theories currently deployed); they are socially useful. These quite traditional notions turn out to be Kuhn's paradigm transcendent criteria of theory choice.

In a 1977 paper 'Objectivity, Value Judgement, and Theory Choice' (Kuhn 1977, chapter 13) re-endorses these values and adds to them: scope (theories ought to apply to new areas beyond those they were designed to explain), and fruitfulness (theories introduce scientists to hitherto unknown facts). Kuhn initially thinks of these as rules of method in the sense introduced in section 3. But owing to the imprecision which can attach to their expression as rules, and the fact that they are open to rival interpretations and can be ambiguous in application or can be fulfilled in different ways, Kuhn prefers to think of them as values to which we could give our general assent.[24] Thus Kuhn adopts a methodology which avoids talk of the rules which we ought to follow as means to realise values, and instead focuses on the values we do, or ought to, adopt in our choice of theories.

Kuhn's list of values does not mention several other important values that have been endorsed by other methodologists. Thus Kuhn does not mention that inductivists and Bayesians put high store on high degree of support of hypotheses by evidence. However in the paper by Worrall in this volume it is argued that it is possible to reconcile this apparent omission with Kuhn's views; the Kuhnian values of scope and fruitfulness are linked to the notion of degree of support, even though Kuhn does not spell this link out. Again constructive empiricists put high value on theories which are empirically adequate; in contrast realists wish to go further and value not only this but also truth, or increased verisimilitude, about non-observational claims. Given what Kuhn says elsewhere[25] we may view him as not endorsing the realists' value of truth, though the constructivists' value of empirical adequacy is one he could adopt. Finally some methodologists' would downplay some of the values Kuhn endorses, such as external consistency or social utility.

The position of Kuhn on methodology after the first edition of *Structure* yields the following picture of a model of weighted values. (1) There is a set of values which can vary over time, and can vary from methodologist to methodologist. A sub-set of these values could comprise a cluster of central values which hold for most sciences and throughout most of their history. Kuhn's own model is more akin to the latter with his set of values forming a tight cluster without containing a larger set. (2) These values are used to guide and inform theory choice across the sciences and within the history of any one science, including its alleged 'paradigm' changes. That is, the cluster of values are science and paradigm transcendent. (3) The model may be either descriptive or normative. Kuhn does not make it clear whether his model is to be understood as a description of how scientists do in fact make their choices, or whether it is to be understood normatively in that it tells us how we ought to make choices. If the latter then there is a need for a justification of the norms it embodies. (4) Kuhn says that the values may be imprecise and be applied by different scientists in different ways. While this is not the case for the value inconsistency (there are fairly precise criteria for internal and external inconsistency for any theory), or for any given degree of accuracy of predictions, some values do exhibit imprecision. Thus accuracy could differ in the required degree. Simplicity might be taken in different ways (simplicity in equations versus simplicity in *ad hoc* assumptions) so that different aspects of a theory might be deemed simple; or the notion of simplicity itself might be taken in different ways or in different degrees. Such imprecision in the interpretation of values can, however, be readily overcome by precisification so that there need not be the wide divergence over the interpretation of values that Kuhn alleges. (5) Different scientists do, as a matter of fact, give different weightings to each of the values.

Following from (4) and (5) there are two aspects in theory choice – an objective aspect in shared values and a subjective aspect in idiosyncratic weightings of values (and interpretation where this arises). In the light of this, Kuhn claims that there is no general 'algorithm' for theory choice – though there is hardly any methodologist who has required that methodological principles should be algorithmic. This allows that different scientists can reach different conclusions about what theory they should choose. First, they might not share the same values; but where they do share the same values they might interpret them differently or give them different weightings. Shared values (with the same interpretation) and shared weightings of these values will be sufficient for sameness of judgement within a community of scientists. However this might not be necessary; it might be possible for scientists to make the same theory choices yet to have adopted different values and/or have given them different weightings. Thus there is the possibility that consensus might be a serendipitous outcome despite lack of shared values and different weightings. However it is more likely that, where values and weightings are not shared, different theory choices will be made, and there is no consensus.

Whether scientists do or do not make theory choices according to Kuhn's model is a factual question to answer. But what does the model say about what we ought to do, and what is its normative/rational basis? In particular why, if T_1 exemplifies some Kuhnian value(s) while T_2 does not, should we adopt T_1 rather than T_2? Kuhn's answer to the last meta-methodological question is often disappointingly social and/or 'intuitionistic' in character. In his 1977 paper Kuhn refers us to his earlier book saying: 'In the absence of criteria able to dictate the choice of each individual, I argued, we do well to trust the collective judgements of scientists trained in this way. "What better criterion could there be", I asked rhetorically, "than the decision of the scientific group"' (Kuhn 1977, pp. 320, 321). As to why we ought to follow the model, Kuhn makes a convenient is-ought leap when he says in reply to a query from Feyerabend: 'scientists behave in the following ways; those modes of behaviour have (here theory enters) the following essential functions; in the absence of an alternative mode that would serve similar functions, scientists should behave essentially as they do if their concern is to improve scientific knowledge' (Lakatos and Musgrave (eds.) 1970, p. 237). The argument is not entirely clear, but it appears to be inductive: in the past certain modes of behaviour (e.g., adopting Kuhn's model of theory choice) have improved scientific knowledge; so in the future one ought to adopt the same modes of behaviour if one wants to improve scientific knowledge. As will be seen a similar meta-inductive argument is at the heart of the meta-methodology of normative naturalism; so Kuhn's model needs to be assessed in the same way advocated by that meta-method.

More recent comments from Kuhn (in a paper entitled 'Rationality and Theory Choice') on the status of his methodology arise in a 1983 symposium on 'The Philosophy of Carl G. Hempel' with Salmon and Hempel. In subsequent reflection on that symposium, Salmon argues that it is possible to reconstrue the features of Kuhn's model of weighted values in terms of subjective Bayesianism (Salmon 1990; see also the paper by Worrall in this volume and Earman 1992, chapter 8 for a further attempt to incorporate Kuhn's model into Bayesianism). Bayes' Theorem is able to account for a large number of our central methodo- logical principles, including accuracy, fruitfulness, scope, and so on (but not social utility unless set in a decision-theoretic context). If Salmon's project in which 'Tom Kuhn meets Tom Bayes' is able to account for the theory choices of Kuhn's model, then the independent status of the model is undercut as it is incorporated into a more wide ranging theory of method.

In his symposium paper Kuhn addresses a point that Hempel had made in an earlier paper about his position, viz., that Kuhnian values are goals at which science aims, and not means to some goal such as puzzle-solving. Here we have followed both Kuhn and Hempel in taking puzzle-solving, accuracy, simplicity, etc. to be values (ends) rather than rules (means), though aspects of them as means do inform their function as ends when employed in theory choice. Given that theories are judged by the values they exemplify, Kuhn takes up a further point that Hempel makes, viz., that rationality in science is achieved through adopting those theories that satisfy these values *better*. Hempel thinks that this criterion of rational justification is near-trivial. However Kuhn turns Hempel's near-triviality into a virtue by proposing that the criterion is analytic, thereby adopting as his meta-methodology a theory of analyticity concerning the term 'science'. In developing his views in this paper Kuhn tells us that he is 'venturing into what is for me new territory' (Kuhn 1983, p. 565). So we can take it that the meta-methodological justification developed here is not one that Kuhn had had in mind before.

Kuhn's account of analyticity is based on what he calls 'local holism'. This is the view that the terms of any science cannot be learned singly but must be learned in a cluster; the terms have associated with them generalisations that must be mastered in the learning process, and the cluster of terms form contrasts with one another that can only be grasped as a whole. If 'learning' is understood as 'understanding the meaning', then analyticity becomes an important part of the doctrine of 'local holism' for the central terms of each sufficiently broad scientific theory. In Kuhn's view the doctrine applies not only to specific theories such as Newtonian Mechanics with its terms 'mass' and 'force' which must be learned together holistically. It also applies to quite broad notions signified by the terms 'art', 'medicine', 'law', 'philosophy', 'theology', and so on; the central terms associated with these notions must also be learned holistically. Importantly 'science' is another such broad notion to be learned holistically since 'science' is in part to be understood in contrast to these terms.

Kuhn recognises that not every science we adopt should possess every value since the values are not necessary and sufficient conditions for theory choice; rather they form a cluster associated with the local holism of the term 'science'. But what he does insist is that claims such as 'the science X is *less* accurate than the non-science Y' is a violation of local holism in that 'statements of that sort place the person who makes them outside of his or her language community' (Kuhn 1983, p. 569). For Kuhn, Y's being more accurate is just one of the things that the local holism of the word 'science' makes Y scientific; Y cannot be non-scientific. For Kuhn, Hempel's near-triviality is not breached because a convention has been violated (this would be the position of the early Popper); nor is a tautology negated. Rather 'what is being set aside is the empirically derived taxonomy of disciplines' (*loc. cit.*) that are associated with terms like 'science'. Like many claims based on an appeal to analyticity, meaning or taxonomic principles, one might feel that the later Kuhn has indulged in theft over honest toil. However in linking his model of weighted values to the alleged local holism of the term 'science', Kuhn comes as close as any to adopting the meta-methodological stance that his theory of method has an analytic justification for its rationality. In this respect Kuhn's position has close affinities with that of Strawson (1952, chapter 9, part II) who tried to justify the rationality of induction in much the same way.

Finally, Kuhn's later position gives no comfort to those sociologists who wish to appeal exclusively to his book:

> The publication of Thomas Kuhn's *The Structure of Scientific Revolutions* in 1962 pointed the way toward the integrated study of history, philosophy and the sociology of science (including technology) known today as science and technology studies (STS). ... It alerted STS practitioners to the mystified ways in which philosophers talked about science, which made the production of knowledge seem qualitatively different from other social practices. In the wake of STS research, philosophical words such as *truth*, *rationality*, *objectivity*, and even *method* are increasingly placed in scare quotes when referring to science – not only by STS practitioners, but also by scientists themselves and the public at large.' (Brante *et al. ibid.*, p. ix)

Kuhn's attempted meta-methodological justification, along with the values he endorses, place his later work firmly within traditional philosophical concerns about scientific method. As many of his critics have noted, his later work is a retreat from many of the claims of his 1962 book.[26]

9. FEYERABEND'S CRITICISM OF METHODOLOGY

For a person who is famous for alleging that the only universal principle of rationality is 'anything goes', or giving his books titles such as *Against Method*, or

Farewell to Reason, it might come as a surprise to some to find that Feyerabend, in his autobiography completed just before his death, makes the following claim on behalf of rationality: 'I never "denigrated reason", whatever that is, only some petrified and tyrannical version of it' (Feyerabend 1995, p. 134). Or, 'science is not "irrational"; every single step can be accounted for (and is now being accounted for by historians ...). These steps, however, taken together, rarely form an overarching pattern that agrees with universal principles, and the cases that do support such principles are no more fundamental than the rest' (*ibid.*, p. 91). Inspecting his earlier career one will find that Feyerabend even proposed some principles of method, such as the Principle of Proliferation, the aim of which is 'maximum testability of our knowledge' and its associated rule is '*Invent, and elaborate theories which are inconsistent with the accepted point of view, even if the latter should happen to be highly confirmed and generally accepted*' (Feyerabend 1981, p. 105).[27]

Feyerabend opposed the following view of scientific method (call it 'Rationalism' with a capital 'R') espoused by Popper and Lakatos (in fact his criticism of Rationalist methodology is almost entirely narrrowly focused upon the principles proposed by the 'Popperian school' and hardly any others):

(I) There is a universal principle (or unified set of principles) of scientific method/rationality R such that for all moves in the game of science as it has historically been played out in all the sciences, the move is an instance of R and R rationally justifies the move.

His opposed position can be easily expressed by shifting the order of the quantifiers in (I) from 'there exists – all' to 'all – there exists', and then adding a qualification about the nature of the principles (call this 'rationalism' with a little 'r').

(IIa) For all moves in the game of science as it has been historically played out in all the sciences, there is some principle (or set of principles) of scientific method/rationality R, such that the move is an instance of R and R rationally justifies the move.[28]

This leaves open two extreme possibilities: that a different rule is needed for each of the moves, or the remote possibility that there is still one universal rule which covers all the moves. Feyerabend's position is close to the first alternative. There might be rules which cover a few moves, but the rules are so contextually limited that they will not apply to a great number of moves and other rules will have to be invoked. Feyerabend also has an account of the nature of these rules:

(IIb) Rules of method are highly sensitive to their context of use and outside these contexts have no application; each rule has a quite restricted domain of application and is defeasible, so that other rules (similarly restricted to their own domain and equally defeasible) will have to apply outside that rule's domain. (Feyerabend often refers to these as rules of thumb.)

Sometimes Feyerabend adopts quite traditional cognitive values for science, such as the rather Popperian Proliferation Principle 'maximum testability of knowledge'. But Feyerabend is also a social critic of science who asks 'what is so great about science with such aims?', and then argues that for some of the moves in the game of science we would have been better off if they had never occurred. For Feyerabend the whole game of science may not be worth playing because science might make a monster of us (Feyerabend 1975, pp. 174, 175). There are better things in life than science, such as acting in plays, singing opera or being a dadaist, and such choices are highly contextual. In sum, for Feyerabend there are a number of broad aspects of science to consider, each of which has its respective values, and these in turn are to be associated with contextual and defeasible rules. By shifting focus from one aspect of science to another, Feyerabend is able to abandon contextual and defeasible methodological rules which are allegedly designed to promote epistemic progress in science, in favour of other goals, for example, dialectical, humanitarian, aesthetic and moral goals, which have little to do with scientific method as standardly understood. It is in respect of these other goals that Feyerabend liked to ask 'What is so great about science?' and then answer by saying 'science is one ideology among many' and can be a threat to the democratic life (Feyerabend 1978, pp. 73, 106).

In taking the position he does Feyerabend is not beyond the pale of rationality; but he is beyond the pale of Rationality. There is much textual support, only a little of which can be cited here, for the view that Feyerabend is not a Rationalist but a rationalist. His opponents are Rationalists who advocate a unique set of universal rules to be applied to all sciences at all times. On occasions he refers to such Rationalising methodologists as 'idealists', or as 'rationalists' (Feyerabend's little 'r' must be read with big 'R' connotations), thereby creating the impression that he must be an irrationalist. But such talk masks Feyerabend's real position.

Direct support for claims (IIa) and (IIb) comes from Feyerabend's response to critics of the 1975 *Against Method* in his restatement of his position in the 1978 *Science in a Free Society*. Feyerabend contrasts two methodological positions: *naive anarchism*, with which his own position should not be confused (presumably Feyerabend is a sophisticated anarchist); and *idealism* in which there are absolute rules – but they are conditional with complex antecedents which spell out the conditions of their application within the context of some universal Rationalist methodology. Of naive anarchism he says:

> A naive anarchist says (a) that both absolute rules and context dependent rules have their limits and infers (b) that all rules and standards are worthless and should be given up. Most reviewers regarded me as a naive anarchist in this sense overlooking the many passages where I show how certain procedures *aided* scientists in their research. For in my studies of Galileo, of Brownian motion, of the Presocratics I not only try to show the *failures* of familiar standards, I also try to show what not so familiar procedures did actually *succeed* [Sic]. I agree with (a) but I do not agree with (b). I argue that all rules have their limits and that there is no comprehensive 'rationality', I do not argue that we should proceed without rules and standards. I also argue for a contextual account but again the contextual rules are not to *replace* the absolute rules, they are to *supplement* them. (Feyerabend 1978, p. 32)

Thesis (b) marks the crucial difference between naive anarchism and Feyerabend's own position; moreover the denial of (b) shows that Feyerabend cannot be an arch irrationalist. There *are* rules worth adopting. Oddly enough, universal rules are not to be replaced but are to be supplemented. What this means is unclear; but it might be understood in the following way. Consider as an example of a universal rule Popper's 'do not adopt *ad hoc* hypotheses'. For Feyerabend this is not to be understood as a universal ban which applies to all sciences and in all circumstances come what may. Sometimes adopting *ad hoc* hypotheses will realise our aims better than not adopting them. Feyerabend seems to suggest that alongside this universal rule are other rules (equally open to supplementation one supposes) about its application. The task of these other rules will be to tell us about the occasions when we should, or should not, adopt this Popperian rule. What Feyerabend needs is the notion of a defeasible rule, but defeasibility is not something he ever discusses. If rules are defeasible, then universalising 'idealists' and Rationalists will not be able to apply rules regardless of their situation. Viewed in this light the passage cited above supports the position outlined in the two parts of (II), as do other similar passages in his writings (Feyerabend 1975, p. 32 and chapter 15; Feyerabend 1978, pp. 98, 99, 163, 164).

In the passages surrounding the last quotation Feyerabend attempts to distinguish between a modified idealism, in which universal rules are said to be conditional and have antecedents which specify the conditions of their application, and his own view in which rules are contextual and defeasible. Presumably the difference is that for Rationalist 'idealists' the conditions of application of the rules are spelled out in the fully specific antecedents of conditional rules. But in Feyerabend's view such conditions cannot be fully set out in some antecedent in advance of all possible applications of the hypothetical rule; at best such conditions are open-ended and never fully specifiable. So Feyerabend opts for a notion of a rule which is not conditional in form but categorical, and is best understood as contextual and defeasible. So even if rules appear to be universal, as in 'do not adopt *ad hoc* hypotheses', there will always be vagueness and imprecision concerning their application. There is also no mention of the conditions under which they can be employed or a time limit imposed on their application; presumably this task is to be left to supplementary rules.

Does Feyerabend adopt as his one and only methodological principle '*Anything goes*' (Feyerabend 1975, p. 28)? No:

> As for the slogan 'anything goes', which certain critics have attributed to me and then attacked: the slogan is not mine and it was not meant to summarise the case studies of *Against Method* . . . (Feyerabend 1987, p. 283) [it] is not a 'principle' I defend, it is a 'principle' forced upon a rationalist who loves principles but who also takes science seriously'. (*ibid.*, p. 284)

Once again Feyerabend uses the term 'rationalist' to name his opponents; but from this it does not follow that his own position is 'irrationalist'. Instead his view is that one cannot have both the complexities of the actual history of science and a universal methodology of the sort loved by those he variously dubs as 'Rationalists' or 'Idealists'.

As he explains in several passages of *Science in a Free Society* (Feyerabend 1978, pp. 32, 39, 40, 188): *'"anything goes" does not express any conviction of mine, it is a jocular summary of the predicament of the rationalist'* (*ibid.*, 188).[29] But the joke has backfired and has been costly in misleading many about Feyerabend's real position. If the Rationalists within the critical tradition want universal rules of method then, given that according to Feyerabend all such rules have counter-examples outside their context of application, the only universal rule left is the 'empty, useless and pretty ridiculous' – 'anything goes' (*loc. cit.*). But this is hardly convincing since the Rationalist need not take up Feyerabend's invitation to adopt 'anything goes'. Any Rationalist will see that from 'anything goes' it follows (by instantiation) that every universal rule of method also 'goes'; but if the Rationalist also accepts Feyerabend's claim that these have counter-examples and are to be rejected as universally applicable then, by *Modus Tollens*, the Rationalists can infer that 'anything goes' cannot be an acceptable rule of Rationalist method – even if jocularly imposed by Feyerabend on serious universalising Rationalists. 'Anything goes' does not mean what it appears to say; it is not even a principle of method that Feyerabend endorses.

Given that one can adopt defeasible principles and remain a rationalist about method, Feyerabend does not appear to be the opponent of theories of scientific method that he is often made out to be, or says that he is. Granted such Feyerabendian principles, what does he say about their justification? There are two approaches. The first would be to take some principle that Feyerabend advocates (such as the Principle of Proliferation or rules of counter-induction) or some principle he criticises (such as the alleged principle of consistency or Popper's anti-*ad hoc* rule) and attempt to evaluate them either on logico-epistemological grounds or on the historical record of the decision context of various scientists. But the latter would take us into a long excursion through episodes in the history of science, and the former has to some extent been carried out.[30] Instead we will look at Feyerabend's meta-methodological considerations in justification for his views on SMs.

If we are to attribute to Feyerabend a meta-methodology concerning his defeasible rules, then it veers between that of a Protagorean relativist and that of a dialectical interactionist in which principles and practice inform one another. In setting out his position he adopts Popper's term 'tradition'[31] to refer not only to mythical and scientific systems of belief, but also to traditions including those of religion, the theatre, music, poetry, and so on. He also speaks of the rational tradition; but unlike Popper he does not privilege it by claiming some special 'second-order' status for it. For Feyerabend all traditions, including the critical or rational tradition, are on an equal par. In resisting the idea that there is a special status to be conferred upon the rules that comprise the tradition, Feyerabend adopts a Protagorean relativism about traditions – at least in the 1978 *Science in a Free Society*. About traditions he makes three claims:

(i) *Traditions are neither good nor bad, they simply are. ....* rationality is not an arbiter of traditions, it is itself a tradition or an aspect of a tradition. ...

(ii) *A tradition assumes desirable or undesirable properties only when compared with some tradition. ...*

(iii) *(i) and (ii) imply a relativism of precisely the kind that seems to have been defended by Protagoras.* (Feyerabend 1978, p. 27)

For at least the Feyerabend of *Science in a Free Society*, there is a rational tradition; and it has contextual defeasible rules which can be used to evaluate claims in other traditions. (We can take the national tradition to include not only the principles of any SM but also any meta-methodology which attempts to justify any SM.) But given his Protagorean relativism about traditions, all such evaluations are from *within* a tradition. There is no absolute tradition, encapsulated in some meta-methodology, which stands *outside* all other traditions and from which we can evaluate them. In this sense no tradition is an absolute 'arbiter' of any other. What does this mean for the contextual defeasible rules of method that Feyerabend endorses? We take this to mean that such rules of method have no further justification other than that they are what we have simply adopted as part of our critical tradition. Their truth, validity or correctness is at best relative to a tradition; there is no further meta-methodological account of their status to be given by appealing to some absolute or privileged tradition of Rationality. It is this relativism that has led some to claim that Feyerabend, even if he admits there are defeasible rules of method, is at heart an irrationalist.

If Feyerabend really adopts a Protagorean relativism about rules of the sort Plato describes in the *Theaetetus*, then at best we can say that there are rules R-relative-to-tradition-T, and rules R*-relative-to-tradition-T*, and not merely rules R and R* which might come into logical relation with one another. Such a version of relativism undercuts the very possibility of rules ever being assessed with respect to one another. But this is often something Feyerabend requires we do. This suggests that Feyerabend is not really a relativist but a pluralist about rules and the traditions they embody (and pluralism need not entail any relativism). The running together of these two notions is evident in the following passage: 'Protagorean relativism is *reasonable* because it pays attention to the pluralism of traditions and values' (*ibid.*, p. 28). What is still excluded by this stance is any attempt to invoke meta-methodology to give an *a priori* or even an empirical justification of his defeasible rules of method. But the pluralism does make possible the critical 'rubbing together' of different traditions, something that Feyerabend would endorse given his principle of proliferation. And it does make possible the following more dialectical view of the interaction between traditions, rules and practices.

There are remnants of the positions of the later Popper and Lakatos with their appeal to the intuitions of a scientific élite in Feyerabend's talk of the interaction between reason and practice, of which he distinguishes three aspects. First, he acknowledges that reason can be an independent authority which guides our practices in science – a position he dubs 'idealistic'. But also 'reason receives both its content and its authority from practice' (*ibid.*, p. 24) – a position he dubs 'naturalism'. Though he does not say how reason gets its authority, his position is one in which a strong role is given to intuitions about good practice; this is reminiscent of the later Popper, Lakatos and the meta-method of reflective equilibrium of section 7. But both naturalism and idealism have their difficulties, says Feyerabend. Idealists have a problem in that too ideal a view of rationality might cease to have any application in our world. And naturalists have a problem in that their practices can decline because they fail to be responsive to new situations and need to critically re-evaluate their practice. He then canvasses the suggestion 'that reason and practice are not two different kinds of entity but *parts of a single dialectical process*' (*ibid.*, p. 25).

But even the talk of reason and practice being separate 'parts' of a single process draws a misleading distinction, and so he concludes: '*What is called 'reason' and 'practice' are therefore two different types of practice*' (*ibid.*, p. 26). The difference between the two *types* of 'practice' is that one is formal, abstract and simple, while the other is non-formal, particular and submerged in complexity. Feyerabend recognises that the conflict between these two types of practices (or 'agencies' as he goes on to call them) recapitulates 'all the "problems of rationality" that have provided philosophers with intellectual . . . nourishment ever since the "Rise of Rationalism in the West"' (*ibid.*, pp. 26, 27). As true as this may be at some level of abstraction, Feyerabend's shift to a position of dialectical interactionism with its additional plea for a Principle of Proliferation with respect to interacting traditions (including those of theories of SM), does have the characteristics of an appeal to some meta-methodological theory. But it tells us nothing more than what we had learned from the intuitionistic approaches of section 7, except that the task of bringing our practices (in science and elsewhere) into line with our theories of those practices (i.e., our theories of SMs) might be harder than we thought. Looked at this way, we can resist the temptation to go relativist by viewing the activity of bringing the rules of 'reason' and the particularity of 'practice' into accord with one another as yet just one more activity on a par with any other activity.

10. NIHILISM ABOUT METHOD: THE SOCIOLOGICAL TURN

Nihilism is the view that there is no legitimation possible for any SM, and in particular there is nothing to be found at the meta-methodological level. Reasons for this are various. For those who take Feyerabend to be opposed to all methodology and to claim that 'anything goes', then he is a nihilist about method. But as we have seen, that is not his position. However he comes close to it when, in one of his moods, he advocates a Protagorean relativism of methods to traditions and epochs of science, rather than a pluralism of methods. Similarly we have seen that at one stage Kuhn claimed that methods are paradigm-relative. But the status of their relativisms remains obscure, unhelpful and in the long run wrong as both recognise in their different ways. Nihilism come in other forms. Thus Lyotard on postmodernism in science:

> But to the extent that science does not restrict itself to stating useful regularities and seeks truth, it is obliged to legitimate the rules of its own game. It then produces a discourse of legitimation with respect to its own status, a discourse called philosophy. I will use the term *modern* to designate any science that legitimates itself with reference to a metadiscourse of this kind making explicit appeal to some metanarrative such as the dialectics of Spirit, the hermeneutics of meaning, the emancipation of the rational or working subject, or the creation of wealth. . . . I define *postmodern* as incredulity towards metanarratives. . . . To the obsolescence of the metanarrative apparatus of legitimation corresponds, most notably, the crisis of metaphysical philosophy and of the university institution which in the past relied on it'. (Lyotard 1984, pp. xxiii, xxiv)

Here talk of 'metanarratives' is akin to what we have called 'meta-methodology' (though Lyotard uses the term 'narrative' in a much wider sense than to refer to only second level SMs or to the first level sciences themselves). But it should be noted that no meta-methodologist we have considered has adopted the first three of the four metanarratives Lyotard mentions in order to legitimate methods in science, and in turn the sciences themselves. What a 'hermeneutics of meaning' might do for any attempt to justify any theory of SM is obscure; in any case it is a topic outside the scope of this 'Selective Survey'.

What of Lyotard's opening suggestion that issues of legitimation arise when we go beyond stating useful generalities and aim for truth? Suppose we were to eschew truth about science's regularities and stay with useful regularities. One aspect of their usefulness must be that generalities remain correct for the next case to which they allegedly apply – that is, their usefulness turns on methodological principles associated with inductive inference in science. But ever since Hume the meta-methodological matter of 'legitimating' inductive inferences has been with us without the advocacy of postmodernism. Moreover, even for the incredulous postmodernist, we need to be given reason to believe, or a 'proof', that there is no legitimation for principles of method (as in the case of the just mentioned inductive methods, or any other methods). But to show that there is no such 'proof' of legitimation can be just as difficult to establish as that there is a 'proof' of its legitimation (whatever 'proof' is taken to mean here). The upshot is that the postmodernist sceptical nihilist about legitimation must indulge in some meta-methodology, just as sceptics who deny that there is any epistemology must also indulge in some epistemology to establish their case.

Such is the alleged urgency of the 'crisis' of legitimation that even the university as an institution is threatened. Whatever crises universities face, or the various sciences either in themselves or in relation to society, it is doubtful that solutions to the methodological problem of legitimation will either relieve such crises, or deepen them if no solution is found. However the fruits of a belief in the failure of legitimation are all too evident in much of the intellectual life of universities, and elsewhere. In Lyotard's book there is little discussion of the sciences and principles of method, even of the sort given by the methodologists mentioned elsewhere in this 'Survey'. And its general orientation towards a Wittgensteinian conception of language and rules does little to establish that science and its philosophy has moved into a postmodern phase. So we will set Lyotardian postmodernism aside, but note that there might be other reasons for its claims about science.[32] Some postmodernists appeal to Kuhn and his sociological turn, and to Feyerabend's anarchism, to provide arguments for their case. But as we have seen neither eschew methodology completely, and both attempt to find some way of legitimating its claims. However a case can be made for the postmodernist position by appeal to sociological studies of science – to which we now turn.

Though sociologists of science often appeal to Kuhn as a precursor, Kuhn resisted any such alliance. Surprisingly he says of his projected treatment of incommensurability in an unfinished book: 'It is needed, that is, to defend notions like truth and knowledge from, for example, the excess of postmodernist movements like the strong program' (Kuhn 1991, pp. 3–4). And elsewhere he adds

'I am amongst those who have found the claims of the strong program absurd; an example of deconstruction gone mad' (Kuhn 1992, p. 9). Kuhn goes on to speak of the manner in which, according to sociological studies, a community of scientists is said to reach a consensus about scientific belief. Kuhn reports that negotiation plays a central role, but little else: '"the strong program" has been widely understood as claiming that power and interest are all there are. Nature itself, whatever that may be, has seemed to have no part in the development of beliefs about it' (*ibid.*, p. 8). Kuhn's last point is important since for the sociologists what scientists believe is constrained only by the negotiations that take place between themselves and not by any role that nature might play in saying 'yes' or 'no' to such beliefs. So what are the claims of the Strong Programme (SP) in the sociology of scientific knowledge?

The sociology of science before the mid-1970s was conservative in that it did not view the very content of scientific belief as a social causal product, and thus open to sociological investigation, as were the funding of science, or its gender biases, or its hierarchies, social organisation and reward system, and so on. A radical shift was made by a number of people such as David Bloor, who gave SP[33] its name. Though he states its four central tenets succinctly it is far from clear what they mean. The first, the Causality Tenet (CT), says of SP: 'It would be causal, that is, concerned with the conditions which bring about belief or states of knowledge. Naturally there will be other types of causes apart from social ones which will cooperate in bringing about belief' (Bloor 1991, p. 7).

Sociologists do not take the care that philosophers do over the distinction between knowledge and belief. So we will go along with Bloor and understand CT as pertaining to *beliefs* (held on the part of some individual x). Since beliefs can be either true or false, then we can immediately incorporate Bloor's second 'Impartiality Tenet' into CT, viz., '[SP] would be impartial with respect to truth or falsity, rationality and irrationality, success or failure' (*loc. cit.*).[34] Thus CT is quite broad with respect to the beliefs within its scope. We will take it to include all scientific beliefs whether they be laws and theories or particular observation statements. Under this heading we can also include the axioms of a systematic presentation of a theory and/or each of the theorems which flow from them (the axioms and theorems being taken either singly or conjointly). CT could also be taken to encompass each of the SMs and meta-methods (including their norms) that philosophers and scientists have proposed, and have been under examination in this 'Survey'; these, too, are grist for the sociological mill of SP understood in its full generality with respect to belief.

What causes a belief p in the mind of some scientist x? It is x's social condition (call this 'S_x') in cooperation with other conditions of x which are not social (for convenience call these 'N_x'). Thus CT, in its full generality with respect to all beliefs and all persons (including scientists), says:

> CT: for all (scientific) beliefs p, and all scientists (and perhaps others) x, there are social conditions S_x, and there are other non-social conditions N_x, which cooperate together such that (S_x & N_x) cause x's belief that p.

What falls within the scope of the non-social causes N_x? It includes items such as our biology, our cognitive structures which we have inherited through the processes of evolution, our similarly inherited sensory apparatus, and our history of sensory inputs (but not our reports of them). Much of the non-social is constant for most of humanity. In contrast the history of our individual sensory input, as well as our individual (or our group's) social condition, does vary from person to person, or from group to group. Thus though the factor N_x must enter into the causal nexus producing our beliefs, it will not explain, apart from appeal to variable sensory input, the variation in our beliefs. What will do this will be the varying social factor S_x. The list of variable social factors is open-ended, but it includes at least: the language we learn and the way we thereby express our beliefs and report our experiences (SP endorses the view that all observation is theory-laden); the beliefs we acquire through acculturation and education; our social circumstance including the class into which we fall. It is the variation in social factors such as these which is alleged to be the main cause of variation in belief. We are to appeal to these factors in offering causal-explanatory accounts of why, say, x believes that p.

Expressed this way, CT can be viewed as a thesis about the causes of belief within a strongly naturalistic programme encompassing a range of mainly social, but also non-social, factors. It is important to note what SP omits from the list of possible causes of belief; in particular, the norms of reason found in logic or methodology (or more properly when these are believed by x) are not admissible causes of x's other scientific beliefs. Their omission is explicitly required by the third Symmetry Tenet which says: '[SP] would be symmetrical in its style of explanation. The same types of cause would explain, say, true and false beliefs' (*loc. cit.*). Much turns on the *types* of causes that produce belief. Less strong sociological theories than SP would allow both of the following: (1) there are false and irrational beliefs in science that could be explained by appeal to some type of factor, e.g., the scientist's political circumstance; (2) a quite different type of factor, e.g., methodological norms, might be invoked to explain true and rationally held beliefs. The symmetry tenet of SP appears to rule out the second quite plausible type of explanation and require that in both cases the causes be of the *same type*, viz., social causes only. This is another sense in which the SP is strong. Only *social* causes are to be admitted in the explanations of belief and its variation. Non-social factors (apart from each person's history of experience) will, of course, also play a causal role, but this can often be relegated to the background of common causal conditions.

Bloor acknowledges that this is one of the grounds on which his critics have resisted SP, saying: 'The problem running throughout most exchanges over the status of the symmetry requirement lies in the clash between a naturalistic and a non-naturalistic perspective. The symmetry requirement is meant to stop the intrusion of a non-naturalistic notion of reason into the causal story' (*ibid.*, p. 177). It is as if the norms of reason and methodology are a *deus ex machina* which intervene in the causal order to produce beliefs; they are not to be admitted as part of the SP's conception of a naturalistic causal order. However Bloor does allow that 'naturalised norms' may enter into the causal order because we are 'natural' reasoners. What this might mean is that, as part of our evolutionary and cultural inheritance, we have acquired beliefs about norms and these 'natural acquisitions' enter in as non-social, rather than social, factors of the causes of belief. (If our natural reasoning capacity is something that arises socially, as does our acquisition of a natural language unlike our Chomskian deep structures, then it enters as a social cause of belief.) But as is well known from research in cognitive psychology, such naturally acquired beliefs about norms offer no grounds for supposing that we use rules of reasoning correctly; given the extent to which we fail to reason according to *Modus Tollens* or commit the gambler's fallacy, the grounds on which we are 'good' 'natural' reasoners are slim indeed.[35] In the long run SP claims about ourselves as 'natural reasoners' comes to the following about the correctness of norms: norm x is correct if and only if the community assents to x. But we are right to be suspicious of sociological formulations which turn merely on communal assent.[36]

If the social factors are quite external to us (e.g., our class status), how do they manage to causally produce beliefs in our minds in a way which is not overtly behaviouristic? It is unclear how external factors, such as class status, manage to penetrate our minds to cause beliefs in something as remote as science without at least our *awareness* of our class interests entering into the casual chain. This suggests that cognitive factors enter into the causal story and that it is cognitive attitudes to social factors, such as beliefs or interests, which are causally active and not the external social factors themselves. If so, CT has been improperly expressed since what has been left out are cognitive attitudes to social factors. This omission is remedied in what might be called the 'interests' version of CT (ICT).

ICT: for all (scientific) beliefs p, and all scientists x, there are interests of x, I_x, such that I_x cause x's belief that p.

Here we can drop reference to the non-social factors since the causal chain is alleged to run independently of them at the cognitive level from interests to beliefs. Now ICT is *not* to be understood as a thesis about, say, the interests that fund, or do not fund, research into some science. This is a legitimate issue for a sociology of science to investigate; but it has no bearing on any sociology of *scientific 'knowledge'*. Rather ICT is a thesis within the sociology of scientific 'knowledge' and as such concerns *beliefs* and their *causes*. In this respect ICT is not a special case of CT since what does the causing is a cognitive attitude to a social factor, and not the social factor itself.

In sections 1 and 2 matters concerning cognitive and non-cognitive interests in science were raised, and it was asked whether such interests could explain why x believes that T (where 'T' is a scientific theory) better than belief in principles of method could. Suppose T does satisfy those interests, including even cognitive interests such as 'predicts a novel fact'. There it was argued that it was quite serendipitous that having some *interest* in a theory, such as the cognitive interest of predicting novel facts, is causally efficacious in choosing T, which does predict a novel fact, rather than some T^* which predicts no novel fact. As we all know, having an interest in some feature does not guarantee that we can always hit upon a thing with that feature. What is wanted, instead, are some methodological principles which, when we use them, are reliable in choosing those theories which exemplify certain values (such as producing novel facts) in which we have an interest. Methods will serve our interests when we use them to choose theories; but merely having an interest without a methodology is no guide to choice.

There is much critical literature on SP.[37] Only one critical point will be raised here. This is the flawed causal methodology adopted by those advocates of SP who undertake historical studies which allegedly exemplify SP. We will consider Forman's study of the emergence of the belief in acausality[38] in physics in Weimar Germany in the decade after the end of World War I. He says of this widespread belief that he 'must insist on a causal analysis' (Forman 1971, p. 3) thus bringing his analysis into conformity with SP. Now CT maintains that for x's belief in acausality there is some social cause S_x of this belief. The position of the quantifier 'there is' in this formulation of CT shows that the social causes of the same belief can (in some cases must) vary for physicists around the world; the alleged social causes of the belief on the part of Weimar physics will clearly be different from the social causes of the belief of, say, some budding Rutherford in the New Zealand of the time, or elsewhere. Those opposed to SP will argue that even if Weimar physicists acquired their belief in acausality socially, those elsewhere might not. Others might initially get their belief by reading the literature in physics (this is in part a social process of information transmission); but they are quite capable of carrying out their own investigations in the light of some theory of SM to get independent evidence for the belief.

So let us restrict the scope of x in CT to the case of the dozen physicists of Weimar Germany that Forman investigates. We will accept unquestioningly his analysis of the cultural milieu of the time in which society was hostile to science and to its notions of causality and embraced the neo-romantic, even 'existentialist', doctrines in Spengler's book *Decline of the West* in which a causal view of the world was also condemned. However it is not sufficient merely to claim that the dozen physicists lived in a milieu of this sort; this could be the case whether they were aware of it or not. For the milieu to have an effect on the physicists' beliefs we need to add that they were *aware* of doctrines that were embraced in their cultural milieu. So CT needs to be reformulated as a thesis not about some S_x causing x's belief in acausality (how could it?), but the significantly different thesis: x's *awareness* of S_x causes x's belief in acausality.

Being socially aware people, most of the dozen physicists Forman investigates were aware of their cultural milieu and its hostility to notions of causality. So was their awareness merely a mental accompaniment to their belief in acausality – this belief being caused in other ways due to the internal development of physics? Or was their awareness of their milieu's hostility the very cause of their belief in acausality? Advocates of SP have to show the latter. And if they can, the result would be rather shocking. Major beliefs in science are just 'social imagery', to quote the title of Bloor's book. They are not acquired on grounds to do with the sciences themselves and their accompanying principles of scientific method; nor are they acquired on grounds to do with how the world is and what our theory says of it, this being one of Kuhn's complaint about SP.

However what is shocking is the failure of sociologists to employ correctly the methodology of causal analysis to show the following claims. Let 'A' be the physicists' *awareness* of their hostile cultural milieu, 'B' their *belief* in acausality, and P an internalist story based in *physics* about how, on the basis of issues in science alone, acausality came to be believed. Then the following needs to be shown about what went on in the mind of each physicist: (1) A accompanies B; (2) P accompanies B; (3) P does not cause B; (4) A causes B. Now (1) and (2) can hold, as Forman shows us; but this does not show that one of (3) or (4) hold.

Of the physicists Forman discusses, perhaps Richard von Mises comes nearest to being his best example as he was a convert to the *Weltschmerz* of Spengler. Did von Mises' Spenglerism cause him to believe in acausality rather than matters internal to the physics with which he was acquainted? Despite all the citations of his work, Forman does not show that it was von Mises' awareness A of his cultural milieu, rather than physics P, which caused this belief B. The closest he gets in his research into von Mises' writings on Quantum Mechanics at the time is a revised address about which Forman makes the following unhelpful remark: 'Admittedly, von Mises has invoked the quantum theory as the occasion for the repudiation of causality' (*ibid.*, p. 81). But even though von Mises might have used the revised address to tell us of his recent change of mind to belief in acausality, this does not show that it was A, his awareness of his cultural milieu, rather than his physical beliefs P, that caused his belief in causality, B. But Forman insinuates that this might be so. But to merely insinuate this, and to invoke other reasons or rationalisations based on Spenglerian considerations, has two failings. First, it does not establish causal claim (4); second, either it attributes a massive amount of self-delusion to von Mises or it tells us that he was lying, given that he mentioned his new belief in acausality in the context of his address on physics. Critics are aware of the methodological shortcomings in the studies alleged to support SP, even those who take the view that scientific belief needs both an externalist (sociological) and internalist (using some SM) explanation:

> when we come down to the content of physics, we must of necessity take into account internal as well as external considerations. ... Forman has succeeded in demonstrating that physicists and mathematicians were generally aware of the values of the milieu .... But when we come to the crucial claims, that there was widespread rejection of causality in physics, and that there were no internal reasons for the rejection of causality, then the weakness in his argument also becomes crucial. For there were strong internal reasons for the rejection of causality ... (Hendry 1980, p. 160)

There are several other ways in which the SP can be interpreted to give an account of theory choice (such as the notion of negotiation) that cannot be explored here. But if SP were true of how we have acquired our scientific beliefs then its nihilistic stance towards meta-methodology would be supported, and its rejection of the role SMs have been alleged to have played in the history of our choice of theories would have been vindicated. If SP were true, it would show that our norms of method cannot, and have not, played any critical role in the evaluation of our theories. At best we are to see our science as a mere reflection of our (believed) social circumstance. Further, postmodernists would have some grounds for their incredulity towards metanarratives. But it is doubtful, given its flawed causal methodology, that SP has any case studies which support it. Despite the popularity of SP, and the many confusions which surround how we are to understand and test its theses, the critical role that methodology can play tells us that we are not rationally obliged to believe SP, whatever our social circumstance.

11. EMPIRICAL APPROACHES II: METHODOLOGY NATURALISED

As far as methodology goes, Quine adopts quite traditional principles such as the standard views on confirmation and disconfirmation, simplicity, conservatism (or familiarity), and so on, without attempting a further elaboration of what such principles might be like.[39] Given that Quine is also an advocate of some version of holism as expressed in the metaphor of Neurath's boat, we can assume that not only the sciences but also their principles of method along with logic are open to revision if there is lack of fit between them and our experiences (or reports of experience). In this respect methodology and the sciences are on a par. However the requirement of 'fit' or overall coherence takes us into the province of meta-methodology in which there is at least some meta-principles of overall consistency and/or degree of coherence across our total web of belief. What is of interest, then, is not so much Quine's views on methodology, but his more radical view that there is no first philosophy on the basis of which we can make a meta-methodological stand in, say, analyticity, or *a priori* knowledge, or whatever (though the requirements of consistency and coherence, along with a hypothetico-deductive model of testing, are necessary if the holistic picture is to be preserved).

The deflation of epistemological, and any meta-methodological, pretensions is well expressed by Quine in a number of places. First, his 'Five Milestones of Empiricism', the fifth being

> naturalism: the abandonment of the goal of first philosophy. It sees natural science as an inquiry into reality, fallible and corrigible but not answerable to any supra-scientific tribunal, and not in need of any justification beyond observation and the hypothetico-deductive method. (Quine 1981, p. 72)

This is a weaker characterisation of naturalism than the one to follow. Pride of place has been given to the H-D method, supplemented with some principles of confirmation and disconfirmation. But one is left wondering whether the H-D method does fulfil the role of a 'supra-scientific tribunal' in needing no justification. Though some methodologists favour the H-D method, many, especially Bayesians, reject it as seriously deficient and paradox ridden.[40] So, what is the justification of any principle of method, H-D or not, within a naturalistic framework?

A stronger version of naturalism occurs in an often-cited passage from Quine's 'Epistemology Naturalized':

> Epistemology ... simply falls into place as a chapter of psychology and hence of natural science. It studies a natural phenomenon, viz., a physical human subject. This human subject is accorded a certain experimentally controlled input ... and in the fullness of time the subject delivers as output a description of the three-dimensional world and its history. The relation between meagre input and the torrential output is a relation that we are prompted to study for somewhat the same reasons that always prompted epistemology; namely, in order to see how evidence related to theory, and in what ways one's theory of nature transcends any available evidence. (Quine 1969, pp. 82, 83)

Quine appeals only to the science of psychology. But if we extend an invitation to all the sciences to join in an investigation into epistemology as a 'natural phenomenon', then we must allow the sociologists of the previous section their say as well. Broadly construed to include all the sciences, there is nothing in Quine's account of naturalism that a supporter of the Strong Programme could not endorse since, as we have seen, they also admit a role for both social and non-social (including psychological) factors in the causation of belief.

But the passage is not without the same difficulties that faced the Strong Programme, viz., its account of the normative within a framework of naturalism. Let us focus on methodology, and in particular the evidence relation. This, we might have initially assumed, is partly a normative or evaluative notion in that we speak of good, less good or bad evidential connections between observation and theory. But this appears to have been replaced by a descriptive account, perhaps containing causal or law-like claims within several sciences from theory of perception to linguistics, about the chain of links from sensory input to linguistic outputs. Is Quine's position eliminativist in that the norms of evidential reasoning are deemed to be non-existent in much the same way as we now deem phlogiston or Zeus' thunderbolts to be non-existent? Or are the norms still there but lead a life in heavy disguise through definition in terms of the predicates of the biological, psychological and linguistic sciences? Or are they supervenient in some way on the properties denoted by such predicates? Whatever the case, we lack an account (if there is one) of the special features of those causal links between sensory input and linguistic output that characterise *justified* or *warranted* or *rational* belief in theory, and the special features which produce unjustified, unwarranted or irrational belief in theory. Moreover which 'human subjects' should be chosen for study: the good or the bad reasoners?

In his recent *Pursuit of Truth* Quine changes tack and emphasises the distinction between the normative and the descriptive saying:

> Within this baffling tangle of relations between our sensory stimulation and our scientific theory of the world, there is a segment that we can gratefully separate out and clarify without pursuing neurology, psychology, psycho-linguistics, genetics and history. It is the part where theory is tested by prediction. It is the relationship of evidential support. (Quine 1992, p. 1)

This appears to suggest that the evidential relation is outside the scope of the naturalists' programme. In his replies in the Quine Schilpp volume there is also a more modulated position:

> A word now about the status, for me, of epistemic values. Naturalism in epistemology does not jettison the normative and settle for indiscriminate description of the on-going process. For me normative epistemology is a branch of engineering. It is the technology of truth seeking, or, in more cautiously epistemic terms, prediction. Like any technology it makes free use of whatever scientific findings may suit its purpose. It draws upon mathematics in computing standards of deviation and probable error in scouting the gambler's fallacy. It draws upon experimental psychology in scouting wishful thinking. It draws upon neurology and physics in a general way in discounting testimony from occult or parapsychological sources. There is no question here of ultimate values, as in morals; it is a matter of efficacy for an ulterior end, truth or prediction. The normative here, as elsewhere in engineering, becomes descriptive when the terminal parameter has been expressed. (Hahn and Schilpp 1986, pp. 664, 665)

Just as Quine once wrote a paper called 'Three Grades of Modal Involvement' so there might be a paper called 'Grades of Naturalistic Involvement', or, if you like in reverse 'Grades of Normative Involvement', in which one investigates a range of possible involvements of the normative with the naturalistic – but neither is the normative totally abandoned, nor is an *a priori* independent 'first philosophy' advocated. In such an investigation Quine's metaphorical talk of the norms of method as engineering, or as 'efficacy for an ulterior end', or as the technology of truth-seeking or of prediction, needs clarification.[41] Concerning methodology, much clarification has already been done by Larry Laudan in his theory of Normative Naturalism (NN) – the next meta-methodology to be investigated.

The account of rules and values given in section 3 accords with that of Laudan who says that methodological rules of the sort 'one ought to do what rule r says' are elliptical and omit reference to epistemic values which are an important, though often unmentioned, axiological aspect of the epistemological context in which rules are to be applied. When values are mentioned explicitly then we have principles of instrumental rationality, or $r–v$ hypothetical imperatives:

(P) if one's cognitive goal is v then one ought to follow r.
Laudan claims that such hypothetical imperatives are warranted if the following universal declarative sentence, or hypothesis, H, is true:
(H) following rule r realises goal v.
A comparative statistical formulation would be:
(H') if one follows rule r then one more frequently realises goal v than by following any alternative r' to r.

The last two kinds of $r–v$ hypotheses will be collectively referred to as 'Hs'.

Claims like (P) are imperatives within some SM. In contrast the Hs are empirical hypotheses about the strategies scientists employ to achieve (or not as the case may be) their scientific values. They are means–ends hypotheses of an instrumental methodology. They can be viewed as sociological hypotheses about a range of actions of scientists and the frequency with which their following particular rules realises particular values. As such the collection of Hs constitute an empirical science about the best strategies we have adopted in constructing the historical sequence of other theories about the world. Thus in parallel to the sciences, there is a history of the development of Hs over time, i.e., a history of the development of our SMs. Moreover the Hs can, like the hypotheses of any science, be proposed, confirmed, rejected and modified in the course of the progressive development of the means–ends science. It is this feature of SMs as a set of Hs that is the focus of NN.

The Hs, we might say, are naturalised versions of principles of SMs. But if we naturalise all of them we have eliminated appeal to any principles of method to decide between the Hs of our ends–means science. One role meta-methodology can have is to provide principles to adjudicate between the various (sets of rival) Hs. To this end Laudan proposes a meta-inductive rule MIR:

If actions of a particular sort, m, have consistently promoted certain cognitive ends, e, in the past, and rival actions, n, have failed to do so, then assume that future actions following the rule 'if your aim is *e*, you ought to do *m*' are more likely to promote those ends than actions based on the rule 'if your aim is *e*, you ought to do *n*'. (Laudan 1987, p. 25; 1996, p. 135)

Sometimes MIR is expressed as an instance of the Straight Rule of enumerative induction, as above, and on other occasions more broadly as a rule about frequencies. Thus Laudan says: 'empirical information about the relative frequencies with which various epistemic means are likely to promote sundry epistemic ends is a crucial desideratum for deciding on the correctness of epistemic rules' (Laudan 1996, p. 156). In the case of the latter, we should be alert to the fact that the meta-methodology of NN is more than MIR: it also carries with it the baggage of statistical methods for the estimation of frequencies. But as will be discussed shortly, many statistical methods have *a priori* aspects; so the meta-methodology of NN cannot be purely empirical in character.

MIR is the core meta-methodological principle of NN. We may use MIR to adjudicate between rival hypotheses about the efficacy of rule *r* compared with any rival *r'* in achieving goal *v*. To illustrate, Laudan points out that while MIR seems quite minimal, it is strong enough to test some commonly accepted rules of SM against the historical record. On this basis he claims that MIR alone is able to eliminate a number of proposed rules such as: 'if you want theories with high predictive reliability, reject *ad hoc* hypotheses' (Laudan 1987, p. 27). Such rules, when supplemented with the value they are supposed to achieve, can be shown to be unreliable with respect to that value. Once a hypothesis *H* has passed its test using MIR, we can formulate it as a hypothetical imperative of method. We can also add it to MIR thereby bootstrapping our way to further meta-principles for adjudicating between rival hypotheses of the means–ends science of naturalised methodology. The warrant for the methodological principle *P* (which is a hypothetical imperative) is just the empirical means–ends hypothesis *H* that has passed its test.

So understood, NN has an important task to perform: given the vast number of methodological principles that scientists and philosophers have proposed, NN advocates an empirically based method, no different from that found in the sciences, for testing just how well the principles perform their task. NN can yield a growing means–ends science of Hs tested by MIR, or MIR supplemented with already confirmed methodological principles; and as this science grows it will in turn yield a body of more refined principles of SM. Importantly NN goes about its task in a manner free from the difficulties that faced the later Popper, Lakatos and the reflective equilibrium principle, discussed in section 7, and also Feyerabend (section 9). No appeal need be made to intuitions; instead episodes from the historical record of the sciences are investigated to uncover the strategies that best realise values. There are a number of case studies of how particular principles of SM have fared under the regime of NN (see the programme outlined in (Laudan *et al.* 1986) and some of the results of testing in (Donovan *et al.* 1992)). However we will focus on questions which arise concerning the meta-methodology of NN.

The programme of NN depends crucially on a meta-rule of induction thereby tying the justification of each methodological principle to the problem of the justification of induction. This raises a number of issues for NN understood in this way. The first has to do with the actual test procedures that advocates of NN have employed in the literature. It turns out that their procedure is not one of collecting a vast number of historical episodes and then performing an induction over, or performing a statistical analysis of, the episodes. Rather some proposed naturalised methodological hypothesis is tested against the historical record using not MIR but principles of instance confirmation or disconfirmation or, as is openly admitted, the H-D method.[42] Though such a deviation from the strict programme of NN might be allowed in the absence of the much harder work of employing MIR alone, none of the three principles are problem-free and at best can only provisionally be used as meta-principles until the more austere programme with MIR alone is launched. For the H-D method is itself highly contested by probability theorists. And principles of instance conformation face problems such as that of grue or the Raven's Paradox, while instance disconfirmation faces the Quine–Duhem problem. These are significant problems in the theory of scientific rationality for which NN assumes an answer rather than provides one.

The second issue is that while Laudan maintains that the principles of SMs are all hypothetical (or categorical with suppressed antecedents), the meta-rule MIR is different in that it is categorical. But it can also be viewed as a hypothetical if its suppressed antecedent contains reference to the value of truth. In making ordinary enumerative inductive inferences about scientific or everyday matters what we want is a reliable inference, i.e., one which, given true premises based on what we have observed in the past, enables us to draw a further truth as conclusion, which is either about the next unobserved case or is a generalisation. Thus the goal of ordinary inductive inference, and of MIR, is truth. We also aim for truth, or high probability of truth, when we test the various Hs, the naturalised form of principles of method. That is, we want to know that the hypothesis 'following r always realises value v' is true, or at least has been highly confirmed. Clearly the value of truth plays a crucial role in meta-methodology and in determining the hypotheses of the means–ends science.

The next set of issues focuses on MIR. First, by itself MIR might be quite unable to perform its task of yielding principles of method when confronted with the massive amount of information provided by episodes in the history of science without the assistance of statistical theory. Some sampling method needs to be employed to determine how many historical cases need to be investigated in order to determine, to a specified degree of accuracy, any proposed means–ends hypothesis.[43] Second, principles of enumerative inference, such as MIR, also obscure two important aspects of inductive inference that Hempel (1981) invites us to separate out. The first is matters to do with rules for determining the probabilities of hypotheses given evidence; the second, formulating rules of acceptance which would tell us what hypothesis we should accept. Pursuing the second aspect takes us in the direction of the theory of epistemic utility which Hempel had earlier developed, or the decision-theoretical approach of others. In a quite standard version of this theory, suppose we are given the probability of some hypothesis H on evidence E. Then there are two 'actions' a scientist can take: either accept H or reject H. Moreover there are two states of the world: either H is true or H is false. This yields a matrix with four outcomes, each of which is assigned some 'utility' by the scientist. Which action should the scientists perform? One standard answer is to accept that hypothesis with the greatest expected utility. Thus problems to do with inductive inference land us squarely in issues central to decision theory and the principles on the basis of which rational decisions are to be made.

What bearing does this have on NN with its meta-principle of MIR? If we were to take as the 'action' following rule r, and if we were to take as the 'state' the realisation of some value v, then providing the scientist can attach utilities to the four outcomes (one of which would be, say, following r but not realising v), then we have a decision-theoretic framework for evaluating principles of method. Viewing matters this way, in the meta-methodology of NN Laudan's MIR has been replaced by decision-theoretic principles. Thus the debate about the status of the meta-methodological principles of NN has been shifted to the debate about the status of the principles of rational decision making which are to be used in determining the more particular principles of the SMs of science. This leaves untouched issues about which principles of SM do, or do not, survive testing. Moreover, even though the materials used in the test procedures are garnered from the historical record of science thereby underlining one way in which NN is empirical, the meta-methodological principles of decision theory by which the test procedures are carried out have a distinctly non-empirical character. Thus NN cannot be, through and through, an empirical science; it is also conceptual, as will be seen at the beginning of the next section.[44,45]

12. PRAGMATISM AND METHODOLOGY

Some approaches to meta-methodology may be characterised as 'pragmatist', though what is intended by that term is not always easy to pin down. If pragmatism is at least characterised as denying that there is a first philosophy which has *a priori* or analytic modes of justification, then Quine's naturalistic account of methodology can be said to be a variety of pragmatism. Laudan endorses such a meta-epistemological claim adding that epistemology, with its methodological principles, is neither a synthetic *a priori* subject, nor is it conventionalist as the early Popper suggested (Laudan 1996, p. 155). For NN, epistemology's hypotheses about inquiry are to be judged like any other hypotheses of science. However Laudan points out that science has both conceptual and empirical elements; and since methodology is to be construed along the lines of a means–ends science, then it too will have both conceptual and empirical elements. In comparing the science of methodology with physics Laudan says: 'Both make extensive use of conceptual analysis as well as empirical methods' (*ibid.*, p. 160). Thus the naturalists' science of methodology will not be wholly empirical but will be in part non-empirical. If so, room will have to be made for a role for conceptual analysis while rejecting the idea of a synthetic *a priori* first philosophy.

In admitting that there are conceptual elements to meta-methodology, Laudan is not so hard-line a pragmatist as Quine. Nor is he so parsimonious as Quine who mentions few goals for science beyond truth and prediction. Though Laudan is a critic of theoretical truth as an unrealisable goal of science (Laudan 1984, chapter 5), he is broadly pluralistic in admitting a wide range of goals for science, both currently and in the past. Insofar as pragmatism is characterised as a pluralism with respect to our epistemic and cognitive goals Laudan is a pragmatist. On a third characterisation of pragmatism in terms of the actual practical success of science, Laudan is also a pragmatist since he requires that methods be judged by their production of successful science. On this broad criterion, the role that Feyerabend assigns to successful practice in assessing methods might make him a pragmatist of sorts; but not obviously the later Popper or Lakatos who base their empirical view of science not on actual practice but on the judgements of a scientific élite about actual practice. One philosopher who puts emphasis on the role of practice in meta-methodological considerations is Rescher – to whom we now turn.

Rescher has developed a systematic account of human knowledge, which eclectically combines elements of scepticism, realism, naturalism, pragmatism and ideas from evolutionary theory. As against the classical pragmatists who construed the truth of individual claims in terms of practical utility, he develops a methodological variant of pragmatism according to which methodological rules are justified by their success in practical application. Rescher accepts a correspondence theory of truth and rejects the classical pragmatist conception of truth applied at the level of individual theses, i.e., propositions; this he calls '*thesis* pragmatism'.[46] This is contrasted with *methodological* pragmatism, which places the locus of pragmatic justification on methodological rules. For Rescher propositional theses about the world are correspondence true or false; and they are justified, not pragmatically, but on the basis of methodological rules. However it is these rules, and not the propositional theses about the world that they help generate, that are justified by successful practical application.

Rescher's methodological pragmatism has much in common with naturalism because it treats methodological principles as subject to empirical evaluation, and hence provides a naturalised account of epistemic warrant. However Rescher recognises a fundamental problem concerning the justification of principles of method which he expresses in terms of the Pyrrhonian sceptic's problem of the *criterion* or '*diallelus*', i.e., circle or wheel (see Rescher 1977, chapter 2 section 2). This is a problem we have met in section 4, viz., that there is no direct or method-independent way of determining that use of a methodological rule yields its goal. Either we use principle M, in which case M presupposes its own correctness, or we use another principle M', in which case there is an infinite regress of principles. We need to see in what way the pragmatist's appeal to practice might overcome this problem.

Rescher conceives of methodological rules as instruments, as tools for the realisation of some end. So understood they are in conformity with the account of rules, and the values that the rules are supposed to realise, given in sections 3 and 11. Rescher also requires rules be evaluated in the same way that instruments are evaluated viz., pragmatically. That is, we need to show that instruments 'work' in that they produce some desired end (the end presumably being a theory which exemplifies some desired epistemic value). However it is not sufficient for a rule to produce its end once, or a few times. Rescher requires that rules regularly realise their ends: 'the instrumental justification of method is inevitably general and, as it were, statistical in its bearing' (*ibid.*, p. 5). Such a requirement is nothing other than Laudan's more explicit account of NN with its appeal to meta-induction or statistical inference. Also akin to NN is Rescher's use of such an 'inductivist pragmatic' method for comparing how well two or more rules realise the same end, or how rules can be refined so as to more regularly realise some end, or how principles once established can be used to supplement the inductivist pragmatic meta-methodology. Rescher, like Laudan, draws out an aspect of the naturalist view of method in which methodological principles of the means–end science, like the hypotheses of any other science, can be revised and improved in the process of examining their degree of pragmatic success.

Induction plays an important role in both the meta-theory of Laudan's NN and Rescher's methodological pragmatism. Rescher extends his pragmatism in an attempt to resolve the problem of the justification of induction (see for example Rescher 1973, chapter 2 and Rescher 1992, chapter 9). We will not pursue the pragmatic justification of induction here. However we will focus on another aspect of Rescher's position that distances his pragmatism from Laudan's NN in that it employs an argument that Laudan finds inadequate on the grounds of NN's own test procedures, viz., a modified version of Inference to the Best Explanation (IBE) which links the success of science produced by the application of methodological principles to the truth-indicativeness of the principles themselves. A similar argument using a simple but direct version of IBE was set out in section 2 to show the superiority of an appeal to methodological principles over non-cognitive interests as an explanation of the instrumental success of the historical sequence of scientific theories. Here some of the promissory notes on which that argument was based will be cashed.

To connect practical success with truth, Rescher deploys a number of broad metaphysical presuppositions which we might refine or abandon in the course of inquiry but which, at the moment, we hold of ourselves and the world in which we act as agents. They are as follows: (a) Activism: we must act in order to survive, hence beliefs have practical relevance since they lead to actions which have consequences. (b) Reasonableness: we act on the basis of our beliefs so that there is a systematic coordination of beliefs, needs and action. (c) Interactionism: humans actively intervene in the natural world and are responsive to feedback from this intervention in both cases of satisfaction and frustration. (d) Uniformity of nature: continued use of a method presupposes a certain constancy in the world. (e) Non-conspiratorial nature of reality: the world is indifferent in the sense that it is neither set to conform to, nor set to conflict with, our actions. On the broad metaphysical picture that emerges, the practical success of our belief-based actions, and the theoretical success of our beliefs, both turn on the correctness of assumptions like (a)–(e). They set out general but contingent presuppositions about the sort of agents we are and the sort of world in which we act and believe (both individually and communally) and the fact that we have achieved a wide degree of success in action and belief.

Consider now principles of method that govern inquiry. They are not merely locally applicable but also apply across a very broad front of inquiry into a wide diversity of matters. They also have a large measure of success in that rules regularly deliver beliefs that instantiate our desired practical goals and theoretical values. Roughly, the idea is that our methods might mislead us some of the time, but it is utterly implausible to hold that most of the time they systematically lead us astray, given our success in the world and the way the world is as presupposed in the general metaphysical picture above. On the basis of this success, what particular epistemic property might we attribute to our methodological principles? Rescher sometimes speaks of this as the 'legitimation problem' for our methods (Rescher 1977, p. 14). A number of terms could be used to name the property possessed by our methods that leads to their success, such as 'validation', 'truthfulness' and 'adequacy'; Rescher uses these terms in various contexts along with 'truth-correlative' and 'truth-indicative' (see Rescher 1977, chapter 6, especially pp. 81, 83). We will settle on the last of these and summarise Rescher's non-deductive plausibility argument as one which shows that the success of our methods is evidence that our methods are truth-indicative.

The argument for this position has much in common with the use of IBE to argue for scientific realism. However Rescher refers to it as a '"deduction" (in Kant's sense)' (*ibid.*, p. 92). But the argument is not strictly deductive (as the shudder quotes allow), but rather an inductive plausibility argument to be distinguished from certain forms of IBE that he wishes to reject. IBE as used by realists is an argument from the instrumental success of our theories to the truth of the theories. Given both realist and various rival non-realist interpretations of some theory, it is alleged that realism offers a much better explanation of, or makes more probable, the theory's instrumental success than any of its rival interpretations (some of which offer no explanation at all). Rescher does not endorse this form of IBE with its strong conclusion that theories are true, or even that they have high truth-likeness. Rather he prefers to say:

> The most we can claim is that the inadequacies of our theories – whatever they may ultimately prove to be – are not such as to manifest themselves within the (inevitably limited) range of phenomena that lie within our observational horizons. . . .
> In general, the most plausible inference from the successful implementation of our factual beliefs is not that they are right, but merely that they (fortunately) do not go wrong in ways that preclude the realization of success within the applicative range of the particular contexts in which we are able to operate at the time at which we adopt them. (Rescher 1987, p. 69)

Let us transfer these considerations to the role that methodological principles play in inquiry. First expand the notion of the 'success' of our theories from merely their instrumental success in pure inquiry to include their success in enabling the domain of human action to be vastly increased (for example, science has investigated the properties of new substances, and this has resulted in new technology such as our having non-stick fry pans with which to cook). And let us shift focus from whether scientific theories are true to what epistemic properties our principles of method must possess if they are to yield, as the result of their application in inquiry, pragmatically successful theories (both practically and cognitively). What we want are principles which not merely yield successful theory on a few occasions, but quite generally; and even though these principles might sometimes fail us, they are not thereby totally invalidated. That is, we want principles which have a high degree of reliability across many sciences and yield success in practical and theoretical contexts while being defeasible. Presenting Rescher's argument this way shows it to be an inductive argument, akin to IBE, from the wide-ranging practical and cognitive success of our theories, to the best explanation of that success, viz., the methodological principles themselves having not gone massively wrong in their application, or having not manifested whatever inadequacy they might possess. This feature we might call the 'truth-indicativeness' of methodological principles. But they are only indicative because they are open to revision and modification in the course of growing inquiry conducted by agents such as ourselves in a world such as the one we inhabit.

Some such modified version of IBE lies at the core of Rescher's meta-methodology of 'methodological pragmatism'. What justificatory basis is there for this form of argument? Rescher's response to the question is to see the re-emergence of the diallelus, or great circle or wheel of justifications (Rescher 1977, chapter 7). Characteristically for a pragmatist, there is no independent standpoint from which a justification can be given for IBE, or any other rock-bottom meta-methodological principle. Rather the same form of argument just given above is used at all points to: (i) validate scientific theories in terms of their success in practical and theoretical application; (ii) validate principles of inquiry (i.e., the level 2 SMs) in terms of their success in yielding the successful science of (i); (iii) validate the metaphysical picture of the world which sets out the presuppositions we must make about the sort of world in which we live and the sort of agents we are that makes such success in (i) and (ii) possible. Some such pragmatist picture, Rescher alleges, closes the circle without making it necessarily vicious. But the pragmatic picture depends on some plausibility argument, or some version of IBE, which is used at different levels and which closes the circle. If the above account of Rescher's argument is correct, it is one which is highly contested by fellow naturalists such as Laudan who restricts his meta-methodology to induction and who explicitly rejects the use of any form of IBE or plausibility argument (Laudan 1984, chapter 5). For those who find that some forms of IBE can be satisfactorily used, this helps draw another distinction between the kinds of pragmatism that Laudan and Rescher espouse.[47]

13. BAYESIANISM AND SCIENTIFIC METHOD

The view that scientific reasoning and scientific method more generally are to be understood in probabilistic terms has recently been given a great deal of attention by its many adherents and is well served by a burgeoning literature.[48] The scope of this book is to explore non-Bayesian approaches to methodology. However in a survey such as this it will not be amiss to review Bayesianism, however briefly, in terms of the framework developed here. Bayesianism is not a unified position, all aspects of which are agreed to by all parties. A brief outline will be given of a standard form of subjective Bayesianism called here 'orthodox Bayesianism', or 'OB' for short. Then we will sketch briefly the position of two philosophers,

Shimony and van Fraassen, who, in quite different ways, deviate from OB. Moreover Bayesianism is not without its critics, one of the more important works in this area being Mayo 1996. Papers in this book which deal with methodology in relation to Bayesianism include those by Fox and Worrall.

The central core of OB involves appeal to the axioms of the probability calculus and the several forms of Bayes' Theorem which follow from these axioms. The various forms of this Theorem provide an account of how probable hypotheses are on given evidence, how rival hypotheses can be compared with respect to (growing) evidence, and how the same hypothesis fairs with respect to different bits of evidence over time. It also provides illumination concerning such matters as the variety of evidence, the Quine–Duhem problem and the various paradoxes that beset theories of confirmation (such as 'grue', the raven's paradox, irrelevant tacking paradoxes, etc.). OB also comes equipped with a number of powerful theorems, such as those which inject a degree of objectivism (or intersubjectivism) through 'convergence of opinion' proofs; these show the conditions under which different persons with widely different initial degrees of belief in some hypothesis can ultimately converge in their relative degree of belief in that hypothesis given evidence, as the evidence comes in over time.

How are probabilities to be understood? By far the most commonly accepted view is that of subjectivism, or personalism, in which probabilities are understood to represent rational degrees of belief on the part of persons. Also central to OB is a principle of updating probabilities in the light of new evidence, the simplest of which is a rule of strict conditionalisation. Understanding the probabilities subjectively, this rule says: if a person acquires evidence E with certainty, then for some hypothesis H the old and new degrees of belief are related as follows: $p_{new}(H) = p_{old}(H, E)$.[49] The above provides a sufficient basis for a wide-ranging theory of confirmation, of the sort set out in Earman (1992). Bayesianism also provides a basis for decision-theoretic approaches to the foundations of scientific inference; this is explored in Savage (1954), Jeffrey (1983) and Maher (1993).

How good an account does OB give of scientific inference? Howson and Urbach explore the connections OB has with the general theory of statistical inference, thereby displaying one advantage it possesses over many other theories of scientific method (outlined in previous sections) which do not always link readily to theories in statistics. Earman also claims that, for all the problems he finds with Bayesianism, it 'provide[s] the best hope for a comprehensive and unified treatment of induction, confirmation and scientific inference' (Earman 1992, p. xi). Further, Salmon argues that, as far as Kuhn's theory of weighted values is concerned, it is possible to give an account of its methodological prescriptions entirely in terms of Bayes' Theorem (Salmon 1990; however see the paper by Worrall in this volume). Earlier Salmon had given his account of how Bayesianism fares with respect to some of its other rivals, such as Popper's theory (Salmon 1967), a matter which Howson and Urbach explore even more fully.

Such comparisons of OB with rival theories of method concerns the virtues of level 2 theories of SM. In judging between such rival SMs, appeal is made to meta-methodological criteria such as greater comprehensiveness, ability to deal with long-standing problems in confirmation theory, and coherence with other mathematical theories (e.g., the advantageous links Bayesianism has to statistical theory). A positive comparative judgement in favour of Bayesianism is made by its adherents despite some of its acknowledged inadequacies (see Earman 1992, chapters 5, 9 and 10). Bayesians also go meta-methodological when they give reasons for why a person's degrees of belief should conform to the probability calculus. Several kinds of justification have been proposed, the most common kind being based on 'Dutch Book' considerations. Ramsey and de Finnetti proposed that there is a connection between degrees of belief and betting behaviour, and used this to justify the conformity of degrees of belief to the axioms of the probability calculus. Commonly in a 'Dutch Book' one makes a set of bets such that whatever the circumstances one cannot win. Thus 'Dutch Book' arguments attempt to show that if one's degrees of belief do not conform to the axioms of the probability calculus then a Dutch book can be made against one in that one's bet on the truth or falsity of hypotheses is such that one can never win. The converse claim is also important, viz., if one's beliefs do conform to the axioms then a Dutch Book cannot be made. These important meta-considerations, which have to do with coherence or consistency conditions for assignments of degrees of belief, have attracted considerable comment but not complete agreement on all details.[50]

The further elaboration and defence of the various Bayesian approaches cannot be made here. Instead mention will be made of two widely differing ways in which some have thought that OB is in need of supplementation, the first being due to Shimony, and the second, only briefly sketched, due to van Fraassen. In a paper[51] which widely ranges over theories of probabilistic method, Shimony set out his version of what he called 'tempered personalism' within the context of a naturalistic view in which we, as beings in the natural world, are capable of reasoning about, and investigating, that world. Shimony argues for at least the elements of the OB position, as already set out above, against other probabilistic approaches to scientific method. But he makes some significant additions; the one discussed here is his 'tempering' condition.

Despite the 'convergence of opinion' theorems, Shimony fears that OB allows people to set their prior degree of belief in some hypothesis at, or close to, the extremes of 0 and 1 so that convergence might not take place in their lifetime of evidence gathering, or converge not at all. The issues of the constraints to be placed on prior probabilities, and the 'swamping of priors', is an important one for Bayesians (see Earman 1992, chapter 6). In order to avoid excessive dogmatism or scepticism on the one hand and excessive credulity on the other, and to encourage a genuinely open-minded approach to all hypotheses that have been seriously proposed, including the catch-all hypothesis,[52] Shimony proposes that the radically subjectivist personalist approach to Bayesianism be tempered by adding the following condition (also suggested by others such as Harold Jeffreys).

Tempering condition (TC): 'the prior probability . . . of each seriously proposed hypothesis must be sufficiently high to allow the possibility that it will be preferred to all rival, seriously proposed hypotheses as a result of the envisaged observations' (Shimony 1993, p. 205).

Shimony also adds that tempered personalism is a contribution towards a 'social theory of inductive logic' of the sort envisaged by Peirce, in contrast to the individualism of the untempered subjectivist interpretation, in that TC applies to all inquirers and all the hypotheses they entertain.

Shimony's TC is a substantive methodological supplement to subjective Bayesian methodological principles (but one that has received little endorsement from strict observers of the tenets of OB). Shimony recognises that there is vagueness in what is meant by 'seriously proposed hypothesis' and that, since there is no way of establishing TC using the axioms of probability, there is an issue about its status. Initially Shimony thought that there might be an *a priori* justification for TC. However in a searching analysis of what TC might mean and what status it might have, Teller shows that there is no *a priori* justification for TC available; nor do other pragmatic justifications work (see Teller (1975) for these arguments). However for Teller, and now for Shimony, the negative conclusion is not problematic since they now advocate empirical rather than *a priori* justifications of principles of scientific method. They also suggest that 'the process of subjecting a method of scientific inference to scientific investigation' (Teller 1975, p. 201) needs to be set in the context of meta-methodological investigations of the sort suggested in section 7 (Goodman's reflective equilibrium) or section 11 (Quinean or other naturalisms).

Shimony in his 'Reconsiderations' of his earlier paper which advocated TC, proposes four further principles of methodology 'chosen to expedite the machinery of Bayesian probability' (Shimony 1993, p. 286; for the principles see sections V and VI); only one principle will be mentioned here. The four principles are proposed as contingent claims which can be shown to fail in some possible worlds; importantly they are alleged to be true of our world and to have some empirical support. In setting out some conditions for the possibility of inquiry for Bayesian agents in a world like ours, they fulfil a similar role to the general metaphysical principles presupposed by Rescher's pragmatic meta-methodology, even though they paint a somewhat different metaphysical picture from that of Rescher. But they also suggest substantive principles of method. Thus Shimony's Principle 2 says: 'A hypothesis that leads to strikingly successful empirical predictions is usually assimilable within a moderate time interval to the class of hypotheses that offer "understanding", possibly by an extension of the concept of understanding beyond earlier prevalent standards' (*ibid.*, 287). Unfortunately this principle is expressed as a descriptive claim rather than a norm, though its transformation to a norm is fairly obvious. As an example Shimony has in mind the prediction of the Balmer series from Bohr's early 1913 theory of the atom. Given this surprising prediction, on Shimony's Principle 2 Bohr's theory ought to be assimilated in whatever way into the prized circle of hypotheses that offer 'understanding'. Any further discussion of Shimony's principle would have to look into the way in which it might expedite Bayesian machinery, or whether Bayesian methods can readily accommodate such a recommendation and that it is otiose. One also needs to look at whether the four principles are all the contingent empirical principles that a fully fledged methodology requires to do justice to our scientific practices.

Another quite different set of modifications to strictly minimal orthodox subjective Bayesianism has been proposed by van Fraassen. His position is finely nuanced with respect to OB, and can be best grasped by considering his own characterisation of traditional sceptical epistemology in terms of the following four theses:[53]

(I) there can be no independent justification to believe what tradition and ourselves of yesterday tell us (i.e., what we find we already believe);
(II) it is irrational to hold unjustified opinion;
(III) there is no independent justification for any *ampliative* extrapolation of the evidence plus previous opinion to the future;
(IV) it is irrational to make such ampliative extrapolations without justification.

Theses (I) and (II) concern belief, with the justification and rationality of belief set out as separate theses. Theses (III) and (IV) concern a special subclass of our beliefs, viz., our ampliative inferences, with the justification and rationality of ampliative inferences and their deliverances set out as separate theses.

It is possible, with a little gentle massaging, to present three other epistemological positions in terms of whether they accept or reject theses (I) to (IV) of the sceptics' epistemology. Let us call 'Traditional Epistemology' (Trad.E) the view that there are justifiable ampliative rules of inductive inference, including even inference to the best explanation; thus Trad.E rejects the sceptic's (III) which holds that, even though there might be ampliative rules, they lack any independent justification. In rejecting (III) traditionalists can then take on board (IV) concerning the rationality of ampliative inference and its deliverances. They also reject (I) in that they hold that all beliefs are open to justification given their rules. That is, traditionalists think that all beliefs are justifiable and that we have at hand all the justifiable ampliative inferences needed to carry out the job of justification. They can now accept (II) since they are armed with the sufficient justifiable beliefs (due to the rejection of (I)) and sufficient justifiable ampliative inferences (due to the rejection of (III)). In sum, Trad.E accepts (II) and (IV) while rejecting (I) and (III).

The position of Orthodox Bayesianism (OB) can be characterised as follows. It agrees with (I) but rejects (II). In accepting (I) OB takes on board what beliefs we currently have and works from there, regardless of what independent justification they may or may not have. But OB rejects (II). Within the context of OB this just means that we are free to assign any prior probability to hypotheses we like and that we are not irrational in so doing; all that is required is that we meet the coherence requirement for the distribution of degrees of belief in so freely distributing. Further, since conditionalisation is not strictly a kind of ampliative inference and is one of the central characteristics of being a Bayesian, then OB, on van Fraassen's characterisation, accepts (III) and (IV) thereby underlining the emphasis on learning from experience and past opinion.

What of the 'New Epistemology' (NE) advocated by van Fraassen? He characterises NE as adopting (I) and (III) but rejecting (II) and (IV). That is, we cannot give an independent justification for either our beliefs or our principles of ampliative inference; but, given the standards of rationality he adopts, it is not irrational to either hold unjustified beliefs or employ ampliative inferences. To characterise his position van Fraassen draws a useful comparison between Prussian law, in which everything is forbidden which is not permitted, and English law in which everything is permitted that is not explicitly forbidden. The Prussian position of Trad.E is well expressed by Bertrand Russell who said: 'Granted that we are certain of our own sense-data, have we any reason for regarding them as signs for anything else . . .?' (van Fraassen 1989, pp. 170, 171). This is to be compared with the English position in which *rationality is only bridled irrationality*' and 'what it is rational to believe includes anything that one is not rationally compelled to disbelieve' (*ibid.*, pp. 171, 172).

The position of van Fraassen's NE with respect to its Trad.E rival is clear. But what of the contrast between NE and OB? Both accept (I) and (III) and reject (II). That is, both NE and OB join hands with the sceptic in holding that nothing, neither beliefs nor ampliative inferences, can be justified in the way Trad.E requires; and they both part company with the sceptic in rejecting the claim that it is irrational to maintain an unjustified belief. However the difference between NE and OB is that NE rejects (IV) while OB accepts (IV). That is, for NE we are permitted to amplify belief even when ampliative inferences are unjustifiable; but for OB this is not permitted and is irrational. In rejecting (IV), the permissive 'English' position of van Fraassen allows one to perform ampliative inferences that lack justification without incurring any epistemic censure for so doing. That is, it is rationally acceptable to believe anything without justification unless one is rationally compelled to disbelieve it. In contrast OB is more restrictive in that it supports Trad.E in rejecting as irrational belief in any claim that transcends the deliverances of ampliative inference.

Van Fraassen's NE is liberal permissive 1960's epistemology while OB wants to claw back some 'law 'n order' in epistemology, but not as much as the regimented adherent to Trad.E. Note that both NE and OB are permissive (on the whole) with respect to what we already believe (both accept (I)). And neither want to reject any of these beliefs on the grounds that, because they lack independent justification of the sort required by Trad.E, they are irrational (both reject (II)). For OB rejecting (II) entails freedom with respect to the assignment of any priors to our beliefs (modulo the coherence requirement). But NE and OB do part company over (IV), one illustrative reason being as follows.

Central to OB are the 'convergence-of-opinion' theorems which say that, even on quite divergent assignment of priors (excluding 1 and 0) to hypotheses, the probabilities given evidence will converge upon continued conditionalisation as that evidence comes in over time. This suggests that Bayesian inquirers should not end up with divergent and irreconcilable beliefs on pain of irrationality; hence a rationale for OB's adherence to (IV). But in rejecting (IV) NE is at odds with OB over the issue of the rationality of divergent irreconcilable beliefs. For OB, if inquirers arrive at divergent irreconcilable beliefs, then someone has been epistemically misbehaving and is to be censured. For NE, that divergent and irreconcilable beliefs have been arrived at by the gentle sway of the rules of NE's permissive epistemology is not necessarily a sign of irrationality, nor a sign of epistemic misbehaviour which is to be censured. Inquirers can have divergent irreconcilable beliefs without epistemic fault. This is one of the lessons to be drawn from the notion of the underdetermination of theory by evidence and the rules whereby theory and evidence are to be assessed.[54]

There are many further considerations surrounding the issues just broached concerning NE which involve van Fraassen's principle of Reflection, the rule of conditionalisation and the role of dynamic Dutch Book arguments. However the above will suffice as an account of one aspect of the nicely differentiated position van Fraassen adopts with respect to OB in opting for NE in its place. His position can be summed up as follows, with a contrast drawn between his position not only with respect to an adherent of OB who looks to unique outcomes from their ampliative inferences but also relativists:

> Like the orthodox Bayesian, though not to the same extent, I regard it as rational and normal to rely on our previous opinion. But I do not have the belief that any rational epistemic progress, in response to the same experience, would have led to our actual opinion as its unique outcome. Relativists light happily upon this, in full agreement. But then they present it as a reason to discount our opinion so far, and to grant it no weight. For surely (they argue) it is an effective critique of present conclusions to show that by equally rational means we could have arrived at their contraries? I do not think it is. ...
>
> So I reject this reasoning that so often supports relativism. But because I have rejected it without retreat to a pretence of secure foundations, the relativist may think that I still end up on his or her side. That is a mistake. Just because rationality is a concept of permission rather than compulsion, and it does not place us under sway of substantive rules, it may be tempting to think that 'anything goes'. But this is not so. (van Fraassen 1989, pp. 179, 180)

14. CONCLUSION

The word 'Selective' in the title is deliberate. There is no way, short of writing a whole book, that could even cursorily mention all those who have made a contribution in support of, or against, the idea of scientific method during the last half-century. Our focus has been on only some authors while others have only been mentioned in passing or not at all. One important omission is any mention of the theory of reliable inquiry which arises out of formal learning theory (see Kelly 1996). However the volume as a whole compensates for this in that one of the workers in the field, Kevin Kelly, has a paper which investigates this approach in relation to normative naturalism.

Other omissions at least include the following: the application of belief revision theories to problems of scientific method; a fuller discussion of the ways in which Bayesianism might be expounded and/or challenged; the work of statisticians such as Fisher, Neyman, Pearson, Jeffreys amongst others that relates to issues in methodology; recent work in AI on modes of inference relevant to methodological matters; and so on. In addition, since the 'Selective Survey' is focused largely on meta-methodological issues, there is not much discussion of particular principles of scientific method such as inference to the best explanation and the like. Nor is there a discussion of particular methodological issues that arise in connection with testing causal claims, the postulation of intervening variables, and so on.

However in focusing on issues in methodology that arise out of Popper, Kuhn and Feyerabend and some of the new approaches to methodology that have developed recently, there is more than enough for one book. What the editors hope to show is that, given some of the anti-methodology trends in philosophy of science which either follow the sociologists of science or postmodernism with its 'incredulity towards metanarratives', there is still much life in the discipline of methodology. Importantly the 'Selective Survey' attempts to put some of the pro- and anti-methodology camps into an overall framework of the sort set out in section 2 in order to show that each occupies some part of the logical space of possible positions, and then to reveal possible lines of critical evaluation of each.

NOTES

[1] Laudan is one methodologist who argues that there are principles of method, which we can suppose can be used to demarcate science, but who argues against several other ways of drawing any demarcation criterion; see Laudan (1983) 'The Demise of the Demarcation Problem'; this is reprinted in Laudan (1996, chapter 11).

[2] Earlier advocates of the three positions would be the inductivist Reichenbach (1949), the probabilist Carnap (1962) and the subjectivist Bayesian Savage (1952).

[3] By far the greatest number of contemporary heirs to the 'inductivist'/'probabilistic' approach are Bayesian. There are a few objectivist Bayesians such as Rosenkranz (1977), but by far the majority are subjectivist Bayesians such as Jeffrey (1983), Earman (1992) and Howson and Urbach (1993).

[4] Such a position is adopted in Earman (1992, chapter 8, section 7). He cautions against the idea of 'The Methodology of Science' of the sort advocated by Popper and Lakatos; he also endorses the Feyerabendian position that there is no such 'Methodology' (but not necessarily for Feyerabend's reasons) and leaves methodological advice of the sort Methodologists might offer to scientists to actions chosen on the basis of maximum expected utility.

[5] Pertinent here are issues to do with the 'personal equation' of astronomers often discussed in texts concerning measurements with optical telescopes to determine the position of a heavenly body at a given time. Under most conditions all observers make systematic misobservations of the time; they are systematically too late. This is an astronomer's 'personal equation' which has to be determined so that recorded temporal observations can be corrected to give true temporal observations.

[6] Sometimes these three terms are given different senses for various purposes; here they will be treated interchangeably.

[7] For a useful introduction and survey of some of the literature on science decision-making see Shrader-Frechette (1991).

[8] The view that there is no formal probability logic of the sort developed for deductive logic is widely held, this being in part one of the lessons to be drawn from Goodman's 'grue' paradox. However the Popperian view that in science we can get by only with deductive logic and no probability logic of any sort is an additional claim which is highly contested. A defence of the Popperian position is given in Musgrave (1999).

[9] Some form of inference to the best explanation, viz., that greater explanatoriness is a guide to truth (or increased truth), is one of the methodological principles used by realists to establish realist claims. Though there are many supporters of the viability of such an inference form, doubts that need to be answered have been raised by Laudan (1984, chapter 5), who argues for its inadequacy on historical grounds, and van Fraassen (1989, chapter 6) who argues that it leads to incoherence.

[10] Aristotle's theory has aspects which accord with quite recent accounts of the function that dreams might play in our physiology: see Gallop (1990).

[11] Accounts of the history of theories of method outside the survey period adopted here can be found, for example, in Losee (1993), Laudan (1981), Oldroyd (1986), Blake et al. (1960) and Gower (1997).

[12] In Donovan et al. (1992) some of these theses are tested against actual pairs of scientific theories such as: the emergence of the theory of plate-tectonics against the background of its rivals,

the theories of Ampère and Biot in electrodynamics and rival theories about nuclear magnetic resonance. See the papers by Frankel, Hofman and Zandvoort in *ibid*. The upshot was that theories are expected to solve some, but not all, of the problems not solved by their rivals or predecessors. Theories are not required to solve all the problems solved by their predecessors; some loss of problem-solving power is allowed to occur in the transition from one theory to another. The first claim is not surprising but the second claim might well be. Methodologists have often claimed that scientists do not, as a matter of fact, accept theories with less content than their predecessors and have also proposed rules prohibiting the acceptance of such theories. Such a rule would seem appropriate if the goal of science is to maximise the content of our theories. But would it be unwise to adopt the strategy prescribed by such a rule if science is to yield new exciting theories? What the above research suggests is that for some episodes in science this goal and its associated rule have not been adopted. Is such behaviour by scientists acceptable, or are they methodologically misbehaving?

[13] In what follows capitals '*R*' and '*V*' stand, respectively, for sets of rules and values while lower case '*r*' and '*v*', with or without subscripts or superscripts, are particular rules and values.

[14] There is no agreed terminology in the literature. What we have called 'methodological principles' (which contain both a rule and a value) are sometimes called 'methods', or methodological 'rules', or 'standards', etc. On our usage, if reference to a value is suppressed then principles will be truncated to rules.

[15] Perhaps Feyerabend would not grant this and would argue that his proliferation principle ought to also hold at the meta-level because proliferation of inconsistent meta-theories is necessary for meta-theoretical advance. Also advocates of para-consistent logics might find adopting a consistency principle at the meta-level is too conservative since they can provide resources for coping with inconsistency.

[16] The conception of rules, values and principles developed in this section has been influenced by that found in Laudan (1996, chapter 7). The idea of meta-methodology in this and the next section has also been influenced by Laudan (*op. cit.*), as well as Hempel (1983) where a distinction is drawn between meta-methodologies which are rationalist in that they have *a priori* elements, and those which are pragmatic and naturalistic in that they emphasise empirical elements. The idea of a meta-methodology can be found in the early Popper; it comes into its own in the work of Lakatos.

[17] Aspects of a transcendental approach are explored in Buchdahl (1980) and in Albert (1984).

[18] It is often complained that Popper failed to recognise that the Quine–Duhem problem stood in the way of falsification and that there is not the asymmetry between verification and falsification that he alleges. However Popper is at least aware of the Quine–Duhem problem and in Popper (1957, section 29, footnote 2) attempts to address it by proposing ways in which the same hypothesis can be tested in different theoretical contexts. Whether his proposal is entirely successful is another matter.

[19] See Popper (1959, section 22) for the difference between falsifiability and falsification and the extra conditions, such as the existence of corroborated falsifying hypotheses, required for falsification. For Popper falsification is not the simple inconsistency of a single observation report and a test consequence of a hypothesis under test, as illustrated in many introductory texts. However there is a further methodological issue, which Popper acknowledges, of how corroborated falsifying hypotheses pass their test. It is also important to note an important feature of Popper's rule about falsification. Once falsified a theory is no longer a candidate for the truth; but this does not mean that one should not still work on a falsified theory. This point is not always made clear in Popper; but it is explicit in Popper (1974, p. 1009).

[20] Logically, Popper's hypothetico-deductivism (H-D) has the same character as Ayer's proposal for a Verification Principle (VP). It is well known that all proposed formulations of the VP have been shown to have counterexamples in that any proposition whatever could be shown to pass the VP test. The same applies to the H-D model as a criterion of demarcation for science; any proposition whatever can be shown to be scientific in that it has testable deductive consequences in the presence of necessary auxiliary claims required by the H-D model. For a recent report on the current state of play concerning VP, and thus the H-D model, see Wright (1989). However the H-D model is not all there is to Popper's demarcation criterion.

[21] Not only are there problems with how such an anti-*ad hoc* rule is to be formulated (Lakatos distinguishes three notions of *ad hoc* in Lakatos 1978, chapter 1), but it is also problematic whether any saving hypothesis can stand in the required increasing content relation given Popper's more formal account of empirical content and degree of falsifiability. This last issue is discussed in Grünbaum (1976).

[22] These are not the only methodological rules Popper proposes. Elsewhere further rules can be found, such as the rules concerning observation statements (*ibid.*, section 22) and rules for the testing of probabilistic statements which are otherwise unfalsifiable (*ibid.*, section 65).

[23] For a fuller account of Popper's early encounters with conventionalism and his, and Lakatos', more empirical approach to meta-methodological matters (discussed in the next section) see Nola (1987). Hempel (1983, section 5) also discusses the status of Popper's meta-methodology saying that despite appearances it is not merely decisionist but has justificatory aspects which are empirical in character.

[24] Some rules fully determine what one is to do and leave little or no choice in action such as 'take your hat off on entering a church'. But other rules leave some choice to the actor; thus 'make a contribution to the collection' does leave open how much one gives in obeying the rule. The methodological rule 'choose theories on the basis of simplicity' is more like the latter than the former sort of rule in that it leaves some options open for the scientist. Because of the openness of some rules, and because Kuhn perhaps thinks of rules as being determinate, he chose to talk of values instead. In order to accommodate Kuhn's position, in the talk of rules R and values V above we need to recognise that they may be fully determinate (or nearly so), or they leave open a range of options as to how one is to fulfil them.

[25] See Kuhn (1970, p. 206) for his rejection of the idea that our theories do approximate to the truth about what is 'really there', and that 'there is, I think, no theory-independent way to reconstruct phrases like "really there"; the notion of a match between the ontology of a theory and its "real" counterpart in nature now seems to me illusive in principle'.

[26] The papers in this collection which deal with Kuhnian themes include those of Pyle, Worrall and Forster.

[27] A list of methodological principles which Feyerabend at one time endorsed is given in Preston (1997, section 7.5). The Principle of Proliferation is criticised in Laudan (1996, pp. 105–110).

[28] Feyerabend often talks of 'rules' or 'standards' rather than 'principles'. Given the terminology adopted here principles become rules if all reference to values is omitted. The varying terminology between ourselves and other writers should cause no problems.

[29] In the 'Preface' to the Revised 1988 edition of *Against Method* Feyerabend again points out: '"anything goes" is not a "principle" I hold' (p. vii), but which he thinks that Rationalists must hold. He explains that he cannot endorse this claim because: 'I do not think that "principles" can be used and fruitfully discussed outside the concrete research situation they are supposed to affect'. This is hardly clear; however it might be taken to be a reference to the contextual character of principles of method.

[30] Earlier criticisms of Feyerabend's position appeared in reviews of *Against Method* to which Feyerabend replied; many of the replies are collected in Feyerabend (1978, part Three). For two recent evaluations of Feyerabend's views on particular principles see Laudan (1996, chapter 5) and Preston (1997, chapter 7).

[31] Popper uses this term, and elaborates his theory of a rational tradition which emerged with the Ancient Greeks and which includes his own critical rationalism, in a paper entitled 'Towards a Rational Theory of Tradition' in Popper (1963).

[32] An alternative line on the ramifications of a postmodernist approach to science that goes beyond the methodological issues discussed here is Rouse (1996, chapters 2–4). Rouse is aware of 'all the analytical deficiencies of Lyotard's essay on the postmodern condition' though he thinks that Lyotard's 'diagnosis ... as one of growing "incredulity towards metanarratives" seems quite accurate' (Rouse 1996, p. 56).

[33] The roots of SP reach back into theories of the sociology of knowledge which have their foundation in the work of Marx. Durkheim's and Mannheim's sociology of knowledge are a more immediate influence. However Mannheim excluded most of science from the scope of the sociology of knowledge while SP extends Mannheim's theory to all of science, thus indicating one sense in which SP is 'strong' and Mannheim's is weak (Bloor calls it the 'teleological model'). Bloor says of Mannheim that his 'nerve failed him when it came to such apparently autonomous subjects as mathematics and natural science' (Bloor 1991, p. 11).

[34] We can also incorporate Bloor's fourth tenet about the reflexive application of SP to itself into the scope of CT; however reflexivity will not be discussed here.

[35] Though there is much in the literature of cognitive psychology on the issue of the biological basis of our competence in reasoning, we will cite one reference to a philosopher who would question our 'natural' rationality: Stich (1985).

[36] Though the matter cannot be discussed here, such community-based notions of correctness raise the following questions: is x correct because the community assents to x?, or does the community assent to x because x is correct? Similar questions were asked in Plato's *Euthyphro*.

[37] See for example 'The Pseudo-Science of Science?' in Laudan (1996, chapter 10). This originally appeared as a symposium with Bloor in *Philosophy of the Social Sciences* 12, 1982. See also Roth (1987, chapters 7 and 8). The second edition of Bloor 1991 'Afterword' contains Bloor's replies to some of his critics.

[38]   Forman is not specific as to what this belief may be; it varies over beliefs such as not every event has a cause, or that there are objective chances, or that there are statistical laws that are not reducible to non-statistical laws, or that physical systems do not evolve deterministically as classically supposed but rather indeterministically, or that not all laws are exceptionless, and so on. Nothing turns on the specific content of the belief during the period of the rise of Quantum mechanics under investigation.

[39]   Quine's most extensive discussion of such principles can be found in Quine and Ullian (1978).

[40]   For criticisms of the H-D method see Reichenbach (1949, pp. 431, 432). Earman (1992, pp. 63–66) discusses a number of difficulties for the H-D method concerning confirmation and the irrelevant tacking paradox.

[41]   For an excellent account of Quine's zig zag path through the issues surrounding norms, naturalism and epistemology, see Haack (1993, chapter 6) and Foley (1994). For a discussion of a different view about the norm/naturalisation issue see the exchange between Hooker (1998) and Siegel (1998).

[42]   In an introduction to a collection of papers which test particular principles of method, Laudan and his co-researchers recognise problems with their test strategy saying:

> Some have asked why we choose to test theoretical claims in the literature rather than to inspect the past of science in an attempt to generate some inductive inferences. After all, the hypothetico-deductive method is not the only scientific method and it is not without serious problems. But our reading of the history of methodology suggests that hypothetical methods have been more successful than inductive ones. Add to that ... that we see no immediate prospect of inductive generalisations emerging from the historical scholarship of the last couple of decades and it will be clear why we have decided on this as the best course of action. Many of the theorists' claims can be couched as universals and hence even single case studies ... can bear decisively on them. Further, given enough cases, evidence bearing on even statistical claims can be compounded. (Donovan et al. 1992, pp. 12, 13)

What this passage shows is that those carrying out research in the light of NN can, and do, abandon Laudan's original meta-inductive rule MIR, or its statistical version, and replace it by H-D methods of test. However this must be provisional since these methods employed at the meta-level must themselves be open to test using meta-principles which are generally acceptable, such as MIR. The H-D method has not achieved comparable acceptance amongst methodologists. Moreover the authors recognise that whatever successes the H-D method has, it cannot be used without some peril. See Donovan et al. (1992, p. xiv ff).

[43]   One important sampling method that could be employed goes by the name of 'Monte Carlo methods', an account of which can be found in many texts on statistics.

[44]   In this brief account of NN we have passed over the problem of the status and role of norms within naturalism. For a discussion of some of these issues and a reply to critics see Laudan (1996, chapters 7 and 9). See also the exchange between Hooker (1998) and Siegel (1998).

[45]   Papers in this book which deal with aspects of NN are those of Kelly and Sankey.

[46]   Rescher holds a correspondence view of truth, as far as the definition of truth is concerned. In contrast, as far as the criterion for the recognition of truth is concerned, Rescher is a coherentist. Pragmatic elements enter into his overall position elsewhere, but not in connection with the definition of truth. For a criticism of the extension of pragmatism to the definition of truth, rather than the rules for recognising truth, see Rescher (1977, chapter IV).

[47]   For an extended and sympathetic account of Rescher's pragmatism with respect to science, see Hooker (1995, chapter 4). Hooker also considers Popper, Piaget and naturalism in the same context of his 'systems theory' of reason and knowledge in science.

[48]   There is a considerable literature starting with the work of Ramsey and De Finnetti in the first half of the twentieth century. More recent accounts include Jeffrey (1983), Earman (1992), Howson and Urbach (1993) and Maher (1993).

[49]   There are more sophisticated forms of updating beliefs, for example where the belief in E is less than certainty; see Jeffrey (1983, chapter 11).

[50]   For considerations largely for and some against synchronic and diachronic forms of the Dutch Book arguments see (Earman, chapter 2), Howson and Urbach (chapters 5 and 6) and Maher (chapters 4 and 5). A critical account can be found in Christensen (1991). See also van Fraassen 1989, chapter 7 and the position of van Fraassen 1995, especially p. 9 where he says 'I will explain why I do not want to rely on Dutch Book arguments any more, nor discuss rationality in the terms they set, though they have a very specific heuristic value'.

[51] The paper 'Scientific Inference' first appeared in a collection in 1970. Reference to the paper will be made to its reprinted version in Shimony (1993); the reprint is followed by a re-assessment of his views called 'Reconsiderations of Inductive Inference'.

[52] Bayes' Theorem standardly applies to a set of exclusive and exhaustive hypotheses. Thus if one considers hypotheses $H_1, H_2, \ldots, H_{n-1}$, then one also has to take into account the catch-all hypothesis H_n which is $\neg [H_1 \vee H_2 \vee \cdots \vee H_{n-1}]$. The hypotheses one actively considers might not contain the correct hypothesis while the catch-all does at least this, even if it is just H_n which simply says 'neither H_1 nor H_2 nor ... nor H_{n-1}'.

[53] The four theses are set out in van Fraassen (1989, p. 178). The brief account of van Fraassen given here is indebted to Kukla (1998, chapter 12), especially sections 12.3–12.6. Kukla provides a useful evaluation of the position set out in van Fraassen (1989, part II, chapter 7, 'The New Epistemology').

[54] It is also one of the grounds on which the dispute between realists and van Fraassen constructive empiricists is held to be both irreconcilable but epistemically blameless. See Kukla (1998, chapter 12.4, pp. 156, 157) for considerations as to why this might not be a compelling way of drawing the difference between OB and NE with respect to the realism/anti-realism issue, since advocates of OB need not be as impermissive as van Fraassen suggests.

REFERENCES

Albert, H., 1984: 'Transcendental Realism and Rational Heuristics: Critical Rationalism and the Problem of Method', in G. Andersson (ed.) *Rationality in Science and Politics*, Reidel, Dordrecht, pp. 29–46.

Black, M., 1954: 'Inductive Support for Inductive Rules', in *Problems of Analysis*, Cornell University Press, Ithaca NY, pp. 191–208.

Blake. R., Ducasse, C. and Madden, E., 1960: *Theories of Scientific Method: The Renaissance Through the Nineteenth Century*, The University of Washington Press, Seattle.

Bloor, D., 1991: *Knowledge and Social Imagery*, The University of Chicago Press, Chicago, second edition (first edition 1976).

Brante, T., Fuller, S. and Lynch, W. (eds): 1993: *Controversial Science: From Content to Contention*, State University of New York Press, Albany NY.

Buchdahl, G., 1980: 'Neo-Transcendental Approaches Towards Scientific Theory Appraisal', in D. H. Mellor (ed), *Science, Belief and Behaviour: Essays in Honour of R.B. Braithwaite*, Cambridge University Press, Cambridge, pp. 1–21.

Carnap. R., 1962: *The Logical Foundations of Probability*, The University of Chicago Press, Chicago, second edition (first edition 1950).

Christensen, D., 1991: 'Clever Bookies and Coherent Beliefs', Philosophical Review 100, 229–247.

Donovan, A., Laudan, L. and Laudan, R. (eds): 1992, *Scrutinizing Science*, The Johns Hopkins University Press, Baltimore, second edition (first edition 1988, Kluwer, Dordrecht).

Earman, J., 1992: *Bayes or Bust?: A Critical Examination of Bayesian Confirmation Theory*, The MIT Press, Cambridge MA.

Feyerabend, P., 1975: *Against Method*, NLB, London.

Feyerabend, P., 1978: *Science in a Free Society*, NLB, London.

Feyerabend, P., 1981: *Realism, Rationalism and Scientific Method: Philosophical Papers Volume 1*, Cambridge University Press, Cambridge.

Feyerabend, P., 1987: *Farewell to Reason*, NLB, London.

Feyerabend, P., 1995: *Killing Time*, Chicago, The University of Chicago Press.

Foley, R., 1994: 'Quine and Naturalized Epistemology', Midwest Studies in Philosophy: Volume XIX: Philosophical Naturalism 19, 243–260.

Forman, P., 1971: 'Weimar Culture, Causality, and Quantum Theory 1918–27: Adaptation by German Physicists and Mathematicians to a Hostile Intellectual Environment', in R. McCormmach (ed.), *Historical Studies in the Physical Sciences*, Vol. 3, University of Pennsylvania Press, Philadelphia, pp. 1–116.

Gallop, D., 1990: *Aristotle on Sleep and Dreams*, Broadview Press, Peterborough Ontario.

Goodman, N., 1965: *Fact, Fiction and Forecast*, Bobbs-Merrill, Indianapolis, second edition.

Gower, B., 1997: *Scientific Method: An Historical and Philosophical Introduction*, Routledge, London.

Grünbaum A., 1976: '*Ad Hoc* Auxiliary Hypotheses and Falsificationism', *The British Journal for the Philosophy of Science* 27, pp. 329–362.

Haack, S., 1993: *Evidence and Inquiry: Towards Reconstruction in Epistemology*, Blackwell, Oxford.

Hahn, L. and Schilpp, P. (eds): *The Philosophy of W.V. Quine*, Open Court, La Salle IL.

Hempel, C., 1981: 'Turns in the Evolution of the Problem of Induction', *Synthese* 46, 389–404.

Hempel, C., 1983: 'Valuation and Objectivity in Science', in R.S. Cohen and L. Laudan (eds.) *Physics, Philosophy and Psychoanalysis* Reidel, Dordrecht, pp. 73–100.

Hendry, J., 1980: 'Weimar Culture and Quantum Causality', *History of Science* 18, 155–180.

Hooker, C., 1995: *Reason, Regulation and Realism: Toward a Systems Theory of Reason and Evolutionary Epistemology*, Albany NY, State University of New York Press.

Hooker, C., 1998: 'Naturalistic Normativity: Siegel's Scepticism Scuppered', *Studies in History and Philosophy of Science* 29A, 623–637.

Howson, C. and Urbach P., 1993: *Scientific Reasoning: The Bayesian Approach*, Open Court, La Salle IL, second edition (first edition 1989).

Jeffrey, R., 1983: *The Logic of Decision*, The University of Chicago Press, Chicago, second edition (first edition 1965).

Kelly, K., 1996: *The Logic of Reliable Inquiry*, Oxford University Press, New York.

Kuhn, T., 1970: *The Structure of Scientific Revolutions*, The Chicago University Press, Chicago, second edition (first edition 1962).

Kuhn, T., 1977: *The Essential Tension*, The Chicago University Press, Chicago.

Kuhn, T., 1983: 'Rationality and Theory Choice', *The Journal of Philosophy* 80, 563–570.

Kuhn, T., 1991: 'The Road Since Structure', in Fine, A., Forbes, M. and Wessels, L. (eds.) *PSA 1990, Volume Two*, Philosophy of Science Association, East Lansing MI, pp. 3–13.

Kuhn, T., 1992: 'The Trouble with the Historical Philosophy of Science', Robert and Maurine Rothschild Distinguished Lecture, Cambridge MA: Department of the History of Science, Harvard University, pp. 3–20.

Kukla, A., 1998: *Studies in Scientific Realism*, New York, Oxford University Press.

Lakatos, I. and Musgrave, A. (eds.): 1970, *Criticism and the Growth of Knowledge*, Cambridge University Press, Cambridge.

Lakatos, I., 1978: *The Methodology of Scientific Research Programmes: Philosophical Papers Volume I*, Cambridge University Press, Cambridge.

Langley, P., Simon H., Bradshaw, G. and Zytkov, J., 1987. *Scientific Discovery: Computational Explorations of the Creative Process*, The MIT Press, Cambridge MA.

Laudan, L., 1977: *Progress and Its Problems: Towards a Theory of Scientific Growth*, Routledge and Kegan Paul, London.

Laudan, L., 1981: *Science and Hypothesis: Historical Essays on Scientific Method*, Reidel, Dordrecht.

Laudan, L., 1983: 'The Demise of the Demarcation Problem', in R. S. Cohen and L. Laudan (eds.) *Physics, Philosophy and Psychoanalysis*, Reidel, Dordrecht, pp. 111–127; reprinted as chapter 11 in Laudan (1996).

Laudan, L., 1984: *Science and Values: The Aims of Science and Their Role in Scientific Debate*, The University of California Press, Berkeley CA.

Laudan, L., 1986: 'Some Problems Facing Intuitionistic Meta-Methodologies', *Synthese* 67, pp. 115–129.

Laudan, L., 1987: 'Progress or Rationality? The Prospects for Normative Naturalism', *American Philosophical Quarterly* 24, 19–31; reprinted in Laudan (1996, pp. 125–141).

Laudan, L., 1996: *Beyond Positivism amd Relativism: Theory, Method and Evidence*, Westview Press, Boulder CO.

Laudan, L. *et al.*: 1986, 'Scientific Change: Philosophical Models and Historical Research', *Synthese* 69, pp. 141–223.

Losee, J., 1993: *An Historical Introduction to the Philosophy of Science*, Oxford University Press, Oxford, third edition.

Lycan, W., 1988: *Judgement and Justification*, Cambridge University Press, Cambridge.

Lyotard, J.-F., 1984: *The Postmodern Condition: A Report on Knowledge*, University of Minneapolis Press, Minnesota.

Maher, P., 1993: *Betting on Theories*, Cambridge, Cambridge University Press.

Mayo, D., 1996: *Error and the Growth of Experimental Knowledge*, University of Chicago Press, Chicago.

Musgrave, A., 1999: 'How to do Without Inductive Logic', *Science and Education* 8, 395–412.

Nola, R., 1987: 'The Status of Popper's Theory of Scientific Method', *The British Journal for the Philosophy of Science* 38, 441–480.

Oldroyd, D., 1986. *The Arch of Knowledge: An Introductory Study of the History of the Philosophy and Methodology of Science*, Methuen, London.

Polanyi, M., 1958: *Personal Knowledge*, Routledge and Kegan Paul, London.

Polanyi, M., 1966: *The Tacit Dimension*, Routledge and Kegan Paul, London.

Popper, K., 1957: *The Poverty of Historicism*, Routledge and Kegan Paul, London.

Popper, K., 1959: *The Logic of Scientific Discovery*, Hutchinson, London.

Popper, K., 1963: *Conjectures and Refutations*, Routledge and Kegan Paul, London.

Popper, K., 1972: *Objective Knowledge*, Clarendon, Oxford.

Popper, K., 1974: 'The Problem of Demarcation', Part II of 'Replies to My Critics', in P.A. Schilpp (ed.), *The Philosophy of Karl Popper*, Open Court, La Salle IL, pp. 976–1013.

Popper, K., 1983: *Realism and the Aim of Science*, Rowman and Littlefield, Totowa NJ.

Preston, J., 1997: *Feyerabend*, Polity, Cambridge.

Quine, W., 1969: *Ontological Relativity and Other Essays*, Columbia University Press, New York.

Quine, W., 1981: *Theories and Things*, Harvard University Press, Cambridge MA.

Quine, W., 1992: *Pursuit of Truth*, Harvard University Press, Cambridge MA., revised edition.

Quine W. and Ullian J., 1978: *The Web of Belief*, Random House, New York, second edition.

Reichenbach, H., 1949: *The Theory of Probability*, University of Califorinia Press, Berkeley.

Rescher, N., 1973: *The Primacy of Practice*, Oxford, Blackwell.

Rescher, N., 1977: *Methodological Pragmatism*, Blackwell, Oxford.

Rescher, N., 1987: *Scientific Realism*, Reidel, Dordrecht.

Rescher, N., 1992: *A System of Pragmatic Idealism Volume I: Human Knowledge in Idealistic Perspective*, Princeton University Press, Princeton NJ.

Rosenkranz, R., 1997: *Inference, Method and Decision*, Reidel, Dordrecht.

Roth, P., 1987: *Meaning and Method in the Social Sciences*, Cornell University Press, Ithaca.

Rouse, J., 1996: *Engaging Science*, Cornell University Press, Ithaca.

Salmon, W., 1967: *The Foundations of Scientific Inference*, University of Pittsburgh Press, Pittsburgh.

Salmon, W., 1990: 'Rationality and Objectivity in Science, *or* Tom Bayes meets Tom Kuhn', in C. Wade Savage (ed.), *Scientific Theories: Minnesota Studies in the Philosophy of Science* Vol. XIV, University of Minnesota Press, Minneapolis.

Savage, L. J., 1954: *The Foundations of Statistics*, John Wiley and Sons, New York.

Shimony, A., 1993: *Search for a Naturalistic World View: Volume I: Scientific Method and Epistemology*, Cambridge, Cambridge University Press.

Shrader-Frechette, K.S., 1991: *Risk and Rationality: Philosophical Foundations for Populist Reforms*, University of California Press, Berkeley CA.

Siegel, H., 1992: 'Justification by Balance', *Philosophy and Phenomenological Research* 52, 27–46.

Siegel, H., 1998: 'Naturalism and Normativity: Hooker's Ragged Reconciliation', *Studies in History and Philosophy of Science* 29A, 623–637.

Skyrms, B., 1975: *Choice and Chance*, Dickenson Publishing Co, Belmont CA, second edition.

Stich, S., 1985: 'Could Man be an Irrational Animal? Some Notes on the Epistemology of Rationality', in Kornblith (ed.) 1994, pp. 337–357.

Stich, S., 1990: *The Fragmentation of Reason*, MIT Press, Cambridge MA.

Strawson, P., 1952: *Introduction to Logical Theory*, Methuen, London.

Teller, P., 1975: 'Shimony's *A Priori* Arguments for Tempered Personalism', in G. Maxwell and R. Anderson (eds.), *Induction, Probability and Confirmation, Minnesota Studies in the Philosophy of Science Volume VI*, Minneapolis, University of Minnesota Press.

van Fraassen, B., 1980: *The Scientific Image*, Oxford University Press, Oxford.

van Fraassen, B., 1989: *Laws and Symmetry*, Clarendon, Oxford.

van Fraassen, B., 1995: 'Belief and the Problem of Ulysses and the Sirens', Philosophical Studies 77, 7–37.

Wright, C., 1989: 'The Verification Principle: Another Puncture – Another Patch', Mind 98, 611–622.

JOHN D. NORTON

HOW WE KNOW ABOUT ELECTRONS

1. INTRODUCTION

In 1997 we celebrated the centenary of Thomson's (1897) 'Cathode Rays' that is conveniently taken as marking the discovery of the electron, our first fundamental particle. The electron is not just our first fundamental particle, but one of the earliest microphysical entities to acquire secure status in modern physics. We see how early electrons acquired this status if we recall the status of the humble atom at this same time. While atoms had been a subject of interest in science and natural philosophy for millennia, their existence and properties remained clouded in debate and clear demonstrations of their existence and properties only emerged in the early part of this century when the same occurred for electrons.

   While the existence and properties of electrons stood at the forefront of physical research at the start of the twentieth century, any doubts about the electron's existence and basic properties soon disappeared. So physicist and historian Edmund Whittaker could review Thomson's investigations of the electron and conclude without apology:

> Since the publication of Thomson's papers, these general conclusions have been abundantly confirmed. It is now certain that electric charge exists in discrete units, vitreous [positive] and resinous [negative], each of magnitude 4.80×10^{-10} electrostatic units or 1.6×10^{-19} coulombs. (Whittaker 1951, p. 365)

Whittaker's confidence reflect a widespread certainty in the physics community about electrons. If the existence and properties of electrons were not assured in 1897, then this assurance arose in the years that followed, so that doubt over the existence of electrons has now moved beyond the realm of normal scientific prudence.

   My concern in this paper is to understand the stratagems used by physicists to arrive at this assurance. I will visit two general argument forms that have been used to affirm the existence and properties of electrons. The first, to be reviewed in section 3, has been brought to the notice of philosophers of science by Wesley Salmon in the corresponding analysis of the reality of atoms. It requires the determination of numerical properties of electrons in many different circumstances. That these properties invariably prove to have the same values – that is, their massive over determination by observation and experiment – is taken as evidence for the existence and properties of electron and that the parameters

67

Robert Nola and Howard Sankey (eds.), After Popper, Kuhn and Feyerabend, 67–97.
© 2000 *Kluwer Academic Publishers. Printed in Great Britain.*

computed are not just accidental artefacts of experiment. In the second, to be discussed in section 4, I will review a strategy for inferring theory *from* phenomena that has entered into the literature under many names including 'demonstrative induction' and 'eliminative induction'. These inferences are especially strong since they proceed from phenomena. Their inductive risk resides principally in the general hypotheses needed to enable the inferences so that security of the inference depends in great measure on our warrant for these general hypotheses. Before proceeding to these two stratagems, in section 2 I will review various traditions of scepticism that might lead us to disavow the existence and properties of electrons in spite of their entrenchment in modern physics. I will indicate why I find each tradition unsuccessful in sustaining scepticism about electrons. Concluding remarks are offered in section 5.

2. VARIETIES OF SCEPTICISM ABOUT MICROPHYSICAL ENTITIES

While the physics community may have harboured no real doubts over electrons for many years, several traditions of criticism have maintained and some continue to maintain that, at best, theories of microphysical entities cannot be taken at face value or, at worst, can in principle never succeed in their goal of revealing the nature of matter on a submicroscopic scale. I have divided these traditions into three classes in increasing order of the severity of their scepticism.

2.1. Evidential Insufficiency

This most modest of sceptical positions merely asserts that there happens to be insufficient evidence to warrant belief in the microphysical entity. While this attitude is so straightforward as to need little elucidation, it is helpful to review one of the most celebrated instances of this sceptical position: Wilhelm Ostwald's rejection of atomism in favour of his energeticism. More precisely, his attitude was that chemical thermodynamics simply did not need the hypothesis of atoms.[1] Those results that were usually thought to require atomism could be secured directly from the phenomena of chemical thermodynamics. For example he could recover stoichiometric laws such as the law of definite proportions. As he explained to an audience assembled in the inner sanctum of British atomism in his (1904, pp. 363, 364) Faraday Lecture, one distinguishes a chemical compound as those solutions for which the 'distinguishing point in the boiling curve'[2] is independent of pressure. He concluded his lecture with a flourish. His suggestions 'are questions put to nature. If she says Yes, then we may follow the same path a little further. If she says No – well, then we must try another path' (p. 522). One might suspect him of a feigned modesty given the history of polemical confrontation with atomists. But subsequent events proved otherwise and showed that Ostwald's scepticism was of the contingent nature appropriate to this category of evidential insufficiency. Famously, in the preface of the 1909 4th edition of his *Grundriss der Allgemeinen Chemie*, he announced:[3]

I have convinced myself that we have recently come into possession of experimental proof of the discrete or grainy nature of matter, for which the atomic hypothesis had vainly sought for centuries, even millennia.

He reflected on J.J. Thomson's successful isolation and counting of gas ions and Perrin's accommodation of Brownian motion to the kinetic theory, concluding

> ... this evidence now justifies even the most cautious scientist in speaking of the *experimental* proof of the atomistic nature of space-filling matter. What has up to now been called the atomistic hypothesis is thereby raised to the level of a well-founded theory, which therefore deserves its place in any textbook intended as an introduction to the scientific subject of general chemistry.

Is scepticism based on evidential insufficiency appropriate in the case of the electron? Whether it is must be decided by an investigation of the evidence available and its interpretation. The material in section 3 and after shows that the existence and basic properties of electrons lies well within the reach of current evidence as long as we are allowed standard stratagems for interpretation of this evidence. Needless to say this sort of scepticism is warranted concerning interactions sustained by electrons in exotic domains for which we have scant or no evidence.

2.2. Programmatic Restrictions

A stronger form of scepticism attempts to avoid entirely the issue of the evidential warrant for microphysical entities. Instead it asserts that the establishment of the existence and properties of microphysical entities is simply not the business of science. Its goals lie elsewhere and suppositions about microphysical entities are perhaps at best an intermediate convenience or a temporary delusion that will pass as the true purpose of science is realised. This attitude is exemplified in positivism or instrumentalism. Ernst Mach is the best known proponent of this attitude, asserting (1882, pp. 206, 207; Mach's emphasis):

> ... it would not become physical science to see in its self-created, changeable, economical tools, molecules and atoms, realities behind phenomena, forgetful of the lately acquired sapience of her older sister, philosophy, in substituting a mechanical mythology for the old animistic or metaphysical scheme, and thus creating no end of suppositious problems. The atom must remain a tool for representing phenomena, like the functions of mathematics. Gradually, however, as the intellect, by contact with its subject-matter, grows in discipline, physical science will give up its mosaic play with stones and will seek out the boundaries and forms of the bed in which the living stream of phenomena flows. The goal which it has set itself is the *simplest* and most *economical* abstract expression of facts.

When this goal is realised, we shall need talk of atoms no more since atoms do not figure in the facts of experience whose simple expression is sought.

This viewpoint promotes a scepticism about micro-entities not by directly casting doubt on our knowledge of them but by suggesting that all consideration of them is, in the last analysis, irrelevant to the true purposes of science. If this tradition of scepticism is taken on its face, it need not concern us here directly, for it amounts to a self-imposed decision not to entertain the existence of entities that are not directly part of the observed phenomena. Why should we impose this on ourselves? There are two cases to consider. Either micro-entities such as electrons lie within the reach of evidence or they do not. In the first case, programmatic scepticism seems wholly unwarranted. We can know about electrons. Why would we choose to know less than we can? Why would we think this a virtue? That

Mach would urge this in the case of atoms suggests that his programmatic restrictions are rooted in a deeper form of scepticism belonging to the second case indicated, in which micro-entities such as electrons lie beyond the reach of evidence. That Mach did harbour the ensuing blanket disbelief in micro-entities is suggested by his dismissal of molecules as things which 'only exist in our imagination' and are 'valueless images'.[4] But now our variety of scepticism is deepened and is seen to rest on presumptions about the methods of science. We must ask how these presumptions can be sustained.

2.3. Methodological Limitations

This deepest form of scepticism asserts that knowledge of micro-entities is something that simply extends beyond the reach of the methods of science. The roots of this form of scepticism can lie in several areas: philosophical, historical and sociological.

In its philosophical form, this version depends on a pessimism concerning the reach of evidence. The underdetermination thesis asserts that no body of evidence, no matter how extensive, will ever be able to determine a unique theory. So no matter how strong the evidential case may appear for some theory, other comparably viable competitors assuredly wait in the wings. The related Duhem–Quine thesis asserts that evidence must confront theory as whole; any particular hypothesis in a theory can be protected from falsification by suitable adjustment of other parts of the theory. For our purposes an immediate corollary is that the empirical success of any theory of micro-entities cannot assure us of the correctness of any particular hypothesis of the theory, so that while we may have an empirically successful theory of electrons, we cannot know as an independent fact that the electron charge is about 1.6×10^{-19} coulombs. I have argued elsewhere (Norton 1993, 1994) that both of these theses are false and that their failure can be shown by looking at the use of a particular strategy of inductive inference, demonstrative induction. I will describe in section 4 below how demonstrative induction was used in the case of the electron.

The historically founded scepticism derives from a recognition of the pervasiveness of change in theories in the history of science. The electron is a clear example. In the course of the century since Thomson's 'Cathode Rays' paper, theories of the electron have undergone near constant revision. Thomson's classical electron is not the electron of Einstein's 1905 relativity theory, which is not the jumping electron of Bohr's old quantum theory, which is not the wave of Schrödinger's wave mechanics, which is not the excitation of a Fermion field in quantum field theory – and so on in multiple subtler variations. The moral, according to the so called 'pessimistic meta-induction,' is that none of the superseded theories was correct, so, by induction, we can have no confidence that our latest theory is correct. Now the existence of a sequence of theories through time may be a manifestation of a pathology: massive, repeated, inexorable error and misconstrual of evidence. Or it may be evidence of great health: a tradition of theory which grows richer by the appropriation of new evidence and in which earlier theories are preserved in limiting form and corrected. In joint work with

Jon Bain (Bain and Norton, forthcoming) we have investigated this meta-induction in the case of the electron. We conclude the latter is the case and that the meta-induction fails for the electron. What the history of the electron reveals is a vigorously growing body of theory concerning the electron, in which the evidential successes of the early theories of the sequence are largely preserved as our understanding of the properties of the electron are refined and expanded. Thus our estimates of the charge and mass of the electron have scarcely altered in over eighty years while we have learned of properties of the electron unanticipated by Thomson: its quantum character, its possession of an intrinsic, quantized spin, that it obeys a Fermi–Dirac statistics and that its electromagnetic interactions may be unified with its interactions in the weak force. Within the sequence is a growing core of stable properties for which physicists have good evidence and which point to the existence of a single stable structure whose existence and nature is revealed in growing detail by the development of theories of the electron. We cannot conclude from this that physicists make no errors or that our latest theory is incorrigible. But optimism and not pessimism is surely licensed, for we are assured that the inevitable errors are sought, found and corrected leaving us with an ever more secure image of the electron.

Finally a sociologically based scepticism seeks to undermine the evidential warrant of scientific theories by examining the social structures and processes that produce the theories. Such seems to be the goal of the 'strong programme in sociology of knowledge' of Bloor (1991) which is intended, apparently, to answer affirmatively the question (p. 3) 'Can the sociology of knowledge investigate and explain the very content and nature of scientific knowledge?'. Insofar as the very content of scientific theories can be explained solely in terms of the social interactions of scientists, then that content can reflect only the conventional agreement of scientists and not an agreement of the theories with nature.[5]

The strong programme embodies a very strong form of scepticism. That the scepticism is justified remains entirely unclear. In evaluating it, we must guard against a simple error. We cannot conclude that a theory only reflects agreement between scientists merely because of the possibility in principle of giving a detailed reconstruction of the social processes that lead to its acceptance. Scientific theories are generated and validated by the communal effort of scientists. Thus it will always be possible to discern and describe the social process that lead to the communal acceptance of this or that theory. But offering such a purely sociological description, as is invited by the strong programme, cannot by itself decide whether a theory agrees with nature or fails to agree or to what degree. The community of astrologers has failed to discern causal influences from sun and moon to the earth; the community of astrophysicists has succeeded in discerning the gravitational influence of sun and moon that raises the earth's tides. That one has failed and one succeeded cannot be revealed merely by noting, even in painstaking detail, the exact course of social interaction that led to acceptance of this or that theory. Such judgement can only be provided by testing the evidential warrant offered by the communities against good epistemic standards, but such comparison has no place in the strong programme. More simply, whether a community has succeeded or failed in its goal of describing nature can only be determined if one is willing to consider what it takes to be successful, but such considerations have been eschewed in the strong programme.[6]

It may well be the case that the manoeuverings of particular scientists are driven by social factors: their needs for wealth or power or the jealous defeat of a rival. But that does not establish that the arguments they mount for the bearing of the theory on evidence are defective. Indeed sound arguments of this type would appear to be the most effective weapons in these battles. Again, as Forman (1971) suggests, the quantum physicists of the 1920s may well have found it expedient to hawk a new physics that emphasised chance and indeterminacy since those characteristics were welcomed by the chaotic society of Weimar Germany. But that would not preclude the possibility that these physicists had in addition good reasons and evidence for the indeterminism of their theory.

Needless to say, there are cases in science in which social factors have illegitimately determined the cognitive content of a scientific theory. A strong candidate is Cyril Burt's investigations of the inheritance of intelligence by means of identical twin studies. The posthumous discovery of anomalies in the statistics of his papers showed that his claims could not be read at face value and raised the question of whether his data has been faked to fit Burt's expectations.[7] Just one such case is needed to refute the view that the scientific endeavour is perfect and invariably offers theories with proper evidential warrant. But refuting that view is of little interest since it is not one that could ever be taken seriously. Rather we need assistance in deciding between two views: the complete scepticism of the strong programme or a more sober view which allows that some scientific theories enjoy proper evidential warrant whereas others do not. Cases such as Burt's do not allow us to distinguish these two views. But what would refute the first view, the complete scepticism of the strong programme, is even one case of a scientific theory with proper evidential warrant. There are many such cases. That of the electron is just one. The nature of the evidential warrant for the existence and properties of electrons will be reviewed below.

While these traditions of scepticism entail that scientists do not, should not or cannot establish the existence and properties of electrons, the broad consensus of physicists is that they long ago succeeded in doing just that. An enormous array of strategies and techniques of great complexity and ingenuity have been employed to this end and, in principle, there may be no common ground between the different arguments used to extract this or that property from the various items of observational evidence. It turns out, however, that we can discern two particular strategies that have been used very effectively to establish the existence and properties of electrons. I will review them in the sections that follow.[8]

3. OVERDETERMINATION OF CONSTANTS

3.1. Reality of Atoms and the Quantum

The great debate over the reality of atoms was resolved with some speed in the first decade of this century. Many contributed to the victory of the atomists, but their undisputed leader was Jean Perrin. His case for atoms was reduced to a single grand argument that was brought to the attention of the modern philosophical community by Salmon (1984, pp. 213–227). Perrin's argument was very simple in concept. Atomism is predicated on the idea that atoms are so small that matter appears continuous on the macroscopic scale. In earlier years atomists were unable to give reliable estimates of the sizes of atoms; they had to content themselves with the assertion that these sizes must be exceedingly small since they had transcended all attempts at measurement. With the coming of the twentieth century this situation changed. Through many different phenomena and experimental techniques, it became possible to estimate the size of atoms. The quantity computed in estimating this size is Avogadro's number N, the number of atoms or molecules in a gram mole.

Perrin himself had worked experimentally on determining the magnitude of N. When this work was drawn together with the work of others, Perrin was able to report roughly a dozen different methods for estimating N and they all gave values of N in close agreement. In the conclusion to *Les Atoms*, Perrin tabulated the resulting estimates of N from methods based on:[9] viscosity of gases (kinetic theory), vertical distribution in dilute emulsions, vertical distribution in concentrated emulsions, Brownian movement (displacement/rotations/diffusion), density fluctuations in concentrated emulsions, critical opalescence, blueness of the sky, diffusion of light in argon, black body spectrum, charge as microscopic particles, radioactivity (projected particles/Helium produced/Radium lost, energy radiated). The methods agreed in giving values of N in the range 60–69×10^{22} (with one exception, critical opalescence, that returned 75×10^{22}). The case for the reality of atoms and molecules lay in this agreement as Perrin explained (p. 215):

> Our wonder is aroused at the very remarkable agreement found between values derived from the considerations of such widely different phenomena. Seeing that not only is the same magnitude obtained by each method when the conditions under which it is applied are varied as much as possible, but that the numbers thus established also agree among themselves, without discrepancy, for all the methods employed, the real existence of the molecule is given a probability bordering on certainty.

The agreement of all these different methods for estimating N is to be expected if matter has atomic constitution. If, however, matter were not to have atomic constitution, then it would be very improbable that all these estimates of a non-existent quantity would turn out to agree.

In his analysis, Salmon (1984, pp. 213–227) has characterized the argument as employing the common cause principle. I do not wish here to pursue the connection to causation and the common cause principle since it seems to me that the essential result is secured already by a simple feature in the logic of the agreement. The agreement between the various estimates of the parameter is expected if the relevant theory is true, but it is very improbable if the theory is false.[10] In this form, the important result resides in an overdetermination of a parameter by many different methods. This overdetermination has been exploited quite frequently in the history of science – more examples follow. Since the parameter determined in these examples is always a constant of a theory, I have called the approach the method of overdetermination of constants.

The method was used by James Jeans when he sought to justify the then emerging quantum theory of the early 1920s in a new chapter added to the 1914 first edition in the 1924 second edition of his *Report on Radiation and the Quantum Theory*. Jeans noted that he had reviewed four phenomena that revealed the failure of quantum theory and the need for a new quantum theory (p. 61) '(i) Black Body Radiation; (ii) The spectra of the elements; (iii) The photoelectric effect; (iv) The specific heats of solids.' In the atomic theory, the size of atoms is set by the magnitude of N with the limit of a continuum theory approached with infinite N. The magnitude of the deviation from classical physics of quantum theory is set by Planck's constant h, with the classical limit arising when h vanishes. So Jeans proceeded to tabulate the values of h derived from the phenomena in these four areas, recovering values in very close agreement; they varied from 6.547×10^{-27} to 6.59×10^{-27}. According to Jeans (p. 61), these concordances demonstrate that the four phenomena 'agree in pointing to the same new system of quantum-dynamics.'

3.2. The Mass to Charge Ratio and the Charge of the Electron

The method of overdetermination of constants, as we shall now see, played an important role in the early history of the electron and much of importance was shown for the electron by demonstrating that the same constant values were recovered for each of the mass to charge ratio of the electron and the charge of the electron in many different circumstances.

When Thomson wrote his 1897 'Cathode Rays' the problem he addressed was not simply the issue of whether there are electrons. The issue was to decide between two theories of the nature of cathode rays. The theory favoured 'according to the almost unanimous opinion of German physicists' is that these rays are 'due to some process in the aether' (1897a, p. 293), that is, 'some kind of ethereal vibration or waves' (1906, p. 145). Thomson, along with his British colleagues, favoured the view that cathode rays consisted of charged corpuscles. More precisely, over the course of the following decade or two, Thomson and others sought to establish a series of properties for cathode rays:

(a) Cathode rays consist of a stream of corpuscles (electrons).
(b) Electrons are negatively charged.
(c) The universality of electrons: all cathode rays consist of electrons of just one type and these electrons are constituents of all forms of matter.
(d) Electrons are much less massive than atoms and molecules.

To make his case, Thomson sought to show that cathode rays had just the properties that would be expected of a stream of negatively charged corpuscles. Thus he recalled that Perrin had shown that cathode rays could impart a negative electric charge to an electroscope and then he reported his own improvement on the experiment.

The bulk of his paper was given over to reporting on two types of experiments: the deflection of cathode rays by magnetic fields and the deflection of cathode rays by electrostatic fields. Qualitatively, these experiments already yielded results indicating that cathode rays consisted of a stream of negative corpuscles. The rays were deflected along the direction of the electric field as expected for negatively charged particles and deflected perpendicular to the direction of the magnetic field as expected for negative charges in motion. The most telling result was that Thomson recovered the same value of m/e, the mass to charge ratio, for both magnetic and electric deflection. For the first case of magnetic deflection, he reported 26 values recovered from three cathode ray tubes operated under different circumstances and they lay in the small range of 0.32–1.0×10^{-7} – although Thomson doubted the accuracy of the tubes that gave the smaller results. For the second case of electric deflection, Thomson reported 7 values of m/e in the range 1.1–1.5×10^{-7}.

The overdetermination of this constant m/e was a strong test of the electron hypothesis. One might imagine that, through some fortuitous agreement of effects, an aetherial wave could be deflected by electric and magnetic fields in directions akin to that of deflected electric particles. But the concordance of the computed values of m/e showed quantitative agreement between the observed deflections and the properties of charged particles that transcends such chance. If a ray is deflected by a magnetic field, one can perhaps choose a value for m/e so that the deflection is compatible with the assumption that the ray consists of charged particles deflected by a magnetic field of strength $\mathbf{H}$ that will deflect the particles with acceleration $\mathbf{a} = -(e/m)\, \mathbf{v} \times \mathbf{H}$. But once this value of m/e is set, no further adjustment is possible to accommodate the deflection due to an electric field strength $\mathbf{E}$. That deflection is just to be measured and it must agree with the acceleration $\mathbf{a} = -(e/m)\mathbf{E}$. That both series of experiments returns the same value of (m/e) assures us that this necessary compatibility has been secured.

While the evidence of this quantitative agreement is strong, Thomson already felt that the qualitative result made the electric nature of cathode rays inescapable. His computation of the ratio m/e was intended to answer further questions. He wrote (1897a, p. 302):

> As the cathode rays carry a charge of negative electricity, are deflected by an electrostatic force as if they were negatively electrified, and are acted on by a magnetic force in just the way in which this force would act on a negatively electrified body moving along the path of these rays, I can see no escape from the conclusion that they are charges of negative electricity carried by particles of matter. The question next arises, What are these particles? are they atoms, or molecules, or matter in a still finer state of subdivision? To throw some light onto this point, I have made a series of measurements of the ratio of the mass of the particles to charge carried by it.

So the result that Thomson emphasised was that the value of m/e recovered was independent of the variation of many factors in his experiment. It did not vary appreciably if he used different gases in his tubes: air, hydrogen, carbonic acid; or if the electrodes were iron, aluminium or platinum. Thomson summarised the agreement in his 1906 Nobel Prize speech (Thomson 1906, p. 148):

> The results of the determinations of the values of e/m made by this method are very interesting, for it is found that, however the cathode rays are produced, we always get the same value for e/m for all the particle in the rays. We may, for example, by altering the shape of the discharge tube and the pressure of the gas in the tube, produce great changes in the velocity of the particles, but unless the velocity of the particles becomes so great that they are moving nearly as fast as light, when other

considerations have to be taken into account, the value of e/m is constant. The value of e/m is not merely independent of the velocity. What is even more remarkable is that it is independent of the kind of electrodes we use and also of the kind of gas in the tube. The particles which form the cathode rays must come either from the gas in the tube or from the electrodes; we may, however, use any kind of substance we please for the electrodes and fill the tube with gas of any kind and yet the value of e/m will remain unaltered.

This invariability of the ratio demonstrated the universality of electrons; they were the same whatever may be the matter from which they were derived. Finally the value of m/e of 10^{-7} was significantly smaller than even the smallest value then known for a charge carrier, the hydrogen ion of electrolysis, whose value in 1897 was estimated as 10^{-4}. From the constancy of m/e and its magnitude, Thomson drew his major conclusion (1897a, p. 312):

> ... we have in the cathode rays matter in a new state, a state in which the subdivision of matter is carried very much further than in the ordinary gaseous state: a state in which all matter – that is, matter derived from different sources such as hydrogen, oxygen, &c. – is of one and the same kind; this matter being the substance from which all the chemical elements are built.

From examination of the ratio m/e for electrons moving freely as cathode rays, Thomson had inferred to the universal presence of electrons in all matter. But are electrons identifiable within matter itself? As it turned out, in 1897, Thomson could report another determination of the ratio m/e from quite a different source. Zeeman (1897) had investigated experimentally the splitting of emission spectra by magnetic fields. As Zeeman explained (p. 232), H.A. Lorentz had communicated to him a theoretical analysis of the splitting. If the emitting atoms were modelled as bound, vibrating ions, the splitting could be accounted for by the magnetically induced alterations in the frequency of vibration. Working back from the magnitude of the shift in the spectral lines, Lorentz's model enabled Zeeman to estimate the charge to mass ratio of the ions: 'It thus appears that e/m is of the order of magnitude 10^7 electromagnetic C.G.S. units.' Thomson (1897b, p. 49) could not resist concluding another briefer treatment of his work on cathode rays by reporting this happy agreement of the value of e/m for charges bound within matter:[11]

> It is interesting to notice that the value of e/m, which we have found from the cathode rays, is of the same order of magnitude as the value 10^{-7} deduced by Zeeman from his experiments on the effect of a magnetic field on the period of the sodium light.

Thomson's demonstration of the constancy of m/e had allowed him to mount a good case for the properties (b)–(d) listed above. But his analyses had not established (a), the corpuscularity of electrons. He could not preclude the possibility that cathode rays and the matter of electrons are a continuous form of matter with a uniform mass and charge distribution so that any portion of the matter would present a constant ratio m/e. The possibility was eliminated by experiments aimed at directly determining the charge of the electron e. The celebrated experiments and analysis is due to Millikan (1913, 1917). But already a decade before Langevin (1904) had assembled an argument for the corpuscularity of the electronic matter that used the overdetermination of constants, that constant being, of course, the electric charge. He reviewed a series of methods then available for estimating electron charge. They were: measurement by ma investigators (including Thomson) of charged water droplets condensed fro supersaturated water vapour; investigations by H.A. Lorentz of the emissive and absorptive power of radiation by metals; and investigation by Townsend of the diffusion of ions in an electric field. These methods all produced values in agreement for the charge of the electron; the values ranged from 3.1×10^{-10} to 4×10^{-10} esu. Townsend's investigations also enabled another deduction of Avogadro's number in agreement with values then accepted. Langevin (p. 202) concluded:

> Here is an important group of concordant indications, all of absolutely distinct origin, which show without doubt the granular structure of electric charges, and consequently the atomic structure of matter itself. The measurements which I have just enumerated allow us to establish, in great security, the hypothesis of the existence of molecular masses.
>
> I seek to point out here this extremely remarkable result, which belongs without doubt to some fundamental property of the ether and of the electrons, that all these electrified centres, whatever may be their origin, are now identical from the point of view of the charge which they carry.

3.3. Limitations

The strength of the method of overdetermination of constants is that it allows comparison and combination of evidence from very diverse domains and, should the evidence disagree, that disagreement will be revealed clearly. The weakness of the method is that the significance of agreement need not always be apparent. We can infer to our intended hypothesis only if we can be assured that the concordance of results is very unlikely to arise if that hypothesis is false. But that assurance may be elusive. For example, the wave theory of light, as expressed in Maxwell's electrodynamics, famously predicts a velocity of prop-agation for light of 3×10^8 m/s. So, if we find numerous independent mea-surements of the speed of light returning this value, are we allowed to infer to the wave theory of light and not to a corpuscular theory? This agreement might well eliminate a Newtonian emission theory of light in which the velocity of light will vary with the velocity of the emitter. But it cannot preclude a theory that merely asserts that light consists of non-quantum, relativistic particles of zero rest mass, for all such particles will propagate at 3×10^8 m/s.

An almost exactly analogous problem arose for Thomson's 1897 argument. Recall that his original purpose was to decide between the corpuscular theory of cathode rays that he favoured and the aetherial wave theory. In introducing his paper (1897, pp. 293, 294), he complained of the difficulty of deciding between the two theories since 'with the aetherial theory it is impossible to predict what will happen under any given circumstances, as on this theory we are dealing with hitherto unobserved phenomena in the aether, of whose laws we are ignorant'. His remarks proved prescient. With the emergence in the 1920s of de Broglie's matter wave hypothesis and then Schrödinger's wave mechanics, it became quite apparent that at least some sort of wave-like theory of the electron would be adequate to the phenomena known to Thomson – although the form of the theory is of a type that we can scarcely fault Thomson for failing to anticipate. As it turns out, however, the bulk of Thomson's conclusions remains unaffected by this development. Electrons are negatively charged systems of just one type that inhere in all ordinary forms of matter at a subatomic level. Thomson also correctly concluded that individual electrons have a definite mass and charge and, at least in the form of cathode rays, do comprise independent systems. These conclusions do not, however, eliminate the possibility that electrons have a wavelike character. That possibility would have appeared remote in 1897, however, when no wavelike form of matter was known to which quite specific, discrete quantities of mass and charge could be assigned. The quantity of energy or momentum assignable to a light wave depended on the intensity and spatial extent of the wave.

4. EVIDENCE AS AN IMAGE OF THEORY

4.1. The Many faces of Demonstrative Induction

These deficiencies of the method of overdetermination of constants can be ameliorated by a stronger technique that gives a far more definitive verdict on the import of evidence. The penalty for this added strength is that situations in which this second method can be used are more contrived and harder to find. In its most general form, it is very simple. One starts with evidence statements, be they observations or experimental reports. From them, with the assistance of some general hypotheses, one deduces a theory or hypothesis within a theory. Several points are important. First, the inference is deductive. So there is no longer any inductive risk associated with the use of an inductive argument form. That risk has been relocated into assertions (the more general hypotheses) and the risk associated with accepting them usually proves easier to assess and control. Second, the direction of the deductive inference is *from* evidence *to* theory. This fact almost immediately de-fangs the underdetermination thesis since the item of evidence is seen to point to a particular theory or even particular hypothesis.

The method has recently been rediscovered by a number of philosophers and it goes under several names.[12] This multiplicity of names is unfortunate since it masks the fact all of these philosophers are discussing essentially the same method. The method was used by Newton in his *Principia* so it is easy to see why it is often called 'Newtonian deduction from the phenomena.' Again, since the arguments employed are deductive (i.e. demonstrative) yet serve a function usually reserved for inductive arguments, it is also natural to label the approach as using 'demonstrative induction'. We might also view the general hypotheses of the arguments extensionally as defining the largest class of theories in which we expect the true theory to be found. The observations or experimental reports then eliminate all but the viable candidates from this universe. In this view, the method employs 'eliminative induction'. Finally, if the universe of theories admits of parameterisation, by far the most common case, then the method has been called 'test theory methodology'.

On first acquaintance, it seems dubious that there might be non-trivial instances of these deductions. A few simple examples dispel this impression. The most straightforward is a simplification of a deduction used repeatedly by Newton in his *System of the World* to recover the inverse square law of gravitational attraction from the phenomena of planetary motion. For simplicity, assume that planetary orbits are circular (as they nearly are) and recall Kepler's third law of planetary motion which relates the period T of a planet's motion to the radius R of its orbit

$$T^2 \propto R^3.$$

This is the phenomenon whose theoretical significance is sought. Newton's laws of motion contribute to the general hypotheses through which this phenomenon will be interpreted. More precisely, his mechanics give us the result that a planet moving at velocity V in a circular orbit of radius R is accelerated toward the centre of the orbit with acceleration $A = V^2/R$. Also we have from simple geometry that the velocity V and period of orbit T are related by $V = 2\pi R/T$. Using these two results we now deduce from Kepler's third law that the planets are accelerated towards the centre of their orbits with an acceleration A that varies with the inverse square of the distance from the centre. For we can write

$$A = V^2/R = (2\pi)^2(R/T)^2(1/R) = (2\pi)^2(R^3/T^2)(1/R^2)$$

and note that (R^3/T^2) is a constant from Kepler's third law so that we have

$$A \propto 1/R^2.$$

In short, we have inferred from the phenomena of planetary orbits to the inverse square law of gravitation, even if only in a special form.

This is a simple example and quite transparent. See the literature cited earlier in this section for more substantial examples drawn from quantum theory, general relativity and other branches of modern physics. Norton (1993), for example, presents a very striking instance. In the ten years following Planck's 1900 analysis of black body radiation, the principal result came to be understood to be a somewhat weak and puzzling one: one could save the phenomena of black body radiation if one presumed some kind of quantum discontinuity, that is, that thermally excited systems could adopt only a discrete set of energy levels. While this result was clearly of some significance, it did not suffice to establish as aberrant a hypothesis as quantum discontinuity. That this hypothesis saved the phenomena did not preclude the possibility that other, more conservative hypotheses might not also suffice. These hopes were dashed in 1911 and 1912, when Ehrenfest and Poincaré showed in a most robust demonstrative induction that one could infer *from* the phenomena of black body radiation *to* quantum discontinuity. They thereby demonstrated the power of evidence to determine theory and, moreover, to force a particular hypothesis and one that was then strenuously resisted by the physics community.

4.2. Bohr's 1913 Atomic Theory

Niels Bohr's (1913) 'On the Constitution of Atoms and Molecules' developed a theory of atomic structure that was as bold as it was successful. Einstein reserved the highest praise for Bohr's achievement when he wrote in his *Autobiographical Notes* (p. 43) that it '. . . appeared to me as a miracle – and appears to me a miracle even today. This is the highest form of musicality in the sphere of thought'. The core of Bohr's theory was an account of the behaviour of electrons bound in orbit around the nuclei of atoms. Famously he supposed that the electrons could persist in a discrete set of stationary states governed by the electrostatic interaction between the electron and the nucleus. When electrons dropped from higher to lower energy states, however, they would emit a quantum of light radiation, thereby enabling Bohr to account for the discrete lines characteristic of atomic emission spectra.

Bohr's first published development of his theory was in the first section of his paper ('Part 1 – Binding of Electrons by Positive Nuclei: Section 1, General Considerations'). In recounting it, I will group and label Bohr's results to aid in later description of his arguments. To begin Bohr laid out the results that govern the orbit of a negatively charged electron around a positively charged nucleus on the assumption that the electron and nucleus interact only electrostatically. These results were standard and comprise:

A: The electrostatic model of electron orbits
An electron of negative charge of magnitude e and mass m much smaller than the nucleus orbits a nucleus of positive charge of magnitude E in a closed elliptical orbit with major semi-axis a and eccentricity ε. The energy W released in forming the orbital state is.[13]

$$W = eE/2a \tag{1}$$

and the frequency ω of the orbit (in cycles per second) is

$$\omega = \frac{\sqrt{2}W^{3/2}}{\pi e E \sqrt{m}}. \tag{2}$$

That orbiting electrons conformed to this electrostatic model was a startling aspect of Bohr's theory – perhaps even as surprising as the quantum discontinuity about to be introduced. Classical electrodynamics was then very well developed. One of its incontrovertible results was that a negatively charged electron in orbit about a positively charged nucleus is accelerated and therefore must radiate its energy and spiral into the nucleus, so that no stable orbit is possible. In considering only an electrostatic interaction, Bohr chose to ignore this prediction.

The next component of Bohr's theory was a restriction to a discrete set of the energy levels admissible for bound electrons. I will express the restriction in two forms:

B: Quantization of energy levels
The stationary electron orbits are restricted to those whose energy W and frequency ω are related by the condition

$$W = \tau h \omega / 2 \tag{3}$$

for τ a positive integer $1, 2, 3, \ldots$ and h Planck's constant. In the context of the electrostatic model, this restriction (3) induces equivalent restrictions on the energy W, frequency ω and major semi-axis a. We recover the first by using (2) to eliminate ω from (3)

$$W = \frac{2\pi^2 m e^2 E^2}{\tau^2 h^2}. \tag{4}$$

Bohr's justification of (3) was somewhat tenuous. He recalled Planck's then latest development of his theory of black body radiation and that it was based on the assumption that an oscillator with natural frequency ν would radiate energy in integral amounts $\tau h \nu$ where as before $\tau = 1, 2, 3, \ldots$ Bohr next considered the process of binding the electron to the atom. In falling from a great distance to a stationary orbit with frequency ω, Bohr simply assumed that Planck's frequency ν would be replaced by *half* the corresponding frequency ω of the orbit so that final energy of the orbit would be given by (3).

Finally, Bohr's (3) proves puzzling to every reader of Bohr's paper if the reader tries to fit the result with the mechanisms proposed in the remainder of the paper. It is easy to interpret (3) as deriving from a sequence of τ emissions as the electron drops to stationary states of successively lower energy. These successive states would differ in energy by the same amount, $h\omega/2 = h\nu$, and the radiation emitted with each transition would be of energy $h\nu$ at frequency ν. The catch is that (4) does not supply such equally spaced energies for the stationary states, so that (3) cannot be justified by the supposition that it represents τ distinct emissions. I will return to this rather unsatisfactory situation below, where we will see that Bohr himself abandoned his justification of (3) in terms of Planck's theory.

In section 2 of his paper, Bohr turned to a more successful application of Planck's notion:

C: Emission of light by quanta
When an electron drops from a stationary state with quantum number τ_1 to a state of lower energy with quantum number τ_2, energy is emitted as homogeneous radiation with frequency ν given by

$$W_{\tau_2} - W_{\tau_1} = h\nu. \tag{5}$$

The combination of (4) and (5) gave Bohr the great success of his theory. It now followed that the emission spectra of an excited atom would contain lines with the frequencies

$$\nu = \frac{2\pi^2 m e^2 E^2}{h^3} \left(\frac{1}{\tau_2^2} - \frac{1}{\tau_1^2} \right). \tag{6}$$

Bohr could now report near perfect agreement with the observed emission spectrum of hydrogen.

D: Observed emission spectrum of hydrogen
The lines of the spectrum are given by the formula

$$\nu = R\left(\frac{1}{\tau_2^2} - \frac{1}{\tau_1^2}\right) \tag{7}$$

where τ_1 and τ_2 are positive integers and the value of the constant R is given as

$$R = 3.290 \times 10^{15}. \tag{8}$$

Bohr's predicted functional form (6) matched exactly the functional form fitted to the observed spectral lines (7). Moreover Bohr could recover the value of the constant R to within plausible experimental error by substituting appropriate values for the hydrogen atom into the constant of his expression (6). For the hydrogen atom, Bohr reported, we have $e = E = 4.7 \times 10^{-10}$, $e/m = 5.31 \times 10^{17}$ and, with Planck's constant $h = 6.5 \times 10^{-27}$, we have

$$\frac{2\pi^2 m e^4}{h^3} = 3.1 \times 10^{15} \tag{9}$$

so that the theory predicts a hydrogen spectrum governed by

$$\nu = \frac{2\pi^2 m e^4}{h^3}\left(\frac{1}{\tau_2^2} - \frac{1}{\tau_1^2}\right). \tag{6'}$$

To summarise, by the close of section 2, the case that Bohr could mount for his theory resided in two arguments. The first, a deductive argument, captures the remarkable fact that the principles of Bohr's theory were able to save the phenomena of the hydrogen emission spectrum:

A: The electrostatic model of electron orbits
B: Quantization of energy levels
C: Emission of light by quanta
_____ (Deduction)
D: Observed emission spectrum of hydrogen.

On the strength of this argument, we can then say that Bohr's entire theory enjoys inductive support by the hypothetico-deductive scheme. I pass over the question of whether the scheme supports one or other of A, B or C in lesser or greater degree and represent this argument as

D: Observed emission spectrum of hydrogen.
_____ (Inductive support
                                                  – Hypothetico-Deduction
                                                    scheme)
A: The electrostatic model of electron orbits
B: Quantization of energy levels
C: Emission of light by quanta

4.3. Bohr's Demonstrative Induction

That so simple a theory should succeed in saving the phenomena of the hydrogen spectrum lends strong support to Bohr's theory. But against it remains the problem that Bohr's suppositions depend on very arbitrary elements. Why are we licensed to revert to simple electrostatics in selecting our stationary orbits? Are the quantum conditions imposed the only ones that will work? Might we not find an account of hydrogen spectra that does not require such wholesale departure from classical physics? The success of the hypothetico-deductive induction sketched above gives us no direct grounds for expecting that other less controversial analyses might not meet with comparable success. Anyone who has worked with Bohr's theory, however, rapidly loses such hopes. One quickly develops an intuitive sense that Bohr's principles, or something of comparable nature, are unavoidable if we are to give an adequate treatment of atomic spectra. His principles are in some sense re-expressions of the information already given us in the discreteness of the spectra. We shall soon see how these intuitions can be put in more precise form.

These intuitions certainly seem to be expressed by Millikan in his summary of Bohr's theory, written a few years after Bohr's paper. Millikan (1917, pp. 207–209) represented Bohr's theory as based on three assumptions, essentially comparable to A, B and C above, with A specialised to the case of circular orbits alone. Apparently wishing to assure the reader that these assumptions were not arbitrary flights of fancy, he announced (p. 209, Millikan's emphasis):

> It is to be noticed that, if circular electronic orbits exist at all, no one of these assumptions is arbitrary. Each of them is merely the statement of the existing *experimental* situation. It is not surprising, therefore, that they predict the sequence of frequencies found in the hydrogen series. They have been purposely made to do so. But they have not been made with any reference whatever to the exact numerical values of these frequencies.

Bohr also clearly sensed the artificiality of his initial development of the theory. His introduction of the condition B: Quantization of energy levels, in form of (3) contained an arbitrary deviation from the theory of Planck, justified only by its success in giving the right result. Planck's theory required emission of light energy in integral multiples of $h\nu$, where ν is the frequency of the emitting oscillator; Bohr based his theory on the supposition that stationary states with frequency ω are formed by the emission of light energy in integral multiples not of $h\omega$ but of $h\omega/2$, when an electron is captured by a nucleus. At this point in his development in his section 1, Bohr promised that all would soon be put right. Reflecting on the assumptions from which he was proceeding, he wrote (p. 5):

> The question, however, of the rigorous validity of both assumptions, and also of the application made of Planck's theory, will be more closely discussed in §3.

In returning to this question in section 3, Bohr (p. 12) immediately retracted one essential element of his argument for (3):

> ... we have assumed that the different stationary states correspond to an emission of a different number of energy-quanta. Considering systems in which the frequency is a function of the energy, this assumption, however, may be regarded as improbable; for as soon as one quantum is sent out the frequency is altered. We shall now see that we can leave the assumption used and still retain the equation [(3)], and thereby the formal analogy with Planck's theory.

Bohr could easily retract this element and with it his earlier justification for the quantum condition (3) for he was about to offer a far stronger derivation. That derivation lay in an inference that proceeded *from* the functional form of the observed hydrogen emission spectrum to the quantum condition (3). In reflecting on the derivation after it was complete, Bohr made explicit that his starting point now lay in observation and the inferences proceeded to theory. He wrote (section 3, p. 14):

> ... taking the starting point in the form of the law of the hydrogen spectrum and assuming that the different lines correspond to a homogeneous radiation emitted during the passing between different stationary states, we shall arrive at exactly the same expression for the constant in question as that given by [(6′)], if only we assume (1) that the radiation is sent out in quanta $h\nu$, and (2) that the frequency of radiation emitted during the passing of the system between successive stationary states will coincide with the frequency of revolution of the electron in the region of slow vibration.

In order to lay out concisely the argument Bohr develops in his section 3, I will again group and label Bohr's results. To begin, Bohr retained A: The electrostatic model of electron orbits, so that stationary electron states are possible. But he made essentially no assumptions about the further character of these stationary states other than:

E: Indexing of stationary electron states
These stationary states are indexed by a parameter τ and governed by the relation

$$W = f(\tau)h\omega \tag{10}$$

where f is an undetermined function.

Since the function f is undetermined, this condition places very little restriction on the stationary states. The explicit presence of $h\omega$ in the formula is unnecessary. It is only there to simplify the final expression for $f(\tau)$, which would otherwise end up containing a factor of $h\omega$. The sole content of Equation (10) is an indexing of the energies W by a parameter τ. The real restriction brought through (10) comes into force when a form for the function f is determined and the parameter τ is shown to admit only integral values. Nothing at this point in Bohr's argument requires τ to be integer valued; it could adopt continuous values. Thus Bohr no longer assumes the quantization of energy levels of the bound electron; he is about to derive it. Combining (10) with Equations (1) and (2) from the electrostatic model, Bohr now inferred that the stationary states satisfy not (4) but the generalised relation

$$W = \frac{\pi^2 m e^2 E^2}{2h^2 f^2(\tau)}. \tag{4′}$$

The assumptions of C: Emission of light by quanta, now enabled Bohr to translate this relation into a condition on the expected emission spectra.

Proceeding as before, these spectra will have lines at frequencies

$$\nu = \frac{\pi^2 m e^2 E^2}{2h^3} \left(\frac{1}{f^2(\tau_2)} - \frac{1}{f^2(\tau_1)} \right). \tag{6''}$$

Comparison with the functional form (7) of D: Observed emission spectrum of hydrogen, now fixed the undetermined function f as

$$f(\tau) = c\tau \tag{11}$$

where c is some constant and τ must be restricted to positive integer values alone. It is this last restriction on the range of values of τ that introduces the quantization of the bound electron's energy levels. That is, the crucial discreteness of the energy levels is now inferred from the observed spectrum and not posited.

All that remains to complete recovery of the quantization of the energy levels is to determine the value of the constant c. Of course that value could be fixed by employing the observed numerical value of R (8) in the formula for the observed spectrum. Bohr, however, proceeded to show that he had no need of this observation to fix the value of c. He could recover it merely by requiring that his theory behave classically in the domain of large quantum numbers.

F: Classical electrodynamics governs emission for large quantum numbers
An electron bound in orbit about a nucleus will emit its energy in light with a frequency equal to the momentary frequency of the orbit. If the electron energy is W, then that frequency is given by (2), $\omega = \sqrt{2} W^{3/2} / (\pi e E \sqrt{m})$.

In the domain of very large quantum numbers – say $\tau = N$ – this classical behaviour is to be imitated by an electron dropping from the $\tau = N$ energy state to the $\tau = N - 1$ state. The transition will generate a quantum of radiation of frequency

$$\nu = \frac{\pi^2 m e^2 E^2}{2h^3 c^2} \left(\frac{1}{(N-1)^2} - \frac{1}{N^2} \right) \approx \frac{\pi^2 m e^2 E^2}{2h^3 N^3} \cdot \frac{2}{c^2} \tag{12}$$

where the approximation introduced is for large N. The condition F requires that there be an emission of radiation at the frequency given by (2). Substituting the expression (4′) for W and the functional form $f(\tau) = c\tau$ into (2), we recover a frequency

$$\nu = \frac{\pi^2 m e^2 E^2}{2h^3 N^3} \cdot \frac{1}{c^3}. \tag{13}$$

Comparison of the two expressions sets the value of c at[14]

$$c = \tfrac{1}{2}$$

so that the expression for the admissible energy levels (4′) reverts to Bohr's original (4) $W = 2\pi^2 m e^2 E^2 / (\tau^2 h^2)$. Bohr's deduction of the condition

B. Quantization of energy levels is now complete. Cast as a demonstrative induction, it can be summarised as:

Observation
    D: Observed emission spectrum of hydrogen
      (functional form (7) only)
General Hypotheses
    A: The electrostatic model of electron orbits
    C: Emission of light by quanta
    E: Indexing of stationary electron states
    F: Classical electrodynamics governs emission
      for large quantum numbers
——————————————————————— (Deduction)
B. Quantization of energy levels

What continues to be noteworthy is that Bohr's argument required only the functional form fitted to the observed spectrum. Bohr's argument allows him to calculate the functional form's constant R and his predicted form is in close agreement with the observed value as we saw in (8) and (9) above.

4.4. A Reduced Form of Bohr's Demonstrative Induction

Impressive as Bohr's argument is, it still retains some features that are troubling to modern readers. The most significant is the continued dependence of the argument and theory on A: The electrostatic model of electron orbits. This model retains entities, such as elliptical orbits of electrons, and quantities such as the orbit's major semi-axis a and frequency ω that have been expunged from the ontology of modern, standard quantum theory. Of course we could not expect Bohr to foresee this. In 1913, with the amazing success of his analysis, it would be entirely reasonable to expect that quantum theory would settle on an ontology of discrete elliptical orbits for electrons with some as yet unknown theoretical element bringing about stochastic jumps between the admissible orbits.

Therefore it is interesting to notice that this electrostatic model is actually inessential to Bohr's demonstrative induction. Essentially the same results can be recovered from a reduced form of Bohr's argument that is compatible with the new quantum theory about to emerge in the 1920s. We will review the reduced version for the special case of an electron bound in the hydrogen atom and infer to the quantization of its energy levels.

The reduced form eschews all talk of elliptical orbits other than in the domain of correspondence with classical theory. Outside this domain, it posits only a stripped down version of the ontology of Bohr's 1913 theory:

A': Existence of stationary electron states
Electrons bound in an atom can persist in a variety of stationary states with energies $W(\tau)$, where τ is an index of these states of undetermined character and W is a strictly decreasing[15] function of τ, so that all states of equal energy are assigned the same value of τ.

No assumption is made or needed that these stationary states are elliptical orbits of some definite size and frequency of localised electrons. What is retained is that these states possess a definite energy. This condition will replace both A and E in Bohr's demonstrative induction. With this replacement, we can proceed as before. We invoke C: Emission of light by quanta, to arrive at the conclusion that the atomic emission spectra will contain frequencies

$$\nu = (W(\tau_2) - W(\tau_1))/h$$

for all admissible values of τ_1 and τ_2, such that $\tau_1 > \tau_2$. Comparing this expression with the functional form (7) of the observed emission spectrum for hydrogen, we conclude that the indices τ_1 and τ_2 adopt only positive integer values, 1, 2, 3, ... and the functional dependence of W is given as

$$W(\tau) = Rh/\tau^2 + \text{constant.} \tag{14}$$

We have now inferred the quantization of energy levels.

The constant R is still undetermined as is the additive constant in (14). Both values are set by invoking F: Classical electrodynamics governs emission for large quantum numbers. We will use a classical electrodynamic analysis in which the energy of an electron spatially very remote from the hydrogen nucleus is set to zero. Such an electron arises in the limit of infinitely large τ, so that we correspondingly set the additive constant of (14) to zero. We now take the case of a large value of $\tau = N$ for the hydrogen atom in which $E = e$ and substitute the simplified expression for $W = Rh/N^2$ into the expression $\omega = \sqrt{2}W^{3/2}/(\pi e^2 \sqrt{m})$ for the orbital frequency of an electron with energy W in the classical analysis. We recover an expression for both orbital frequency ω and frequency of emitted radiation ν

$$\nu = \omega = \frac{\sqrt{2}}{\pi e^2 \sqrt{m}} \cdot \frac{R^{3/2}h^{3/2}}{N^3}. \tag{15}$$

This classical process will be imitated by the light emitted in the transition from the state with $\tau = N$ to $\tau = N - 1$. We have from (7) that the frequency of light emitted in this process will be

$$\nu = R\left(\frac{1}{(N-1)^2} - \frac{1}{N^2}\right) \approx R \cdot \frac{2}{N^3}. \tag{16}$$

Setting the two frequencies of (15) and (16) equal we solve for an expression for $R = 2\pi^2 me^4/h^3$ which is just the expression for R in Bohr's theory. That is, we have recovered B: Quantization of energy levels, expression (4) restricted to the special case of the hydrogen atom for which $e = E$:

$$W(\tau) = \frac{2\pi^2 me^4}{h^2\tau^2}. \tag{4''}$$

We can summarise this reduced demonstrative induction as follows:[16]

Observation
   D: Observed emission spectrum of hydrogen.
      (functional form (7) only)
General Hypotheses
   A': Existence of stationary electron states
   C: Emission of light by quanta
   F: Classical electrodynamics governs emission
      for large quantum numbers
   ——————————————————————————————— (deduction)
   B: Quantization of energy levels (for electron
      in hydrogen atom, (4″))

This reduced demonstrative induction recovers the quantization of electron energy levels from a strict subset of Bohr's commitments. Bohr's A and E have been replaced by A' which is itself entailed by A and E. The inference now only returns results for the hydrogen atom. The argument can be readily modified to allow recovery of the corresponding results for other atoms if we are able to affirm in D that the emission spectra of these other atoms are also governed by the functional form (7).

It is also noteworthy that all the assumptions of this reduced demonstrative induction are compatible with the new quantum mechanics that emerged in the 1920s. The stationary states of A', for example, would simply correspond to the energy eigenstates of a bound electron. Therefore we would expect the conclusion to remain valid in the new quantum mechanics. And it does, of course. The energies of (4″) are simply the energy eigenvalues of an electron bound in a hydrogen atom.

This reduced demonstrative induction also gives us some insight into the much discussed logical inconsistency of Bohr's theory. That inconsistency lay in the presumption of the electrostatic model for electron orbits. That model provided for no radiation and thus had to be arbitrarily suspended as expedience required. In addition, one needed to ignore the massive body of evidence in other domains that showed that the behaviour of electrons was governed not merely by electrostatics but by electrodynamics. The reduced demonstrative induction shows us that Bohr's use of the electrostatic model was inessential for his celebrated account of atomic spectra. A subset of his commitments, free of manifest inconsistency, suffices for recovery of atomic spectra. This resolution is compatible with my earlier analysis (Norton 1987) of the logical inconsistency of the old quantum theory of black body radiation. There I urged that the viability of the theory depended on the existence of a consistent subtheory from which the essential results of the theory could still be recovered. We have now seen that the same strategy succeeds with Bohr's 1913 theory.

4.5. *The Strength of Bohr's Demonstrative Induction*

Bohr's theory provided a greatly deepened understanding of the properties of electrons bound in atoms. How strong was Bohr's evidence for these properties?

We have seen that Bohr's theory did not merely depend on its success in saving the phenomena of atomic spectra. There was a sense in which Bohr's theory was inferred from that phenomena. Thus Bohr's results are as secure as the reports of the phenomena and the demonstrative inductions that take us from them to the theory. In this section, I will assess the strength of the demonstrative induction. Since the argument itself is deductive, we need not torment ourselves with an evaluation of the degree of inductive risk introduced by an inductive argument form.[17] We have relocated all our inductive risk in the premises of the arguments. I will take the reports of atomic spectra as unproblematic and consider the general hypotheses that allow us to translate them into Bohr's theory. Bohr's case for his theory is made insofar as we can establish these general hypotheses.

There is very strong evidence for these general hypotheses. The evidence for them is of two types. The first is external and stems from other results in physics. The second is internal and derives from the way that the demonstrative inductions succeed.

To begin, we can review the external evidence by considering the general hypotheses individually. The condition C: Emission by light quanta, as Bohr makes clear, is imported from Planck's treatment of black body radiation. The general result – that systems of atomic size would emit energy in quanta of magnitude $h\nu$ – had become a fixture of the physics of the preceding decade. The result was difficult to interpret for there was no classical account of it, but it had repeatedly proved its utility . It was the core of Planck's original 1900 analysis of black body radiation and continued to feature in his more recent theories. Einstein has also developed analogous notions extensively commencing with his celebrated 1905 introduction of the notion of the light quantum. Finally the analyses of Ehrenfest and Poincaré of 1911 and 1912 (see Norton 1993) had shown the unavoidability in treatments of black body radiation of energy discontinuities associated with quanta of energy of size $h\nu$.

Soon after Bohr's investigations, Einstein's (1916a,b) celebrated 'A and B coefficients' papers gave even more secure foundation to C: Emission of light by quanta. Einstein pictured molecules with discrete energy levels in thermal equilibrium with radiation. He supposed that energy exchanges were governed by just three probabilistic processes: spontaneous and induced emission and absorption. From this extraordinarily simple foundation, he recovered Planck's formula for the distribution of energy in heat radiation. In the recovery, he compared his formula with that of the Wien displacement law and *concluded* that the frequency of light ν emitted or absorbed when the molecule alters its energy between energies ε_m and ε_n is given by the formula $\varepsilon_m - \varepsilon_n = h\nu$, remarking immediately that this result is just 'the second rule in Bohr's theory of spectra' (Einstein 1916b, p. 69). That is, the formula (5) of C: Emission by light quanta, was derived by Einstein along with the Planck formula.

The condition F: Classical electrodynamics governs emission for large quantum numbers, was easier to understand and virtually impossible to avoid. It required only that the behaviour of electrons revert to classical behaviour when they are no longer closely bound to atomic nuclei. A full and very secure account of the behaviour of such free electrons was provided by the crown jewel of ninteenth century physics, Maxwell's electrodynamics, as perfected by the turn of the century.

Most troublesome is A: The electrostatic model of electron orbits, which provides for the existence of stationary electron orbits governed by electrostatics.[18] To begin, Rutherford's experiments had shown that atoms consisted of very small positively charged nuclei and associated negatively charged electrons. Thus something like the electrostatic model with electrons orbiting a small nucleus was suggested. It was clear that the model could not be governed by classical electrodynamics in its entirety, for then the electron must radiate its energy and spiral into the nucleus. As Bohr (1913, p. 4) observed, this prediction of classical electrodynamics was not in accord with observation:

> A simple calculation shows that the energy radiated out during the process considered will be enormously great compared with that radiated out by ordinary molecular processes.

The simplest response is just to switch off that component of the classical theory that leads to radiation, that is, to revert to electrostatics. But how can we be assured that we have preserved the correct component of electrodynamics? As it turns out, according to the new quantum mechanics developed in the 1920s, Bohr preserved too much of the classical theory in continuing to represent bound electrons as possessing definite positions, elliptical orbits and the like. The reduced demonstrative induction of the preceding section shows, however, why this excess ontology was not fatal to Bohr's theory: it was simply superfluous to the treatment of atomic spectra. We would now locate the essential component of A merely in the supposition of the existence of stationary states. As the reduced demonstrative induction shows, the electrostatic quantities Bohr introduces through the model A can be introduced instead through the condition F: Classical electrodynamics governs emission for large quantum numbers.

In spite of the obviously problematic character of a theory that embodies the electrostatic model A, Bohr could be assured that there was still something very right about the theory. This assurance would come in the way that his demonstrative induction succeeded. This yields the internal evidence for the general hypotheses foreshadowed above. In brief, the demonstrative induction's result is massively overdetermined. Just as the overdetermination of constants gives inductive support for the theory in which they arise, so this overdetermination gives inductive support for the soundness of the demonstrative induction and the general hypotheses in particular.

The way in which this inductive support arises is strongly analogous to a more familiar circumstance, which I will use to elucidate the support. This analogy can be shown by introducing yet another way of describing demonstrative induction. As the title of section 4 indicates, in a demonstrative induction, we can conceive of the evidence as an image of theory, much as cameras and other optical instruments provide images of objects. Some of the structural properties of the objects are encoded within the image and, by suitable analysis, we can recover these properties from the image. For example, by stereoscopic analysis of aerial photographs, we can determine the heights of objects on the ground, although these heights might not be apparent on a casual scan of the photographs. Correspondingly, the evidence of Kepler's third law is a kind of image of the law of gravitation and encodes within it information about the structure of the law. The demonstrative induction sketched above allows us to extract that structure. Similarly, observed atomic emission spectra are images of the theoretical structure that interests us, the energy spectrum of the bound electron. We read the image and recover that energy spectrum with a demonstrative induction.

When we interpret an image produced by an optical instrument, we are inferring from a two dimensional image to aspects of the full structure of the three dimensional object that produced the image. To begin, we have some confidence in the interpretation if any simple reading at all of the image is possible; that is, if the image is not just noise. Correspondingly, we have some initial confidence in Bohr's demonstrative induction simply because it is possible at all and as simply as it is. But the mere fact that an interpretation of the image is possible, cannot give final assurance of its correctness. Optical systems are typically troubled by aberrations. How do we know that we are not mistaking such an aberration for a real feature of the original object, much as Galileo misinterpreted the distorted images of the rings of Saturn and inferred that the planet had ears?

That assurance comes when we procure multiple images of the same object, taken, for example, from many different angles. If we reconstruct the same object from each image, we become very confident of our interpretation. The multiplicity of images overdetermines the character of the object; each image provides a test of the interpretation of the other images. Correspondingly, Bohr's observation report on atomic spectra massively overdetermine his resulting theory. Only a small part of the spectral observations catalogued by (7) are needed to complete his demonstrative induction. These spectral observations are customarily divided into series with frequencies

$$R\left(1 - \frac{1}{n^2}\right) \quad \text{with } n = 2, 3, 4, \ldots \quad \text{(Lyman series)},$$

$$R\left(\frac{1}{2^2} - \frac{1}{n^2}\right) \quad \text{with } n = 3, 4, 5, \ldots \quad \text{(Balmer series)},$$

$$R\left(\frac{1}{3^2} - \frac{1}{n^2}\right) \quad \text{with } n = 4, 5, 6, \ldots \quad \text{(Paschen series)},$$

$$R\left(\frac{1}{4^2} - \frac{1}{n^2}\right) \quad \text{with } n = 5, 6, 7, \ldots \quad \text{(Brackett series), etc.}$$

It is easy to see that just the first of these alone, the Lyman series, provides an observational premise rich enough to support Bohr's complete demonstrative induction. That is, with Bohr's other premises, it is sufficient to force the functional form $f(\tau)$ to adopt the value (11) with integer values for τ. Thus the remaining series serve as tests, akin to the image of the same object from a different angle. If the theory were baseless, we would not expect each series to yield results concordant with the others when interpreted through the demonstrative induction. But it is already a part of the Bohr's induction that they all yield the same result.[19] The spectra of substances other than hydrogen could in principle supply more observation-images that would serve to overdetermine the theory further. This possibility was hard to realise in 1913. Bohr's (1913) discussion of the helium spectrum (pp. 10, 11) and of other substances (pp. 11, 12) was too hesitant to admit this possibility. Again, Bohr's treatment had concerned emission spectra only. In principle similar determinations would be possible for absorption spectra, but Bohr's (1913) discussion (pp. 15, 16) shows that such efforts would then have been premature.

There was another way in which the observations overdetermine the resulting theory. Bohr clearly took some pride in the power of his theory to give a definite value for the constant R of the formula (7), even though the value of the constant was already known from observation. In the analogy of the optical images, the value of the constant is another image of the energy spectrum of the hydrogen atom. On the basis of earlier images (the functional form of (7)) we predict what that new image must be. When the prediction matches the observation, we are assured again of our interpretation of the images.

5. CONCLUSION

Our present knowledge of the electron is the result of a century of vigorous investigation. While electrons are almost unimaginably small and abstruse in character, we can come to know of their existence and properties at the highest level of confidence. This paper has illustrated two of the stratagems used to reach this level of confidence. It also illustrates the utility of history of science in philosophy of science. That utility has become a commonplace of the last few decades of research. However I believe that it has often been misused. With talk of revolution and incommensurability widespread, it has been used to emphasise the irrational and the accidental in the history of science. While we must never lose sight of the highly contingent and often erratic character of science, it is all too easy to see nothing but this character in the history of science. One result is the pessimistic meta-induction discussed in section 2.3, which erroneously purports to establish the failure of all scientific theories without any explanation of how the failure arises.

When one approaches a speculative theory as bold as Bohr's 1913 theory, we add an easy drama to our histories if we overemphasise the irrational and accidental. Even the best of historians of science can be lured to do so. Thus Pais (1991, p. 148) calls Bohr's derivation of his expression for the constant in the spectral law (7) '... the most important equation that Bohr derived in his life. It represented a triumph over logic'. Pais found this notion so congenial that this section of his text is entitled 'Triumph over logic: the hydrogen atom'. If this is all we see in Bohr's achievement and others like it, then we end up with an image of science as a collection of imaginative, speculative leaps, untempered by prudence or reason. It is hard to have confidence that such an endeavour can supply us with an accurate and stable picture of physical reality.

What the analysis of this paper shows is that there was quite another side to Bohr's achievement of 1913. Once Bohr had conceived of the notion of accounting for atomic spectra through the transition of electrons between stationary states, the observed spectra and the requirement of concordance with then current physics drove him to a unique result, a particular energy spectrum for the electron. The passage to this result is laid out by Bohr himself in the demonstrative induction recounted in section 4.3. This shows that, in addition to any irrationality in Bohr's work, there is a core of sober theorising, firmly anchored in evidence. The reduced demonstrative induction of section 4.4 above shows us what this core is, how it is anchored in observation and, finally, that the success of Bohr's theory was quite independent of the much noticed inconsistency of his use of electrostatics in an electrodynamic system. While we admire Bohr for his brilliant, speculative leap, it is this sober core of his theory that survived the 1910s and was preserved in later theories of atoms and the electron. If we want to revel in the heroics of science, we should ask our history of science to report on these grand leaps. But if we want to understand how science succeeds in developing an ever more perfect picture of the physical world, we should ask our history of science about these stable cores of sound theorising that survive from one day to the next.

University of Pittsburgh

NOTES

[1] In Ostwald's (1904, p. 508) words: 'Chemical dynamics has, therefore, made the atomic hypothesis unnecessary for this purpose [of deducing stoichiometric laws] and has put the theory of the stoichiometric laws on more secure ground than that furnished by a mere hypothesis.'
[2] If one plots the boiling point of a solution against the composition of the solution, the distinguishing point is a maximum or minimum of the curve. A solution boiling at this temperature does not change its composition.
[3] Translation from Brush (1976, Vol. 2, p. 699). Ostwald's emphasis. This demonstration of the atomic hypothesis represented a serious threat to the Second Law of Thermodynamics which is apparently violated by fluctuation phenomena such as Brownian motion. That the Law could be retained for at least macroscopic purposes required careful analysis and the resulting literature mutated and evolved into a quite surprising direction. See Earman and Norton 1998–1999.
[4] As quoted in Brush (1976, Vol. 1, p. 286) from Mach's *The History and Root of the Principle of Conservation of Energy* (Chicago: Open Court, 1911), p. 86.
[5] Although Bloor protests that such scepticism is not the goal of the strong programme (e.g. p. 166), it is hard to see how that goal can be disavowed when Bloor looks to sociology to 'explain the very content ... of scientific knowledge.' Certainly sufficient of his critics have supposedly misunderstood Bloor's intentions in this way for him to need an Afterword (pp. 165–170) devoted to correcting them. Among them is the eminent sociologist Ben-David.
[6] Such considerations do play a role in Collins' (1982, 1985, Ch. 4) analysis of the experimenter's regress. He describes the failure of scientists to achieve agreement with nature through experimental methods because of a fatal circularity in their use of experimental apparatus: correct results can only be obtained from good experiments; but good experiments are just those that produce correct results. Not even Collins holds that this regress supports a blanket scepticism about all experiment – some experiments are not defeated by it (see Collins 1985, p. 84). At best it suggests that some experimental claims are not well founded. In general the regress is broken by the independent calibration of the apparatus: we know which are the good experiments because we check that they give correct results in cases in which we know independently what the correct results are. Indeed it remains open as to whether there are any interesting cases of the regress. That Collins' example concerning the detection of gravitational radiation fails to illustrate the regress is shown by

Franklin (1994) and he also gives further general discussion of the role of calibration in experiment. (For Collins' rather odd reply, see Collins 1994.)

[7] See for example Dorfman (1978). The debate over Burt's work continues. See Mackintosh (1995).

[8] I will not consider Hacking's (1982) analysis in which he urges that the electron made the transition from a theoretical entity to one that was realistically construed when physicists began to manipulate electrons as part of further investigations. This is because Hacking's analysis does not reveal the basis for physicists belief in the reality of atoms; rather it displays evidence of that belief, their willingness to think of electrons as something that can be manipulated for other ends.

[9] The list quotes the row headings from Perrin's (1913) table, p. 215 from the translation volume.

[10] As an example of the latter, consider someone trying to mount a case for an aether in the nineteenth century. They may succeed in finding different methods of estimating the earth's (non-vanishing) velocity through the aether. But exactly because there is no such velocity, we would have no expectation that truly independent methods could yield concordant estimates.

[11] Thomson's remark contains a trivial, possibly even typographical error. The ratio e/m computed by Zeeman has the approximate value 10^{+7}; whereas its inverse m/e routinely computed by Thomson has the approximate value 10^{-7}. Millikan (1917, pp. 40, 41) was sufficiently impressed with Thomson's overall argument that he used it in his text to answer the question 'Do all atoms possess similar constituents? In other words, is there a primordial subatom out of which atoms are made?'. His answer came in recalling the magnitude and constancy of e/m recovered by Thomson and Wiechert for cathode rays and Zeeman's 1897 discovery of the same ratio for charges within atoms.

[12] See for example Bain (1998), DiSalle et al. (1994), Dorling (1973, 1990, 1995), Gunn (1997), Harper (1990, 1997), Harper and Smith (1995), and Norton (1993, 1994, 1995). For a critical response see Bonk (1997) and Hudson (1997).

[13] Bohr does not adhere to the modern practice of presenting the binding energy W of the electron as a negative number. His energy W is the positive energy released during formation of the atom. Thus lower energy electrons correspond to higher values of W. I will adhere to Bohr's sign convention.

[14] Bohr also alludes to a slightly more general analysis that would give the same result. He asserts that classical electrodynamic analysis of an emitting electron in an elliptical orbit will yield radiation at frequencies $n\omega$, where $n = 1, 2, 3, \ldots$ so that the emitted spectrum has frequencies $\nu = \omega n$. This is returned in his theory by taking the case of an electron with a large quantum number N and considering emissions associated with transitions from state N to state $(N - n)$. For large N and small n, the expressions (12) and (13) are now replaced by analogous expressions

$$\nu = \frac{\pi^2 m e^2 E^2}{2h^3 c^2} \left(\frac{1}{(N-n)^2} - \frac{1}{N^2} \right) \approx \frac{\pi^2 m e^2 E^2}{2h^3 N^3} \cdot \frac{2}{c^2} \cdot n$$

and

$$\nu = \frac{\pi^2 m e^2 E^2}{2h^3 N^3} \cdot \frac{1}{c^3} \cdot n.$$

Comparison of the two expressions yields the same result, $c = 1/2$.

[15] Recall Bohr's sign convention for energy: W is the positive energy released on binding the electron to the nucleus, so the deeper the binding and the smaller the index, the greater the positive energy.

[16] While Bohr's 1913 theory is commonly presented in terms of the quantization of the orbital angular momentum of the electron, I have not cast the demonstrative induction in terms of angular momentum, because the theory's treatment of orbital angular momentum is not entirely satisfactory. Classical analysis shows that the angular momentum l of an electron with energy W is given by $l^2 = m e^2 E^2 (1 - \varepsilon^2)/2W$, where ε is the orbit's eccentricity. If we now substitute W with (4), the expression for quantized energy levels, we recover only a partial statement of the quantization of angular momentum: $l = \sqrt{1 - \varepsilon^2}(h/2\pi)\tau$. This does not yet give us the quantization of orbital angular momentum into multiples of $h/2\pi$ since nothing yet precludes the eccentricity ε adopting a continuous range of values. The further condition needed to achieve this arose first in Sommerfeld's (1923, Chapter 2) elaboration of Bohr's theory in which he quantized both degrees of freedom of the two dimensional electron orbit, introducing a radial quantum number n_r and an azimuthal quantum number n_φ. Their sum $n = n_r + n_\varphi$ is the principal quantum number and corresponds to the τ of (4). In this scheme, ε is restricted to a discrete set of values by the condition $\sqrt{1 - \varepsilon^2} = n_\varphi/n$. Substitution

into the above expression for angular momentum now returns the expected quantization of angular momentum $l = (h/2\pi)n_\varphi$. Since this quantization is expressed in terms of the azimuthal quantum number, Bohr was in no position to recover the result from his emission spectra. In a well known degeneracy in Sommerfeld's theory, the energy of a bound electron depends only on the principal quantum number and is $W = 2\pi^2 me^2 E^2/(h^2(n_r + n_\varphi)^2)$, so that an examination of this energy spectrum (4) alone could not enable Bohr to discern the two quantum numbers that comprise the principal quantum number. Bohr was still able to report the quantization of angular momentum, but only by the artifice of momentarily restricting himself to circular orbits (Bohr 1913, p. 15). In that case the radial quantum number vanishes and the principal and azimuthal quantum numbers become equal.

[17] This is a notoriously difficult problem if we do not embed the inference in a richer framework, such as in a Bayesian analysis (and that introduces further problems). How many instances are needed to give a high degree of certainty in instance confirmation? One or two cyanide fatalities may convince us that large doses of cyanide are always fatal. But one or two dry summers may not convince us that all summers are dry. How much certainty accrues to an hypothesis when it makes a single successful prediction? How much with a second successful prediction?

[18] This condition couples naturally with E: Indexing of stationary states. But I need say little about E. since it adds essentially nothing to the suppositions of A. It functions more as a definition. Its equation (10) supplies the definition of an index τ of these stationary states, without restricting the character of these states.

[19] The qualification, of course, is that the range of τ varies in each case: The Lyman series only delivers the full range $\tau = 1, 2, 3, \ldots$; the Balmer $\tau = 2, 3, 4, \ldots$; the Paschen $\tau = 3, 4, \ldots$; etc. At the time of writing Bohr (1913), he knew only of the Balmer and Paschen series but anticipated the existence of the others as series (p. 9) 'which are not observed, but the existence of which may be expected.' These two then known series are already sufficient to give the overdetermination under discussion.

REFERENCES

Bain, J., 1998: "Weinberg on QFT: Demonstrative Induction and Underdetermination," *Synthese* 117, 1–30.

Bain, J. and Norton, J.D., forthcoming: 'What Philosophers of Science Should Learn from the History of the Electron,' in A. Warwick and J.Z. Buchwald (eds.) *The Electron and the Birth of Microphysics. Dibner Institute Studies in the History of Science and Technology.* MIT Press, Cambridge, MA.

Bloor, D., 1991: *Knowledge and Social Imagery* (2nd edn.), University of Chicago Press, Chicago.

Bohr, N., 1913: 'On the Constitution of Atoms and Molecules,' *Philosophical Magazine*, 26, 1–25.

Bonk, T., 1997: 'Newtonian Gravity, Quantum Discontinuity and the Determination of Theory by Evidence,' *Synthese* 112, 53–73.

Brush, S., 1976: *The Kind of Motion We Call Heat*, North-Holland, Amsterdam.

Collins, H.M., 1982: 'The Replication of Experiments in Physics,' in Barnes, B. and Edge, D. (eds.), *Science in Context: Readings in the Sociology of Science*, Open University Press, Milton Keynes, pp. 94–116.

Collins, H.M., 1985: *Changing Order: Replication and Induction in Scientific Practice.* Sage Publications, London.

Collins, H.M., 1994: 'A Strong Confirmation of the Experimenters' Regress,' *Studies in History and Philosophy of Science* 25, 493–503.

DiSalle, R., Harper, W.L., Valluri, S.R., 1994: 'General Relativity and Empirical Success,' in Jantzen, R.T. and Mac Keiser, G. (eds.), *The Seventh Marcel Grossmann Meeting on recent developments in theoretical and experimental general relativity, gravitation, and relativistic field theories: Proceedings of the Meeting held at Stanford University, 24–30 July 1994*, World Scientific, Singapore, pp. 470–471.

Dorfman, D., 1978: 'The Cyril Burt Question: New Findings,' *Science* 201, 1177–1186.

Dorling, J., 1973: 'Demonstrative Induction: Its Significant Role in the History of Science,' *Philosophy of Science* 40, 360–372.

Dorling, J., 1990: 'Reasoning from Phenomena: Lessons from Newton,' *PSA 1990*, Vol. 2, 1991, 197–208.

96 JOHN D. NORTON

Dorling, J., 1995: 'Einstein's Methodology of Discovery was Newtonian Deduction from the Phenomena.' in Leplin, J. (ed.) *The Creation of Ideas in Physics*, Kluwer, Dordrecht, pp. 97–111.

Earman, J. and Norton, J.D., 1998–1999: 'Exorcist XIV: The Wrath of Maxwell's Demon,' Parts I and II, Studies *in the History and Philosophy of Modern Physics* 29B, 435–471 and 30B, 1–40.

Einstein, A., 1916a: 'Strahlungs- Emission und -Absorption nach der Quantentheorie,' *Deutsche Physikalische Gesellschaft, Verhandlungen* 18, 318–323.

Einstein, A., 1916b: 'Zur Quantentheorie der Strahlung,' *Physikalische Gesellschaft Zürich, Mitteilungen* 18, 47–62; *Physikalische Zeitschrift* 18, 121–128; translated G. Field, 'On the Quantum Theory of Radiation,' in van der Waerden, B.L. (ed.), *Sources of Quantum Mechanics*, Dover, New York, 1968, pp. 63–77.

Einstein, A., 1979: *Autobiographical Notes*, Open Court, La Salle and Chicago.

Forman, P., 1971: 'Weimar Culture, causality and quantum theory, 1918–1927' *Historical Studies in the Physical Sciences*, 3, 1–115.

Franklin, A., 1994: 'How to Avoid the Experimenters' Regress', *Studies in History and Philosophy of Science* 25, 463–491.

Gunn, D., 1997: Test-theory Methodology in Physics, Ph.D. Dissertation, University of Canterbury, New Zealand.

Hacking, I., 1982: 'Experimentation and Scientific Realism,' *Philosophical Topics*, 13 , 71–87; reprinted in Boyd, R. *et al.* (eds.) *The Philosophy of Science*, Bradford, Cambridge MA, 1991.

Harper, W., 1990: 'Newton's Classic deductions from Phenomena,' *PSA 1990*, Vol. 2, 1991, 183–196.

Harper, W., 1997: 'Isaac Newton on Empirical Success and Scientific Method,' in Earman, J. and Norton, J.D. (eds.), *The Cosmos of Science: Essays of Exploration*, University of Pittsburgh Press, Pittsburgh; Universitätsverlag Konstanz, Konstanz, pp. 55–86.

Harper, W. and Smith, G.E., 1995: 'Newton's New Way of Inquiry,' in Leplin, J. (ed.) *The Creation of Ideas in Physics*, Kluwer, Dordrecht, pp. 113–166.

Hudson, R.G., 1997: 'Classical Physics and Early Quantum Theory: A Legitimate Case of Theoretical Underdetermination,' *Synthese* 110, 217–256.

Jeans, J., 1924: *Report on Radiation and the Quantum Theory* (2nd edn.) Fleetway Press, London.

Langevin, P., 1904: 'The Relation of Physics of Electrons to Other Branches of Science,' in *Physics for a New Century: Papers Presented at the 1904 St. Louis Congress*. Tomash Publishers/American Institute of Physics, 1986, pp. 195–230.

Mach, E., 1882: 'The Economical Nature of Physical Inquiry,' in McCormack, T.J. (trans.), *Popular Scientific Lectures* (3rd edn. with additional new lectures, 1898; 5th edn., 1943), Open Court, La Salle, Illinois, 1986.

Mackintosh, N.J. (ed.), 1995: *Cyril Burt: Fraud or Framed?*, Oxford University Press, Oxford.

Millikan, R.A., 1913: 'On the Elementary Electrical Charge and the Avogadro Constant,' *Physical Review* 2, 109–124, 133, 136–143.

Millikan, R.A., 1917: *The Electron: Its Isolation and Measurement and the Determination of some of its Properties*, University of Chicago Press, Chicago.

Norton, J.D., 1987: 'The Logical Inconsistency of the Old Quantum Theory of Black Body Radiation,' *Philosophy of Science* 54, 327–350.

Norton, J.D., 1993: 'The Determination of Theory by Evidence: The Case for Quantum Discontinuity 1900–1915,' *Synthese* 97, 1–31.

Norton, J.D., 1994: 'Science and Certainty,' *Synthese* 99, 3–22.

Norton, J.D., 1995: 'Eliminative Induction as a Method of Discovery: How Einstein Discovered General Relativity,' in Leplin, J. (ed.) *The Creation of Ideas in Physics*, Kluwer, Dordrecht, pp. 29–69.

Ostwald, W., 1904: 'Faraday Lecture: Elements and Compounds' reproduced in Knight, D.M. (ed.), *Classical Scientific Papers: Chemistry*, American Elsevier, New York, 1968.

Pais, A., 1991: *Niels Bohr's Times, In Physics, Philosophy, and Polity*, Clarendon, Oxford.

Perrin, J., 1913: *Les Atoms*, Libraire Felix Alcan. Trans. D.Ll. Hamick, *Atoms*, Ox Bow Press, Woodbridge, CT, 1990.

Salmon, W.C., 1984: *Scientific Explanation and the Causal Structure of the World*, Princeton University Press, Princeton.

Sommerfeld, A., 1923: *Atomic Structure and Spectral Lines* (3rd edn., rev., 1934), trans. H.L. Brose, Methuen, London.

Thomson, J.J., 1897a: 'Cathode Rays,' *Philosophical Magazine* 44, 293–316.

Thomson, J.J., 1897b: 'Cathode Rays,' Discourse in Physical Science, Royal Institution, Friday April 30, 1897 in Bragg, W.L. and Porter, G. (eds.), *Royal Institution Library of Science. Physical Sciences.* Vol. 5, Applied Sciences Publishers, London.

Thomson, J.J., 1906: 'Carriers of Negative Electricity' Nobel Lecture, December 11, 1906, in *Nobel Lectures: Physics, 1901–1921*, Elsevier, Amsterdam, 1967, pp. 145–53.

Whittaker, E., 1951: *A History of the Theories of Aether and Electricity*, Vol. 1: *The Classical Theories*, Nelson, London. Reprinted, Dover, 1989.

Zeeman, P., 1897: 'On the Influence of Magnetism on the Nature of the Light emitted by a Substance,' *Philosophical Magazine* 53, 226–39.

ANDREW PYLE

THE RATIONALITY OF THE CHEMICAL REVOLUTION

1. HOW NOT TO CRITICIZE KUHN

Kuhn's account of science, in his *Structure of Scientific Revolutions*[1] (SSR) attracted a lot of hostile criticism from philosophers of science.[2] Kuhn was accused of such terrible sins as idealism, relativism, and irrationalism, of casting doubt on the rationality and objectivity of science. Kuhn himself, in a series of replies to his critics, made a number of concessions, and qualified or toned down many of his more extravagant claims.[3] But a more robust response can be found among some of Kuhn's disciples in the history and especially in the sociology of science.[4] They insist that Kuhn, in SSR, was simply telling it as it is, i.e., describing accurately the way that science was (and is) done. *Prima facie*, of course, such an account would be merely descriptive, and would have no implications for normative methodology.[5] If we were to find out that scientists are stubbornly irrational individuals, perversely hanging on to old theories even in the face of powerful counter-evidence, we could simply dismiss such a finding as regrettable evidence of human frailty, but of no epistemological significance. A normative element might, however, creep in by the back door, by way of a historical study of the heroes of science. Can we imagine discovering that Copernicus, Galileo, Newton, Lavoisier, Darwin and Einstein were all bad scientists? Perhaps the norms of science are already implicit in the practice of scientific enquiry, rather than waiting to be discovered by the *a priori* methods of philosophers. If real science is as Kuhn describes it, his disciples can argue, so much the worse for the methodological canons of inductivists and Popperians. To count as philosophy of science at all, accounts of methodology must be sensitive to what actually goes on in physics, chemistry, biology, etc.

Against someone who takes this line, the usual philosophical criticisms of Kuhn will have no effect whatsoever. If one wishes to take issue with such Kuhnians, one must question Kuhn's reading of the History of Science, and especially his reliance on a few key examples of scientific revolutions. One of Kuhn's central examples in SSR is the antiphlogistic revolution in chemistry at the end of the eighteenth century, associated with the name of Lavoisier. But was the Chemical Revolution a Kuhnian paradigm-shift? With the honourable exceptions of a fine paper by Alan Musgrave[6] and an intriguing discussion by Howard Margolis,[7] the question has been surprisingly neglected by philosophers.[8]

Robert Nola and Howard Sankey (eds.), After Popper, Kuhn and Feyerabend, 99–124.
© 2000 *Kluwer Academic Publishers. Printed in Great Britain.*

The answer, as we shall see in due course, is a definite 'no'. Of itself, this result would do little to shake the confidence of the Kuhnians. But if similar accounts can be given of the other revolutionary episodes in the History of Science, the cumulative effects of such an argument should be overwhelming. If the argument of SSR is a generalisation based on a few key cases from the History of Science, it matters that Kuhn gets those cases right.

2. HOW NOT TO WRITE THE HISTORY OF SCIENCE

During the course of the twentieth century, the historiography of science has itself suffered a revolutionary change. One might say that historians of science have escaped from the clutches of Scylla (Whiggism) only to fall into Charybdis (relativism). Any adequate history of science must seek to find a way between these twin perils. Let us first characterise the errors to be avoided, then see if this enables us to see the way ahead.

Much early writing of the History of Science is unashamedly Whiggish. Just as Whig historians of politics re-wrote constitutional history to make it seem as if it were all leading up to them, so the Whig historian of science assesses past scientific theories from the viewpoint of those currently accepted, and evaluates them accordingly. Whig historians also tended to accept the Baconian myth that scientific knowledge is acquired by some sort of quasi-algorithmic Method. Now if this were so – if the natural sciences were indeed in possession of such a Method – it would follow that false theories are the result of mistakes. (If you set me a sum in elementary arithmetic, and I get the wrong answer, I am guilty of a mistake.) It is no longer controversial, however, that the sciences possess no such Method: all our theories are hypothetical and fallible.

The Whig's assimilation of all theoretical falsehoods to the category of errors leads him to adopt an extremely dismissive attitude towards 'errors' (i.e., theories no longer accepted), often portraying them as mere products of prejudice and superstition. This is especially the case when the past theories in question (e.g., Aristotelian physics, Ptolemaic astronomy, Stahlian chemistry) differ radically, and even conceptually, from our own. The theories of his own day appear, to the Whig historian, so obviously right that all those who initially opposed them must, it seems, have been victims of blindness or prejudice. The other side of the same coin is the elevation to the lofty status of 'precursors' of scientists whose views – however arbitrary and ill-founded they may have been when they were first proposed – happened to anticipate currently-accepted theories. For example, the half-baked evolutionary ideas of some Presocratic philosophers make them 'precursors of Darwin', or the random atomic swerve of Epicurus becomes an anticipation of the quantum jump.

The vicissitudes of Whiggish history of science can at times border on the comic. For example, nineteenth century historians of optics felt obliged to apologise for Newton's acceptance of the 'wrong' (i.e., corpuscular) theory of light. But then, with the discovery of the photon, Newton's theory (in which corpuscles of light 'ride' on aether-waves) was suddenly hailed as a brilliant anticipation of modern views. The moral to be drawn from this story should be obvious. The quality of Newton's reasoning in the *Opticks* did not change during the course of subsequent history, for the simple reason that this quality must be assessed in relation to the evidence available at the time, not two centuries later.

Much modern writing in the History of Science can be seen as a reaction against Whiggism and (naive forms of) inductivism. Against such inductivism, the tentative and hypothetical nature of scientific thinking, and the non-existence of any infallible Method, are duly emphasised: our own theories are no different in this respect from those of the past. Against Whiggism, one will hear the 'pessimistic induction': All past theories have proved false; ours are no different in principle; therefore ours are – in all probability – also false. Doubtless, we are told, Whiggish historians of the twenty-second century will poke fun at us for accepting such absurd theories as Quantum Mechanics and Special Relativity.

Taken to its limits, this viewpoint can lead to an extreme form of irrationalism. Denying that scientific theories can be proved, rendered highly probable, or even conclusively refuted by empirical evidence, some modern historians of science have abandoned traditional epistemology altogether. In place of the usual normative concepts of evidence and rational belief, they appeal to social factors to explain the *de facto* acceptance and rejection of scientific theories. Instead of asking 'What ought one to believe, on the basis of such-and-such evidence?' they ask, 'What counts as rational here, given the norms of this society?'. The relativism implicit in this approach has been explicitly endorsed by some of its more extreme spokesmen.[9] What ought to be believed, such people will contend, is relative not just to evidence but to social context: criteria of rational belief will vary from one society to another, and there is no higher-order meta-criterion to settle the issue between different criteria.

Historians of science of this persuasion often refuse – in reaction against Whiggism – to evaluate past science at all. They prefer to elucidate the internal coherence and rationality (each according to its own lights, of course) of past belief-systems; in sharp contrast to the Whig, they show no interest in awarding points. Even if the philosophy behind this viewpoint is indefensible (and I think it is), their sympathetic reconstruals of old ideas have often been of great hermeneutic value. The historian interested in seeing, e.g., alchemical or astrological world-views from the inside can provide illumination regarding what it was like to inhabit such a thought-world; the Whig historian, dismissing all such theories as folly, prejudice or superstition cannot.

Much modern History of Science focuses on so-called 'Scientific Revolutions', i.e., periods of radical discontinuity when one world-view or global theory replaces another, e.g., the transition from Ptolemaic to Copernican astronomy or from phlogistic to antiphlogistic chemistry. If relativism and irrationalism are defensible anywhere, it would seem, it will be at these abrupt transitions from one world-view to another diametrically opposed – or even conceptually disparate – one.[10]

But, some will be asking, is not the *via media* between the Scylla of Whiggism and the Charybdis of relativism just plain obvious? Let us assess a past theory not according to its resemblance to our own (the error of Whiggism) but rather by its relation to the evidence available at the time. This enables us to evaluate the science of the past without laying ourselves open to the accusation of Whiggism. This is surely a step in the right direction, but it cannot be the whole story. A pair of related objections still stand in our way.

Relativist Objection 1. Are you not just assuming, the sophisticated relativist will respond, that the theory–evidence relation is timeless, ahistorical, and indifferent to social context, i.e., that criteria of theory-appraisal do not alter from age to age, or from one society to another?[11] And is it not just another form of Whiggism to seek to evaluate past theories by our criteria, mistakenly assumed to be timeless and ahistorical? Perhaps eighteenth-century chemists, for example, had different epistemic aims from modern chemists, and hence employed different canons of theory-appraisal. If so, there simply is no timeless theory–evidence relation, and evidence which we perceive as lending strong support to a theory need not have been so perceived at the time. One might try to meet this objection in two very different ways.

Reply 1. One might just assert, either as a normative philosophical demand, or as a generalisation from the recorded history of science, that there is sufficient common ground, sufficient sharing of cognitive ideals, to make our judgement of past theories non-anachronistic. A philosopher like Karl Popper, for example, simply characterises the scientific enterprise in terms of adherence to certain regulative maxims ('seek always to maximise information-content', 'do not have recourse to *ad hoc* shifts to "save" refuted theories', etc.): anyone failing to act according to these maxims is, for Popper, not a genuine scientist.[12] If this sounds somewhat too legislative a solution to our problem, one could instead run a similar argument as a historical generalisation. We could assess, according to our lights, Aristotle's arguments for the spherical figure of the Earth, or Tycho's critique of Copernicus: if our assessment of strengths and weaknesses coincides with that of contemporary critics, we have at least *prima facie* evidence of the historical invariance of criteria of assessment.

Reply 2. The other, and weaker, response is to relativise one's epistemology to contemporary standards, i.e., to ask, in any particular case, which of the rival theories was better supported by the available evidence, not in some timeless sense of 'better supported', but in the sense employed by the protagonists in the debate. If in fact there turns out to be a common value-system running throughout the History of Science (as some of us are still inclined to suspect[13]) this will turn out to be no concession at all. We will use eighteenth-century standards in assessing, say, the arguments of the Chemical Revolution but – lo and behold – these standards will turn out not to be significantly different from our own. We will find that the participants in the debate are doing exactly what we would do – looking for the simplest and most coherent account capable of doing justice to the phenomena.

Relativist Objection 2. In cases of scientific revolution, it is alleged, canons of scientific acceptability, and even of what counts as a scientific explanation, are themselves subject to abrupt change. This, according to some writers, is precisely what distinguishes a revolution from everyday varieties of theory-change. But if this is the case, eighteenth-century chemists might not have shared any common set of values, in which case our recommendation (above) to 'use the currently-accepted epistemic values' becomes worthless, because there are two (or more) competing sets of values at stake.

This is one way into the notorious problem of incommensurability, which has plagued a whole generation of philosophers of science since Kuhn and Feyerabend first proposed the idea[14] in 1962. I do not pretend to have a definitive solution to all the problems of incommensurability, but a little sorting-out of issues may be in order.

3. INCOMMENSURABILITY THESES

In the first place, it is essential to distinguish different incommensurability theses. Kuhn often suggests that the History of Science manifests incommensurability of problems and values. Since we evaluate a theory according to its problem-solving capacity, these two factors will prove conceptually inseparable – if we assign great importance to the solution of a particular problem, we are committed to assigning high marks to any theory which solves it. Stronger incommensurability theses invoke Gestalt psychology, suggest that partisans of rival theories inhabit different 'worlds', and allege the existence of communication-barriers between the inhabitants of these different worlds. Conceptual incongruity, it is claimed, ensures mutual incomprehensibility in such cases.[15]

But how much of this rhetoric is true? One has to concede, I think, that incommensurability of problems and values will be found everywhere in the History of Science – it will be manifest in every case of theory-choice, not merely in revolutionary episodes. Any pair of genuinely competing theories will have their respective strengths and weaknesses: one theory will cope with problems A-S but fail with T-Z; its rival will cope with E-Z but fail with A-D; which theory we favour will depend on the relative weights we attach to the various problems we are hoping to solve. One theory may be simpler, its rival more accurate, or more coherent with other established beliefs. Incommensurability of values is of course the norm in moral and political debate: the politician is faced every day with choices between, e.g., lives vs. civil liberties (drink-driving laws) or jobs vs. forests (logging disputes). Value-incommensurability appears in such cases as the inevitable consequence of the admission of a plurality of irreducible values. If we value theories in science for a variety of distinct reasons, we may have to learn to live with this sort of incommensurability.

One can admit incommensurability of problems and values, however, without surrendering to Gestalt-psychology, talk of different 'worlds', or mutual incomprehensibility. In the case of the Chemical Revolution, no such problems arose. Kuhn tells us that Priestley and Lavoisier see different things when they collect the 'air' given off in the reduction without addition of *mercurius calcinatus per se*.[16] But why should we not say that they see the same thing, not just in the obvious sense that the referent of Priestley's 'dephlogisticated air' is the same as that of Lavoisier's 'oxygen gas', but in the stricter sense that they both see the metallic mercury being restored and a colourless 'air' being evolved? There may be examples of scientific revolutions where the visual experience itself changes, but there seems to be no reason whatever to suppose that this is one of them.

As for mutual incomprehensibility and communication failure, it is conspicuous only by its absence. The clearest of all accounts of the phlogiston theory is given by Lavoisier in his *Réflexions*[17] – his later self does not have to struggle to understand his own earlier views. Nor does he have any problem regarding the *reference* of terms such as 'phlogisticated air' and 'dephlogisticated air', as found in the writings of his predecessors and his opponents.[18] The shared craft tradition of chemistry guarantees the reference of such terms by way of providing a recipe (couched in neutral terms) for their production. Nor do opponents of the new chemistry such as Kirwan and Priestley manifest any great difficulties in understanding the writings of Lavoisier and his disciples. In fact, both Kirwan and Priestley show the clearest possible signs of having a rather good grasp of the antiphlogistic theory: they anticipate how Lavoisier *et al.* are going to respond to their objections.[19] There was, of course, vigorous opposition to the new nomenclature of 1787,[20] but the objection was not that the new terminology was incomprehensible but that it was question-begging.[21] Opponents of the new chemistry objected – not unreasonably – to a terminology which was committed to a theory of composition they did not accept.

On our view, then, incommensurability of problems and values is sure to be present in the Chemical Revolution (as in every other case of choice between competing theories), but, unless further evidence appears to the contrary, we can dismiss Kuhn's talk of Gestalt-shifts, different worlds, and mutual incomprehensibility as so much empty rhetoric. Let us suppose, then, that we are faced with two competing theories, T1 and T2, each with its accredited domain of solved problems and its complementary domain of unsolved problems or anomalies. A rational choice between two such theories will require a weighting of the various problems and (still more fundamental) a decision as to what problems the theories ought to solve. Now if the partisans of rival theories assign different weights to the various problems at stake, one may easily find that this incommensurability of problems and values involves, as a corollary, rational undecidability.

Does this entail either relativism or irrationalism? Not at all. In the first place, value-incommensurability merely makes undecidability possible; it certainly does not entail that it will be present in any given instance. Assignment of different weights to the various problems is perfectly compatible with consensus: T1 might outscore T2 on all problems, or on a sufficient majority to gain higher marks on all the judges' score cards. To 'save' T2, in such a case, a staunch defender will find herself obliged to assign enormous weight to the small group of problems it can solve and T1 cannot. The extension of T1 to deal with these outstanding anomalies (anomalies, of course, only from its point of view) then forces either surrender or retreat to some other anomaly for T1 – to which, now, still greater weight must be assigned by the remaining defenders of T2. Another offensive move open to the advocates of T1 is to show that T2 suffers from a perfectly parallel anomaly: shared anomalies cancel out and lend no net support to either theory.

All these moves, it is easy to show, took place in the debates of the Chemical Revolution. The partisans of phlogiston (Φ), men such as Macquer, Kirwan, and

Priestley, were forced gradually onto the defensive, obliged to attribute ever-greater weight to the dwindling number of problems where Φ-theory was still seen to advantage. The attackers, Lavoisier and his disciples, extended their anti-Φ-theory to more and more new territory, and dismissed remaining diffi-culties (e.g., over affinities) as shared anomalies, and thus as not counting par-ticularly against them. New empirical discoveries (the composition of water, the existence of carbon monoxide) reduced the grounds on which Φ-theory could be defended. By 1800, all its defences had effectively been swept away.

What of the alleged role of simplicity as a factor militating in favour of the new chemistry? The relativist will be suspicious of any such claim, and will argue (a) that it is merely a contingent and socially relative fact that we prefer simple theories at all; and (b) that the concept of simplicity has no absolute sense, but must be relativised to time, society, and world-view. For the relativist, it may make sense to speak of 'simple' (classical Greece), 'simple' (medieval Christen-dom) or 'simple' (phlogistic chemistry), but not of 'simple' *simpliciter*. In answer to these two objections, I would contend that (a) any intelligent being with limited memory-space, fallible recall, and finite computational speed must place some positive epistemic value on simplicity (only God could entirely dispense with it). Exactly how much weight should be placed on simplicity as a factor in theory-appraisal is, of course, a much more delicate business: once again, we can expect value-incommensurability to arise here. The really difficult philosophical ques-tion about simplicity is whether it is merely a pragmatic virtue, or whether we have any grounds for thinking simpler theories more likely to be true.[22] (For-tunately, this is a question we do not have to answer here.) As for objection (b), there are some contexts where 'simple' can be given a straightforward numerical sense, which seems not to be relative to period, social background, or theoretical allegiance. For example, if Lavoisier represents a given reaction as a single dis-solution involving three terms $(AC + B \rightarrow AB + C)$, where the phlogistic chem-ist represents the same reaction as a double dissolution involving four terms $(AC + BF \rightarrow AB + CF)$, it seems permissible to claim that the former account is objectively simpler. Indeed, Lavoisier's opponents sometimes admitted that his theory was simpler than theirs,[23] while maintaining that its extra simplicity is achieved only by over-simplification and distortion of the facts.

4. THE CHEMICAL REVOLUTION

To understand the arguments of the Chemical Revolution, one must first under-stand the phlogiston theory of Becher, Stahl, and their followers. The central idea behind this theory is the notion of chemical 'principles' whose presence in a compound explain certain salient properties. Phlogiston is the principle of combustibility, the substance whose presence in a compound confers the prop-erty of combustibility. In traditional alchemy, 'Sulphur' played this role; it was the German chemist Stahl who coined the new term 'phlogiston', and who was largely responsible for the teaching and dissemination of the theory.[24] Stahl's Φ-theory is firmly rooted in metallurgical practice, in particular, in the use of charcoal in the reduction of metal ores. Charcoal being itself combustible, is a source of Φ; during the smelting of an ore, it imparts its Φ to the calx and thus 'revivifies' the metal. Phlogiston, then, is the principle of reduction, of 'reducing power'; the chemical phenomena which the theory was designed to handle are those involving the transfer of this power from one substance to another.[25] Let us run briefly through orthodox phlogistic accounts of some common chemical reactions – one always understands a theory better for seeing it in action.

(1) *Simple Combustion and Calcination.* In combustion, a base B loses its Φ to the air, which becomes phlogisticated. The eventual saturation of the air with Φ stops the combustion – hence the need for a continual supply of fresh air. The Φ emitted explains the heat and light characteristic of combustion: it thus deserves Macquer's name of 'feu fixé', fixed or chemically combined fire.[26] Expressed in symbols, simple combustion becomes:

$$(B + \Phi) \rightarrow B + \Phi\uparrow \quad (= \text{heat, light})$$

The calcination of metals was seen as fundamentally the same process, albeit slower and less spectacular.

(2) *Metallic Replacement Reactions.* When one metal (iron, say) displaces another (copper, say) from solution, this occurs in virtue of the different Φ-affinities of the two metals. The copper, having a stronger affinity for the Φ, can seize it from the iron: thus metallic copper is precipitated and the iron goes into solution. On the basis of such reactions, the Swedish chemist Torbern Bergman was to draw up his affinity table for Φ.[27]

(3) *The Dissolution of Metals in Acids.* When a metal is dissolved in an acid, its Φ is released; sometimes this Φ will be given off in the form of 'light inflammable air' (our hydrogen gas); sometimes it will 'phlogisticate' the acid, yielding perhaps 'nitrous gas' (our nitric oxide, NO) or 'volatile sulphureous acid' (our sulphur dioxide, SO_2), i.e., the phlogisticated forms of nitric and sulphuric acids respectively. Not surprisingly, the latter type of reaction is more likely to occur when the acid is concentrated.

In the dissolution of a metal in a weak acid, a given weight of metal will release a constant quantity of inflammable air, irrespective of the acid used. This fact was seen by Henry Cavendish as powerful evidence that the inflammable air comes from the metal, not from the acid.[28] That this 'inflammable air from metals' just is pure Φ was suggested by Cavendish, endorsed by the Irish chemist Richard Kirwan, and taken seriously by Bergman and Priestley.[29] It must, after all, contain Φ (since it is inflammable); and it is, clearly, very simple in nature (an 'air' of very low specific gravity), so perhaps it is just phlogiston pure and simple.

(4) *Metallic Reductions.* The reduction of metallic calces by charcoal was, as we have seen, one of the foundations of the Φ-theory. Priestley's discovery that the calces of some metals (e.g., lead) could be reduced by 'inflammable air' was seen by him as a crucial experiment establishing the Φ-theory[30] – and, incidentally, lending considerable weight to Kirwan's identification of Φ with inflammable air. The calx, says Priestley, simply absorbs the inflammable air to form the metal – almost, he felt, ocular proof of the correctness of Φ-theory.

(Unfortunately for Kirwan's identification, however, this light inflammable air proved incompetent to reduce other calces – pure Φ, surely, should reduce anything.)

(5) *The Composition of the Non-Metals (Charcoal, Phosphorous, Sulphur, etc.)*. Each of these non-metals consists, according to the Φ-theory, of a specific acid, plus a full complement of Φ. Some acids (e.g., nitric, sulphuric) also admit of partial phlogistication, yielding compounds (our NO, SO_2) intermediate between the original acid and its fully phlogisticated compound. This pattern of analysis was extended to arsenic, antimony, and even into organic chemistry by the research of Bergman and Scheele.[31]

It should now be easy to see why Φ-theory was so widely accepted and – on the whole – such a success: it was the first great theory of *redox* reactions, the first chemical theory to place *redox* reactions firmly at the centre of the chemical stage. It also gave rise to a clear picture of the fundamental identity of the processes of combustion, calcination, and respiration – all involve essentially the emission of Φ. Thus we find the Scot Adair Crawford articulating a phlogistic account of respiration (venous blood, he suggests, is more phlogisticated than arterial; the lungs function to discharge this excess Φ from the body) which was to prove the natural precursor for that of Lavoisier.[32] Meanwhile, another of the phlogistic chemists, Joseph Priestley, was investigating another aspect of the Φ-cycle. If animals are continually phlogisticating (and hence spoiling) the air, how is it that our atmosphere remains respirable? Plants, Priestley discovered, have the power to dephlogisticate the air, i.e., to take in foul and unbreathable air and give out 'dephlogisticated air' (our oxygen) in its place.[33] Nature observes a Φ-cycle: plants take in Φ from foul air (with the aid of sunlight), and build up phlogistic matter; animals eat the plants and burn this phlogistic matter, releasing its Φ back into the atmosphere. One simple transformation, and it all begins to sound very modern!

And yet, almost incredibly, one Whiggish historian of chemistry (White) denies that the Φ-theory was of value to eighteenth-century chemistry.[34] He portrays Φ-theory as a mass of errors and confusions, of no value to the practising chemists. This seems to be a gross error, the sort of error we might expect from a Whig historian.[35] Because Φ-theory subsequently turned out to be false, it could not have been of heuristic value at the time? I hope I no longer need to point out that this would be a complete *non-sequitur*.

But was Φ-theory only of heuristic and instrumental value? Was it merely of use as a guide to the production of new experiments? I want to defend a stronger claim, to say that, although Φ-theory was not itself true, it taught chemists a lot of important and abiding truths, e.g., the fundamental identity of combustion, calcination, and respiration, the existence of a chemical balance in the biosphere between plants and animals, the possibility of transferring reducing power from one substance to another, and so on.[36] In an important sense, Φ-theory carved Nature at the joints.[37] Science, it would appear, is cumulative after all: even if the axioms of the Φ-theory were false, much of its middle-level theory (the part that guided actual chemical research) was true. Moreover, these truths or insights into Nature were abiding: they did not disappear during the Chemical Revolution.

Before going on to discuss the antiphlogistic theory of Lavoisier, which was eventually to supplant Φ-theory, let us mention two of the anomalies which were to prove instrumental in its ultimate downfall. In his *Digressions Académiques* of 1772, Guyton de Morveau (a good phlogistonist before his 'conversion' in 1787) attempted to patch up the Φ-theory, but only succeeded in exposing its weaknesses to critics of the acumen of Bayen and Lavoisier.[38] The first of these two anomalies is the gain in weight of metals during calcination. The importance of this phenomenon has been grossly exaggerated by historians. Historians of a Whiggish persuasion have even claimed that it provides a crucial experiment, refuting Φ-theory and establishing in its place the rival anti-Φ-theory of Lavoisier.[39] This is doubly false: the anomaly did not refute Φ-theory, nor could it establish the rival theory of Lavoisier, for the very simple reason that Lavoisier had, at this time, no anti-Φ-theory to offer, no global theory of chemistry of anything like the breadth and power needed to rival Φ-theory.

The weight-gain anomaly forced phlogistic chemists to address the question of the weight of phlogiston. Stahl himself seems to have been indecisive on this point; his disciples were left to make their own choices.[40] Does the mere loss of Φ make the calx outweigh the original metal? If so, two distinct explanations are possible:

(a) Φ has negative weight, or absolute levity (Black, for a while[41]);
(b) Φ has very low specific gravity; its loss, therefore, involves a gain in net weight as measured in air (Guyton, Chardenon[42]).

Neither of these accounts will withstand much scrutiny. To refute Guyton's theory, one would need only to weigh the metal and its calx *in vacuo* instead of in the air. As for the hypothesis of 'absolute levity', it falls foul of Newtonian physics: if Φ has mass but negative weight, a pendulum with a calx bob should swing faster than one with a metal bob.[43]

Most phlogistic chemists did not, however, endorse such positions. Priestley, for example, admits that absolutely light Φ would solve some problems, but rejects the notion outright;[44] Scheele and Bergman argue quite explicitly that Φ is material and therefore has weight;[45] Kirwan follows Cavendish in identifying Φ with light inflammable air (hydrogen), which has measurable weight. So if a metal, on calcination, loses something ponderable, but still gains weight, it must simultaneously gain something heavier. This gives rise to a new set of variants on Φ-theory.

(c) Calcination involves the 'fixing' of ponderable igneous corpuscles (Boyle,[46] Baumé[47] or 'saline' corpuscles from the flame. This theory falls foul of Lavoisier's painstaking quantitative work, aimed at showing that (i) ponderable matter is not transformed into imponderable, nor vice versa; and (ii) ponderable matter does not pass through the walls of glass vessels – *pace* Boyle, calcination will not occur in sealed vessels.

(d) The Φ in the metal reacts with dephlogisticated air (our oxygen) to form 'fixed air' (our carbon dioxide) which then remains in the calx. This theory, proposed by Kirwan,[48] will cope easily enough with the weight-gain, and will not be without empirical support (most metal oxides, after all, do contain some portion of carbonate). Unfortunately, this version of Φ-theory sacrifices some of its original explanatory power: on this account, the Φ emitted during calcination is not actually given off into the surrounding air. Apply the same account to combustion, e.g., of non-metals such as sulphur and phosphorous, and one cannot cite the emission of Φ as the cause of the heat and light of combustion.

(e) Calcination, say Cavendish and Priestley,[49] is the substitution of Φ by water; the observed weight-gain is simply the weight of the water added, less that of the Φ given off in the following reaction (A = Air, W = Water):

$$(E + \Phi) + AW \rightarrow (E + W) + A\Phi \uparrow$$

(f) Calcination is a substitution-reaction (Macquer, Richter[50]), in which dephlogisticated air replaces Φ, thus:

$$(E + \Phi) + (O + Cal) \rightarrow (E + O) + (\Phi + Cal) \uparrow$$

This sort of compromise-theory remained a viable alternative to Lavoisier's antiphlogistic theory as late as 1800, and proved popular among German chemists, reluctant to forsake completely their great compatriot Stahl.[51]

What can we conclude from this quick survey of the options facing the phlogistic chemist? Only that the weight-gain phenomenon posed a genuine difficulty, and one which generated a number of very different responses. It could not, however, be described as a knock-down refutation.

More significant, in many ways, was the reduction *without addition* of some metal calces, e.g., that of mercury. If phlogiston is the source of reducing power, how can a reduction be achieved without the addition of any Φ-source such as charcoal? Once again, the reactions of phlogistic chemists showed wide variation:[52]

(a) According to Priestley, 'red precipitate of mercury' is not a true calx, precisely because it can be 'revivified' without addition.[53] It is produced, he claims, merely by a mechanical rearrangement of the particles of the mercury, thus possessing all the Φ of the metal. It is therefore merely an 'apparent' calx, an exceptional substance, hence not a sound basis for any generalisation about the nature of calcination. (Priestley is remarkably casual about questions of chemical composition – in this case, as in others, he seems prepared to allow two substances with strikingly different properties to have the same chemical composition.[54])

(b) Mercury metal, says Macquer, has a 'superabundance' of Φ. All that is lost in the formation of the red precipitate is this excess; what remains is sufficient to restore the metallic lustre. But, Guyton soon showed, the process (successive calcination and reduction without addition) can be repeated indefinitely. Macquer's 'excess' Φ proves inexhaustible![55]

(c) Phlogiston, according to Scheele, can be derived from the decomposition of heat, which is a compound of Φ and 'fire air' (oxygen).[56] The decomposition of heat thus releases both the Φ needed to restore the metal, and

the 'fire air' collected. Unfortunately, however, Scheele's theory falls foul of Lavoisier's strictures about the conservation of ponderable matter and the impossibility of ponderable matter passing through barriers such as glass.[57]
(d) Φ, says Macquer in his later works, can be identified with the matter of light: this explains the successful reduction of red precipitate under a burning glass.[58]

Why can the calces of some metals, but not those of others, be reduced without addition? The answer, for the phlogistic chemist, is easy: the bases of metals such as gold, silver, and mercury have very powerful Φ-affinities, and can therefore take Φ from sources that the bases of other metals cannot. In the table of Φ-affinities, the bases of the noble metals will be at the top, those of iron and zinc near the bottom.[59] (Bergman's table of Φ-affinities only needs to be turned upside-down to yield the table of oxygen-affinities Lavoisier needed for his antiphlogistic theory.[60])

What conclusions can we draw from this discussion? Guyton's *Digressions* of 1772 certainly raised some difficult questions for phlogistic chemists. While attempting to resolve the anomalies, Guyton only succeeded in drawing attention to them: none of the 'solutions' we have canvassed is entirely satisfying. As early as 1774, the Φ-theory was attacked in two anonymous articles in Rozier's *Journal de Physique*.[61] The author (Bayen?) stressed just these two problems – weight-gain during calcination, and reduction without addition – in issuing a serious challenge to the credibility of the Φ-theory. Phlogistic chemistry, the author argues, is a mess, full of internal contradictions and arbitrary assumptions.[62]

At this time, however, this was all that anyone could do. Throughout the 1770s, there existed no adequate global rival to the Φ-theory. Lavoisier, for example, could explain the weight-gain in combustion and calcination in terms of the fixation of the ponderable base of oxygen gas,[63] and could account for the heat and light of combustion in terms of the disengagement of its 'caloric' or heat-stuff, but had no account of the evolution of an inflammable air from metal–acid reactions, nor from charcoal and steam, both of which were perfectly straightforward for the phlogistonist. In the *Avertissement* to the 1774 *Opuscules*, Lavoisier tells us that he had planned to include a discussion of 'calcination the wet way', but eventually decided to omit it.[64] Around 1780 he was planning a 'proto-Traité', but abandoned the project because he could not explain metal–acid reactions and the evolution of inflammable air.[65] Lavoisier thus held his fire,[66] and delayed his attack on Φ-theory until the last crucial piece was in place. In his memoirs of this period, i.e., around 1780, he snipes at phlogiston, reminding his readers that it is merely a hypothetical substance and hinting that chemical theory could dispense with it,[67] but he goes no further. The revolution in chemistry dates not from 1773 but from 1783, when Lavoisier first heard of the composition of water.[68] Only then was he in a position to launch his attack. That Lavoisier had no idea, prior to hearing from Blagden the news of Cavendish's experiment, of the composition of water, is quite clear: during his collaboration with Bucquet in 1777, he expected the combustion-product of inflammable air to be vitriolic acid (our sulphuric acid) while Bucquet expected fixed air (our carbon dioxide).[69]

Lavoisier's theory of oxidation, developed during the hectic period between Priestley's visit of 1774 and the *Mémoire sur la Combustion en général*[70] (1777) was firmly based on the reduction without addition of red precipitate of Mercury – a reaction which, he felt, provided a key to unlock the secrets of chemistry. It took some time for Lavoisier to recognise that the 'air' evolved in the reaction (Priestley's 'dephlogisticated air', Scheele's 'fire air'), was not just a modification of atmospheric air but a distinct component of it, but once he had made this step, other things began to fall into place. If (*pace* Priestley) 'red precipitate' is a true calx, the two anomalies haunting the Φ-theory can be laid to rest. The weight-gain on calcination of metals is simply the weight of the 'pure' or 'vital' air absorbed – there is no need to postulate a simultaneous loss of anything. As for reduction without addition, all one needs to say is that mercury metal has a low affinity for the ponderable base of this 'pure' air, and this affinity decreases with temperature, making it relatively easy to reduce mercury calx without addition of a Φ-source such as charcoal.

As for carbon, sulphur, phosphorous, etc. they all become simples capable of forming acids when combined with a sufficient amount of 'pure air'. Hence the new name of 'oxygen' or 'acid-former' for this special air, which turns out to be the principle of acidity. All acids, Lavoisier thought, contain oxygen, and any substance can be converted into an acid by combination with enough oxygen.[71] Instead of degrees of phlogistication, starting with an acid and ending up with sulphur (or carbon, or phosphorous), one has degrees of oxygenation, starting with sulphur (carbon, phosphorous) and ending up with their respective acids, thus:

Φ-*theory*:   Addition of Φ: Vitriolic Acid → Sulphureous Acid → Sulphur

O-*theory*:   Addition of O: Sulphur → Sulphureous acid → Sulphuric Acid

The crucial point here, of course, is reversal of the order of composition: what Φ-theory represents as compound, O-theory represents as simple, and vice versa.[72]

Cavendish's famous experiment provided Lavoisier, in 1783, with the last vital clue, almost the final missing piece in the jigsaw puzzle. Now, at last, he could see how the pieces would fit together. Cavendish had sparked together 'light inflammable air' (our hydrogen) and 'dephlogisticated air' (our oxygen) and obtained an equal weight of water. To Lavoisier, the conclusion was obvious: water is a compound of oxygen and the light inflammable air (now renamed 'hydrogen' or 'water-former'). This, however, was not Cavendish's conclusion. Instead of concluding that water is a compound of the two airs, he concluded that each of the airs is a 'modification' of water, with either an excess of Φ (inflammable air) or a deficiency of it (dephlogisticated air). Sparking them together thus restores ordinary water. The reaction thus looks like this:

$$(W + \Delta\Phi) + (W - \Delta\Phi) \rightarrow 2W$$

As Cavendish himself pointed out, this new version of Φ-theory begins to look empirically equivalent to Lavoisier's theory.[73]

Priestley too repeated Cavendish's experiment, but arrived at a very different result. Water is deposited, Priestley grants, but this is not the true reaction-product of the combustion. This, he claims, is 'nitrous acid'.[74] The French chemists, hearing of Cavendish's startling results, and of the very different interpretations of these results by Lavoisier, Priestley, and Cavendish himself, proceeded to repeat the experiments under a variety of conditions. Priestley's acid, they showed, is obtained only when the 'vital air' (oxygen) is in excess of the 1 : 2 ratio needed for complete combustion: with excess inflammable air, and with slow combustion instead of a spark, no acid is formed, and the product is pure water. The acid, they concluded, is merely the result of an impurity, of the presence of some 'azote' (nitrogen) in one or other of the original gaseous reagents.[75]

Lavoisier was now, for the first time, in a position to explain the evolution of light inflammable air (hydrogen) in metal–acid reactions. The metal, he says, must take up oxygen to go into solution; this oxygen may be taken *either* from the acid (releasing fumes of 'volatile sulphureous acid' or 'nitrous gas'), *or* from the water (releasing hydrogen gas). In a concentrated acid, the metal is more likely to take its oxygen from the acid; in a dilute acid, water is more abundantly available as a source of oxygen. Some acids, e.g., 'marine acid' (our hydrochloric acid) retain their oxygen so tightly that only the water is decomposed, never the acid.[76]

The oxygen theory is now, i.e., in the mid-1780s, almost complete, and Lavoisier is able to propose a system of antiphlogistic chemistry capable of doing battle on all fronts against Φ-theory. Let us quickly run through the central points of the new doctrine.

(1) Combustion (calcination, respiration) involves a decomposition, by the fuel, of oxygen gas. The ponderable base of the gas combines with the fuel (augmenting its weight); the imponderable 'caloric' (essential to the gaseous state) is released, accounting for the evolution of heat and (in some cases) light.

(2) Metallic replacement reactions involve a simple competition between metals for oxygen: a metal with a high oxygen-affinity (iron, say) will take oxygen from the salts of other metals (copper, say), thus precipitating the less reactive metal out of solution.

(3) Metal–acid reactions have been dealt with above: the dissolving metal takes oxygen either from the acid or from the water, to pass into solution. Cavendish's discovery that a given quantity of metal releases only a given quantity of 'inflammable air', irrespective of the acid used, can also be accommodated. So much metal needs so much oxygen, so decomposes only so much water, and thus releases so much hydrogen.

(4) Reduction of metal calces by charcoal is essentially similar to a metallic replacement reaction: the carbon has a higher affinity for oxygen than the familiar metals, and can therefore rob their calces of oxygen, yielding the revivified metal and 'fixed air' or 'aerial acid' (our carbon dioxide). The earths

lime and magnesia may, Lavoisier suggests,[77] be oxides of metals with higher oxygen-affinities than carbon, and hence irreducible with charcoal – a hint which would be followed up by Humphry Davy. As for the reduction of some (and only some) calces by hydrogen, this too results from a simple competition for oxygen – water will be formed in all such cases.

(5) The composition of Sulphur, Carbon, Phosphorous, etc. is denied: as far as our analyses reach, says Lavoisier, these are simple substances, capable of combining with oxygen to form their respective acids.

One cannot help being impressed by the enormous amount of common ground that links the two rival theories, by their shared domain of problems and their shared explanatory ideals. Both are 'principle' theories, i.e., account for a common feature of a group of related compounds in terms of the presence of some common 'principle'. The Φ-theory can be transformed into its rival by replacing 'loss of Φ' with 'gain of O' and vice versa. Once one has treated one or two reactions this way, one can run through orthodox phlogistic explanations of chemical processes, generating their antiphlogistic equivalents by a simple transformation. Lavoisier did not have to work out the O-theory in all its details: the phlogistonists had already done most of the work for him.

Before going on to discuss the weaknesses of Lavoisier's theory, we must at least mention some of the compromise-theories that proved so popular before the final triumph of the new chemistry. (The very existence of such compromise theories counts against some of Kuhn's more extreme claims.) Lavoisier's discovery that 'pure', 'vital', or 'dephlogisticated' (de-Φed) air enters into calces could not long be resisted, but could readily be incorporated into a variety of double-decomposition theories of combustion. Such theories were championed by Macquer, Guyton (before his conversion), Richter, Gren, and others.[78] On this theory, calcination is best represented as:

$$(E + \Phi) + (O + Cal) \rightarrow (E + O) + (\Phi + Cal) \uparrow$$

Alternatively, one could adopt Kirwan's theory:

$$(E + \Phi) + (O + Cal) \rightarrow (E + (\Phi + O)) + Cal \uparrow$$
$$\text{fixed air}$$

Or that of Priestley:

$$(E + \Phi) + (A + W) \rightarrow (E + W) + (A + \Phi)$$
$$\text{de-}\Phi\text{ed air} \qquad\qquad \Phi\text{ed air}$$

By 1800, the old phlogiston theory was dead, and the outstanding dispute was between Lavoisier's theory and a spectrum of compromise-theories. How might such a debate be settled? Here the factor of simplicity comes into play on the side of Lavoisier. His theory of combustion is objectively simpler than compromise theories in that it represents combustion in terms of 3 factors rather than 4.[79] Furthermore, it allows the chemist greater access to simples: on the compromise versions of Φ-theory, there is no access to Φ itself (except on Kirwan's version), nor to the supposed 'earths' that are the bases of the metals, which remain hypothetical. The simplicity of Lavoisier's theory was emphasised by its defenders, by converts such as Black,[80] and was granted by opponents such as Kirwan.[81] It is not the case that the opposed theories or paradigms carried with them their own, internal notions of simplicity: even Φ-theorists freely granted the extra simplicity of Lavoisier's theory. How much weight to attach to it *vis-à-vis* other criteria of theory appraisal is, of course, a much more difficult and delicate matter.

Another factor urged by Lavoisier in his famous *Réflexions sur le phlogistique* (1785) was the superior clarity and coherence of his theory. By the time the *Réflexions* were written, there simply was no monolithic Φ-theory: one had to compare the new oxygen-theory against the quite distinct Φ-theories of Macquer, Scheele, Kirwan, Priestley and others. Each of these Φ-theories would have its characteristic strengths, i.e., would be competent to stand up to Lavoisier on some problems; but each would also suffer from its particular Achilles heel. And the respective explanatory strengths of the various Φ-theories could not be pooled, simply because they involved incompatible assumptions about the properties of phlogiston: one chemist attributes weight to Φ, another denies it; one allows Φ to pass through glass, another denies it; one accepts the composition of water, another denies it; and so on.[82] Since one or other of the various Φ-theories can provide a solution for any given problem, one may experience the illusion that 'Φ-theory' is holding its ground when in fact it is falling apart. This explains the persistent accusations that phlogiston is a 'Proteus' capable of changing its nature from one phlogistic chemist to the next, or even within the pages of the same chemist.[83]

Disciples of Kuhn might seize on this as a piece of favourable evidence. Kuhn, after all, speaks of 'proliferation' of theories, by the defenders of the old paradigm, as a characteristic response to crisis.[84] But Kuhn's other marks of crisis are simply absent: there is no explicit discontent, no debate over fundamentals, no recourse to philosophy. Phlogistic chemistry was a successful and progressive research programme when it was overthrown. As for the proliferation of versions of Φ-theory, there seems no reason to regard it as a response to crisis – it was there all the time. In both its social and its intellectual dimensions, phlogistic chemistry was a loose and ramshackle affair from the start.[85] If one must impose a Kuhnian interpretation on the Chemical Revolution, one might do better to say that Lavoisier imposed the first paradigm on the science of chemistry, and that the various versions of phlogistic chemistry count as pre-paradigmatic.[86]

Lavoisier's *Réflexions* posed a challenge to the defenders of phlogiston: come up with a clearly articulated version of your theory adequate to the phenomena, or give it up. The challenge was never adequately met. This does not mean, however, that the triumph of the new chemistry was all plain sailing, that Lavoisier's own theory was free from problems and anomalies. Let us run through a few of the more prominent difficulties.

(1) The oxygen-theory of acidity, seen by Lavoisier himself as an integral part of the new chemistry, was never universally accepted. It was doubted not just by

Cavendish, but by Lavoisier's own followers Berthollet and Fourcroy, before finally being laid to rest by the electrolytic experiments of Humphry Davy.[87] Critics pointed out the inconsistency between Lavoisier's operational definition of 'element' and his insistence that 'marine' or 'muriatic' acid (our HCl) must be compound (although as yet undecomposed) because as an acid it must contain oxygen.[88]

(2) There was persistent confusion, on both sides of the controversy, about inflammable airs, in particular, about the distinction between hydrogen and carbon monoxide. In a water–gas experiment ($C + H_2O \rightarrow CO + H_2$) the 'heavy' inflammable air is derived, say Priestley and Kirwan, from the charcoal. Not so, replies Lavoisier: it comes from the decomposition of the water. To the end, however, he could not get his experimental results to fit his theory. In the *Traité*, he actually admits to suppressing some experimental details.[89] The hydrogen gas, he suggests, may 'dissolve' some of the carbon, producing an inflammable air of high specific gravity which will burn with a characteristic blue flame, leaving fixed air (our CO_2) as its combustion-product. But this simply will not do. In one of Kirwan's experiments, he showed that dry charcoal, heated in a crucible with a small hole at the top, will emit an 'inflammable air'.[90] Here we seem to have incontrovertible proof that an inflammable air is being given off from the charcoal, in the absence of any water to decompose. Only the discovery of carbon monoxide in 1800 cleared up this outstanding anomaly for the new chemistry. Until then, Priestley and others could still score points with their experimental proof that charcoal contains inflammable air.

(3) Kirwan, in his *Essay on Phlogiston* pointed out a number of anomalies in Lavoisier's table of oxygen-affinities, i.e., cases where, it seems, $X > Y$ and $Y > X$ simultaneously.[91] Iron, for example, will decompose water, thus taking oxygen from hydrogen, yet hydrogen will reduce a higher oxide of iron to a lower, and thus take oxygen from iron. In their replies to Kirwan, Lavoisier and his followers were able to reply that such anomalies should not count against the new chemistry.[92] In the first place, all affinity tables suffer from such anomalies and exceptions, due to a host of complicating factors such as temperature, solubility, and (crucial in the above case) degree of saturation. Furthermore, Lavoisier retorts, exactly the same objections could be advanced against Φ-theory, since high O-affinity = low Φ-affinity and vice versa. Shared anomalies cancel out.

(4) The oxygen-theory, it could be alleged, leaves O-affinities as irreducible primitives, and can therefore provide no account of why all metals, or all combustibles, are akin. But it is easy to see that this objection, too, counts equally against the phlogistic and anti-phlogistic theories. The Φ-theory leaves Φ-affinities as primitive and irreducible, and thus cannot explain the qualitative likeness of Φ-hungry substances like earths and acids. Thus the Φ-theory can explain combustibility and metallicity but not acidity; the O-theory can explain acidity (presence of sufficient oxygen) but leaves combustibility (high O-affinity) unexplained. If one allows the *absence* of a 'principle' to have explanatory power, these lacunae could be filled – but filled symmetrically, for both theories. There seem to be no grounds here for preferring one theory to the other.[93]

(5) Defenders of phlogiston sometimes claimed that the antiphlogistic theory could not account for the chemical roles of light and electricity. But this was a feeble argument: no chemical theory of the time could do more than acknowledge the existence of some puzzling phenomena in these areas. There is no reason to believe that Lavoisier's theory was at a disadvantage in this respect.

5. CONCLUSIONS

The following points sum up the gist of the historical story.

(1) Phlogistic chemistry was a highly successful theory, supported by a host of plausible problem-solutions and facing, c. 1770, one or two minor anomalies which seemed to pose no real threat. It was already undergoing proliferation, but not as a response to perceived crisis.

(2) The attempts of Guyton (1772) and others to resolve those problems (weight-gain, reduction without addition) served merely to exacerbate them, and to bring them to the attention of men like Bayen and Lavoisier.

(3) On the basis of his famous experiment with red precipitate of mercury, followed by a series of experiments on the combustion of non-metals, Lavoisier developed his theory of oxidation, which could account for combustion, calcination, and the composition of acids. By the mid-late 1770s this account of combustion was in place, but Lavoisier did not have a global theory competent to take on Φ-theory. As for the phlogistic chemists, they accommodated his findings (in ways that were more or less *ad hoc*) into a variety of compromise-theories.

(4) At this time, Lavoisier could not account for other phenomena (e.g., calcination 'the wet way', and the reduction of metal calces by inflammable air) that could easily be accounted for in terms of phlogistic chemistry. Lavoisier therefore held his fire, and was praised by Macquer for his restraint.

(5) When news came to him in 1783 of Cavendish's synthesis of water, the crucial piece of the puzzle fell into place. The *Réflexions* were read in 1785, and converts flocked to his banner, e.g., Berthollet[94] (1785), Fourcroy[95] (1786), Guyton[96] (1787), and a little later Black.[97] William Higgins tells us, in his *Comparative View* (1789) that he had converted in 1784.[98] Kirwan gave up the unequal struggle in 1791, and even Priestley wavered. The 'old guard' did not (*pace* Kuhn) just gradually die out; they were, for the most part, converted to the new chemistry.[99] They converted at the right time, i.e., after the water-controversy, and for the right reasons, citing the gain in explanatory power accruing to the new theory as a result of the discovery of the composition of water.

(6) Though a few outstanding problems still remained (e.g., 'heavy inflammable air'), the weight of the evidence was now firmly behind the new chemistry; defenders of phlogiston were forced to adopt cumbersome and *ad hoc* compromise theories. The discovery of carbon monoxide (1800) undermined the last major objection of Priestley.

From 1785, defenders of phlogiston found themselves forever on the retreat, conceding one piece of ground after another. (To illustrate this, one need only glance at the three phlogiston-theories of the German chemist Gren – the first is pure Stahl; the last is almost pure Lavoisier, with Φ tagged on almost as an afterthought, and doing no real work.[100]) They are obliged to attribute ever-greater weight to the ever-dwindling number of issues on which they can still score points. There is no sharp cut-off point where 'rational doubt' gives way to 'pig-headed obstinacy' or 'perversity', yet it is transparently clear that in his last defences of Φ-theory Priestley was on very weak ground, having to attribute enormous weight to one or two minor difficulties to muster any credibility at all.[101] By this time, of course, he was a marginal figure, exiled in America, and cut off from the main stream of European science.

What philosophical morals can we derive from this historical story?

(1) There is some evidence of incommensurability of problems and values in the Chemical Revolution, but there is no reason at all to suppose that this entails any relativistic or anti-rationalistic conclusions. It shows only that the appraisal of scientific theories is not algorithmic, but leaves room for *judgement*.[102]

(2) There is no evidence at all of Gestalt-switches or of mutual incomprehensibility: the level of communication during the debate is uniformly high, in both directions. The fact that 'phlogiston' does not translate into the language of the new chemistry simply does not matter. Translation-failure is perfectly compatible with mutual understanding and with rational appraisal.[103]

(3) As for Kuhn's later explanation of paradigm-shifts in terms of changes in similarity-relations, and thus a re-classifying of the same set of objects,[104] we need to make some careful distinctions. In one sense this re-classification does occur, in another sense it does not. If you think of classification in chemistry as theory-driven and top-down, you will think of the distinction between simples and compounds as primary, and other taxonomic categories as subdivisions of the categories of simples and compounds. From this point of view, Lavoisier's revolution is truly a transformation of chemistry. But if you think that classification in chemistry is practice-driven and bottom-up, the supposed transformation is all-but invisible. The chemists' familiar categories of metals, earths, non-metals, acids and salts survive the revolution intact.

What I want to say is that we can, quite legitimately, assess the rationality of episodes in the History of Science, even revolutionary episodes such as the Chemical Revolution. We can say, meaningfully and I think truly, that Lavoisier's anti-Φ-theory had the balance of the evidence in its favour by the late 1780s, and certainly by 1800.[105] And we can do this without Whiggism: the best supported theory at a given time is the rational one to accept at that time. 'Best' here refers, of course, to contemporary standards of assessment: to avoid the charge of 'philosophic' or 'methodological' Whiggism I deliberately allowed historians to relativise their judgements of rationality to conform to the norms of the period they are studying. I am in fact sympathetic to the notion of a core minimal rationality running through the whole History of Science,[106] and continuous with common-sense notions of rational belief, but I have not chosen to defend such a reactionary view in this article.[107] What I have sought to show is that, even if we allow ourselves to relativise methodology to historical context, we can still say that the Chemical Revolution was a rational process, i.e., that the chemical community behaved rationally in abandoning Φ-theory when they did, and for the reasons it did, and embracing the new chemistry.

A disciple of Kuhn might, of course, respond by saying that Kuhn had already retracted all the anti-rationalist rhetoric about Gestalt-switches and 'conversion experiences', and had explained the talk of mutual incomprehensibility and different 'worlds' in terms respectively of local translation failure and re-classification of a fixed set of objects. Perhaps my 'rationalistic' account of the Chemical Revolution is now perfectly consistent with the more sophisticated views of the later Kuhn. Perhaps Kuhn was only calling – as he himself insisted – for a more nuanced and historically sensitive account of scientific rationality than that provided by inductivists and Popperians.[108] If so, I am happy to greet the older and wiser version of Thomas Kuhn as an ally. My foes are those historians and sociologists of science who took, and still take, the errors of SSR as their gospel.

University of Bristol

NOTES

[1] Kuhn (1962, 2nd edition 1970).
[2] See, for example, Lakatos and Musgrave (1970) and Scheffler (1967).
[3] The Postscript to the second edition of SSR (Kuhn 1970) begins this process, which continues in 'Second Thoughts on Paradigms' in Kuhn (1977, pp. 293–319), in the 'Afterwords' in Horwich (1993, pp. 311–341), and in a number of other late papers.
[4] Barnes (1983), Barnes and Bloor (1982).
[5] See Popper's response, 'Normal Science and its Dangers' to Kuhn in Lakatos and Musgrave (1970, pp. 51–58). Popper admits that Kuhn's 'normal science' is real, but denies that it is normal in either sense of the word.
[6] Musgrave's Lakatosian account of the Chemical Revolution (in Howson 1976, pp. 181–209) is admirable in many respects, but over-emphasises the role of the time-factor, i.e., the progressive nature of the new chemistry. Phlogistic chemistry was also generating lots of new, confirmed predictions in the late eighteenth century – e.g., in Scheele's work on acids.
[7] Margolis (1993, chapters 4 and 5, pp. 43–67). Margolis' account contains significant insights (e.g., he dates the Revolution in Chemistry to 1783 instead of 1773) but, to my mind, over-emphasises the non-rational role of 'cognitive barriers' and thus preserves too much of what is wrong in Kuhn.
[8] There is one sociological study, McCann (1978), which treats the Chemical Revolution as a Kuhnian paradigm-shift, but the work reads like an unintended *reductio ad absurdam* of the author's approach. Anyone who thinks he can provide insight into the content of a scientific revolution by counting the articles produced by English and French chemists deserves no more than a dismissive footnote.
[9] Laudan (1990b) makes one of the characters in his dialogue an explicit champion of relativism. It is easier, in many ways, to invent a spokesperson for relativism than to deal with real philosophers of relativist leanings. But see Barnes and Bloor (1982) for an explicit defence of relativism.
[10] Kuhn has often been criticised for saying, of two paradigms, that they are both incompatible and incommensurable. The objection is that if they are incompatible, they must contradict one another, i.e., one must assert some proposition p which the other denies, so they must share concepts. More sympathetic critics have attempted to rescue Kuhn from this objection by saying that where two paradigms are incommensurable, the propositions held true under one paradigm cannot even be formulated under the other. It is not that p is true-in-paradigm-P1 and false-in-paradigm-P2, but that p is true-in-paradigm-P1 and unstatable-in-paradigm-P2. See Sankey (1994).
[11] The most prominent advocate of the historical variation of scientific methodology is of course Larry Laudan; see Laudan (1990a).
[12] Popper (1959). See also his 'Normal Science and its Dangers' in Lakatos and Musgrave (1970, pp. 51–58).
[13] Laudan seems at times to confuse two very different questions. If we ask, for example, whether doctors in the Middle Ages had explicitly formulated the methodology of the double-blind test, the answer is obviously no. But if we ask whether they could have appreciated the rationale for such

a test, the answer is surely yes. The power of the imagination in medicine was already a commonplace, spelt out in such authorities as Avicenna's *Canon*. And the rationale for the experiment designed to exclude any causal role for the imagination (whether of physician or of patient) is surely transparent.

[14] See Kuhn (1970, chapter 9, pp. 92–110); Feyerabend, 'Explanation, Reduction, and Empiricism' in his (1981, pp. 44–96). Sankey's *The Incommensurability Thesis* (1994) is almost exclusively concerned with semantic aspects of incommensurability, and is thus not very helpful to us here. Chapters 7 and 8 of his collection, *Rationality, Relativism, and Incommensurability* (1997) do deal with value-incommensurability.

[15] See the Postscript to Kuhn (1970, p. 202) for such 'communication breakdown'.

[16] See Kuhn (1970, pp. 118, 120).

[17] Lavoisier (1862, pp. 623–655).

[18] For helpful accounts of reference-fixing, see Kitcher (1978) and Nola (1980). We need a fundamentally causal theory with a descriptive element to enable us to say that 'phlogiston' does not refer, although 'dephlogisticated air' does.

[19] Kirwan (1789, pp. 67, 83, 84, and 126, 127). For Priestley, see his letter to Mitchill of June 14th 1798 (in Schofield 1966, pp. 292–294) and Priestley (1796, pp. 292–294).

[20] See de Morveau *et al.* (1787).

[21] Both Kirwan and Priestley make this objection – that the names of substances in Chemistry ought to be neutral between opposed theories. See Priestley (1796, pp. 40, 41).

[22] For an explicit denial that the pragmatic virtues are guides to truth, see Van Fraassen, especially his (1980). For a less sceptical account, see McMullin in Horwich (1993, pp. 55–78).

[23] Kirwan (1789), admits the greater simplicity of Lavoisier's theory, but warns his readers against 'a false shew of simplicity' (pp. 7, 8).

[24] Stahl (1730) . For commentary, see Metzger (1926, 1927) and Oldroyd (1973).

[25] If there is a single 'problem-solution' that is central to phlogistic chemistry, it is the reaction of charcoal with 'vitriols' (sulphate salts) to release sulphur. See Metzger (1927). For perceptive and sympathetic commentary, see Lavoisier's *Réflexions* (Lavoisier 1862, p. 494).

[26] Macquer's use of 'feu fixé' for Φ seems to have fixed the usage in French chemistry. For discussion of Macquer, see Partington (1962, p. 80 ff).

[27] Bergman (1785 p. 219).

[28] Cavendish (1766).

[29] Kirwan (1789, pp. 4, 5). An advantage of my theory, Kirwan argues, is that it renders Φ isolable by experiment.

[30] Priestley to Wedgwood, 6th March 1782; Priestley to Franklin, 24th June 1782 (in Schofield 1966, pp. 205, 209). For comments, see Partington (1962, p. 268), and Toulmin (1957).

[31] A glance at Scheele (1786) is sufficient to see how progressive the research programme of phlogistic chemistry could be. Scheele used the Φ-theory to predict the existence of whole families of hitherto unknown acids which could be produced simply by dephlogistication of non-metals, both inorganic and organic. He then went on to produce and isolate these new acids.

[32] Crawford (1788). For Lavoisier's abiding concern for the chemistry of life, and much incidental insight into Lavoisier's working methods, see Holmes (1985).

[33] Priestley to Franklin, 1st July 1772, in Schofield (1966, p. 104). For comments, see McEvoy (1978).

[34] White (1932, pp. 11, 183).

[35] For a survey of Whig historiography, as applied to the Chemical Revolution, see MacEvoy (1997).

[36] Lavoisier begins his *Réflexions* by praising Stahl for the discovery of two truths: the identity of combustion with calcination, and the possibility of transferring 'inflammability', e.g., from charcoal to sulphur. See Lavoisier (1862, p. 494).

[37] This, I suspect, is the heart of my disagreement with Kuhn. He doesn't believe in natural kinds; I think Φ-theory hit on some real natural kinds, and thus began the process of developing a natural taxonomy for chemistry. For Kuhn's 'revolutionary transcendental nominalism' see Sankey, 'Kuhn's Ontological Relativism' in his (1997, pp. 42–65).

[38] Morveau (1772). For comments, see Smeaton (1957, pp. 21, 22).

[39] For an early warning against such Whiggism, see Toulmin (1957).

[40] See Partington and McKie (1937, pp. 370, 371).

[41] Perrin (1983).

[42] See Partington and McKie (1937, p. 373 ff (Chardenon) and p. 389 ff (Guyton)).

[43] Joseph Black seems to have taken the 'negative weight' hypothesis seriously for a number of years, but was eventually persuaded that it was untenable. His friend the physicist Robison stressed its incompatibility with Newtonian physics. For this story, see Perrin (1983).

[44]   Priestley (1774, p. 267). For Priestley's dismissal of the 'negative weight' hypothesis, see Partington and McKie (1937, pp. 403, 404).

[45]   Bergman (1785, pp. xvii–xviii). For the identification (almost) of Φ with 'light inflammable air', see p. 211.

[46]   Boyle (1744, Vol. 3, p. 706ff).

[47]   For Baumé's account of weight-gain in his *Chymie experimentale et raisonnee* (1773), see White (1932, pp. 83, 84).

[48]   Kirwan (1789, pp. 38, 168).

[49]   See Priestley (1796, p. 33).

[50]   Macquer (1778), articles 'phlogiston' and 'calcination'. For Richter, see Partington (1962, pp. 630, 631).

[51]   For an interesting account of the reception of antiphlogistic chemistry in Germany, see Nordmann (1986).

[52]   Perrin (1969) gives a careful account of these variations.

[53]   Priestley to Franklin, 24th June 1782 in Schofield (1966, p. 209). Priestley clung to this bizarre view of 'red precipitate' until the end. See Priestley (1796, pp. 24, 25).

[54]   See Basu (1992) for further discussion.

[55]   See Perrin (1969, pp. 143, 144).

[56]   Scheele (1777).

[57]   See Lavoisier (1862, pp. 391–402) for his careful critique of Scheele's ideas.

[58]   For Macquer's identification of Φ with the matter of light, see Macquer (1778, article 'phlogistique').

[59]   Bergman (1785, p. 219). Bergman comes close to accepting the identification of Φ with light inflammable air, and consistently insists that Φ is ponderable.

[60]   See Lavoisier (1790, pp. 159, 160).

[61]   See Perrin (1970–1971).

[62]   Perrin (1970–1971, p. 131).

[63]   Lavoisier (1776).

[64]   Lavoisier (1776, *Avertissement*, pp. xxviii, xxix).

[65]   See Siegfried (1982, pp. 33, 34).

[66]   In a letter of 1778, Macquer praises Lavoisier's restraint, and is pleased to see how easily Lavoisier's results can be accommodated within Φ-theory. See Partington (1962, p. 83).

[67]   The 'Mémoire sur la Combustion en général' of 1777 (Lavoisier 1862, pp. 227, 228) reminds us that the supposed existence of Φ in charcoal, phosphorous, and sulphur is only a hypothesis. The 'Mémoire sur un Procedé Particulier pour convertir le Phosphore en acide phosphorique' of 1780 (*ibid.*, p. 282) goes a little further, rejecting phlogistic explanations as resting on unproven assumptions and suggesting that Φ is dispensable.

[68]   Philosophers' accounts of the Chemical Revolution often go wrong by over-emphasising the weight-gain experiments of 1772–1773, and then asking why phlogistic chemists were not converted to Lavoisier's new theory in 1773. Honourable exceptions include Alan Musgrave, and Howard Margolis whose work (1993) came to my attention long after I had written the first version of this paper (1984). Margolis locates the Chemical Revolution – rightly, in my view – in the water-controversy of the 1780s.

[69]   See Lavoisier (1862, pp. 335, 336). Here the oxygen-theory of acidity may have led Lavoisier astray.

[70]   Lavoisier (1862, pp. 225–233).

[71]   See the 'Considérations générales sur la nature des Acides' (Lavoisier 1862, pp. 248–260), and Le Grand (1972).

[72]   The importance of this point was stressed in a paper by Siegfried and Dobbs (1968).

[73]   Cavendish (1784). For comments, see Partington (1962, pp. 334, 335).

[74]   Priestley to Wedgwood of 8th January, 1788 (Schofield 1966, p. 249). He held on to this view until the end – see Priestley (1796, pp. 33–37).

[75]   See Daumas and Duveen (1959), Snelders (1979).

[76]   Lavoisier (1862, pp. 509–527).

[77]   Lavoisier (1790, pp. 159, p 177).

[78]   For Guyton's compromise theory in his *Elémens de Chymie* of 1777–1778, see Smeaton (1974). For Gren's three versions of Φ-theory, see Partington (1962, pp. 620–622, 633, 634).

[79]   Some historians have denied that Lavoisier's theory was simpler, arguing that both theories require a heat-stuff, and the only difference is whether it is found in the fuel or in the air. But the caloric theory of heat is so well grounded in the phenomena of latent and specific heats that it was becoming shared property in the late eighteenth century. Phlogistic chemists too would need some

account of the permanent elasticity of gases, and of the phenomena of latent heat discovered by Black. So they would end up requiring four terms where Lavoisier's theory requires only three.
[80]   Black's letter to Lavoisier of 24th October 1790 is quoted in Perrin (1982, pp. 162, 163). Black refers to the new chemistry as 'more simple and plain' and as 'so simple and intelligible'.
[81]   Kirwan (1789 pp. 7, 8).
[82]   See Lavoisier's *Réflexions* (1862, p. 640) and his response to Φ-theory in Kirwan (1789, p. 15).
[83]   Lavoisier (1862, p. 640).
[84]   Kuhn (1970, chapter 8), 'The Response to Crisis', pp. 77–91.
[85]   See Metzger (1926, p. 457) for the loose and open-ended nature of Stahl's programme for chemistry.
[86]   Anyone whose prime concern is with the sociological dimension of Kuhn's thought will find this line of thought especially compelling. After all, the close-knit group clustered around Lavoisier, with their own nomenclature, textbook and house journal do look like a group of rebels plotting a coup – and establishing a much more centralised and authoritarian system than had existed before. See Court (1972).
[87]   See Le Grand (1972).
[88]   Lavoisier seeks to present his oxygen-theory of acidity as a simple induction: all known acids contain oxygen; therefore (probably) all acids contain oxygen. See the *Traité* (Lavoisier 1790, pp. 64, 65) and his discussion of acids in Kirwan (1789, pp. 334, 335). For commentary, see Perrin (1973) and Siegfried (1982).
[89]   Lavoisier (1790, p. 87). See also Fourcroy in Kirwan (1789, p. 226). This *ad hoc* hypothesis of the 'dissolution' of carbon in hydrogen gas became the standard antiphlogistic account.
[90]   Kirwan (1789, pp. 182, 183).
[91]   Kirwan (1789, p. 41).
[92]   See Lavoisier's response in Kirwan (1789, p. 45 ff), and Fourcroy's careful discussion of the oxides of iron (*ibid.*, pp. 214–218). Lavoisier admits frankly in the *Traité* (Lavoisier 1790, 'Preface,' pp. xx, xxi and pp. 177, 178) that the theory of affinities is still in a messy and unsettled state.
[93]   It has sometimes been said that Lavoisier puts an end to the old chemical tradition of quality-bearing 'principles'. This is simply an error. For corrective remarks, stressing Lavoisier's continued allegiance to the chemistry of quality-bearing principles, see Perrin (1973) and Le Grand (1972).
[94]   For Berthollet's conversion, see Le Grand (1975).
[95]   For Fourcroy's conversion, see Smeaton (1962).
[96]   For Guyton's conversion, see Smeaton (1974).
[97]   For Black's conversion, see Perrin (1982). Black had been teaching the new chemistry for some years before his famous letter to Lavoisier of 1790.
[98]   Higgins (1789, 'Preface').
[99]   Perrin (1988b) provides the best documentation of this important point.
[100]   For Gren's three Φ-theories, see Partington (1962, pp. 620–622, 633, 634).
[101]   See Priestley (1796).
[102]   Sankey (1997, pp. 145, 146), 'Judgment and Rational Theory Choice'.
[103]   Sankey (1991, 1994) has consistently defended translation-failure, but insists that such failure does not entail communication-breakdown. One can sometimes understand a passage one cannot readily translate.
[104]   See Kuhn (1983), Hoyningen-Huene (1990).
[105]   Kuhn of course offers, in the *Postscript* to SSR itself (Kuhn 1970, p. 199), just such a 'rationalised' account of revolutions. But he thinks that such accounts are just *post hoc* rationalisations. I think that what Kuhn offers as *post hoc* rationalisation represents the real history better than his own story of a paradigm-shift.
[106]   See Worrall (1988), Siegel (1985).
[107]   Case studies such as this could, however, be used as a source of evidence for an invariant scientific rationality – our judgements of what is simpler, more coherent, and more faithful to the facts do not seem to diverge significantly from those of the participants in the debate.
[108]   Kuhn (1970, p. 199).

REFERENCES

Barnes, B., 1983: *Thomas Kuhn and Social Science*, Columbia University Press, New York.
Barnes, B and Bloor, D., 1982: 'Relativism, Rationalism, and the Sociology of Knowledge', in M. Hollis and S. Lukes (eds.), *Rationality and Relativism*, Blackwell, Oxford, pp. 21–47.
Basu, P.J., 1992: 'Similarities and Dissimilarities between Joseph Priestley's and Antoine Lavoisier's Chemical Beliefs, *Studies in the History and Philosophy of Science* 23, 445–469.

Bergman, T., 1785: *A Dissertation on Elective Attractions*, trans. T. Beddoes, reprinted by Frank Cass and Co., London, 1970.

Boyle, R., 1744: *Works*, T. Birch (ed.), 6 Volumes, London.

Cavendish, H., 1766: 'Experiments on Factitious Airs', *Philosophical Transactions of the Royal Society of London*, Vol. LVI.

Cavendish, H., 1784: 'Experiments on Air', *Philosophical Transactions of the Royal Society of London*, Vol. LXXIV, pp. 119–153.

Chalmers, A., 1982: *What is this thing called Science?*, 2nd edition, Open University Press, Milton Keynes.

Court, S., 1972: 'The Annales de Chimie, 1789–1815', *Ambix* 19, 113–128.

Crawford, A., 1788: *Experiments and Observations on Animal Heat and the Inflammation of Combustible Bodies*, 2nd edition, London.

Crosland, M., 1973: 'Lavoisier's Theory of Acidity', *Isis* 64, 306–325.

Daumas, M. and Duveen, D., 1959: 'Lavoisier's relatively unknown large-scale decomposition and synthesis of water, February 27 and 28, 1785', *Chymia* 5, 113–129.

Donovan, A.L., 1975: *Philosophical Chemistry in the Scottish Enlightenment*, The University Press, Edinburgh.

Donovan, A.L., (ed.), 1988: *The Chemical Revolution: Essays in Reinterpretation*, Philadelphia; special volume of *Osiris* (2nd series).

Feyerabend, P.K., 1981: *Philosophical Papers*, Vol. 1, Cambridge University Press, Cambridge.

Fichman, M., 1971: 'French Stahlism and Chemical Studies of Air', *Ambix* 18, 94–122.

Fine, A., 1975: 'How to Compare Theories: Reference and Change', *Nous* 9, 17–32.

Guerlac, H., 1976: 'Chemistry as a Branch of Physics: Laplace's Collaboration with Lavoisier', *Historical Studies in the Physical Sciences* 7, 193–276.

Guerlac, H., 1977: *Essays and Papers in the History of Modern Science*, Johns Hopkins University Press, Baltimore.

Higgins, W., 1789: *A Comparative View of the Phlogistic and Antiphlogistic Theories*, London.

Holmes, F.L., 1985: *Lavoisier and the Chemistry of Life: An Exploration of Scientific Creativity*, University of Wisconsin Press, Madison, Wisconsin.

Horwich, P. (ed.), 1993: *World Changes: Thomas Kuhn and the Nature of Science*, MIT Press, Cambridge MA.

Howson, C. (ed.), 1976: *Method and Appraisal in the Physical Sciences*, Cambridge University Press, Cambridge.

Hoyningen-Huene, P., 1990: 'Kuhn's Conception of Incommensurability', *Studies in the History and Philosophy of Science* 21, 481–492.

Hoyningen-Huene, P., 1993: *Reconstructing Scientific Revolutions: Thomas S. Kuhn's Philosophy of Science*, University of Chicago Press, Chicago.

Kirwan, R., 1789: *An Essay on Phlogiston and the Composition of Acids*, 2nd edition, reprinted by Frank Cass and Co., London, 1968.

Kitcher, P., 1978: 'Theories, Theorists and Theoretical Change', *The Philosophical Review*, 87, 519–547.

Kuhn, T.S., 1970: *The Structure of Scientific Revolutions*, 2nd edition, University of Chicago Press, Chicago (first edition 1962).

Kuhn, T.S., 1977: *The Essential Tension*, University of Chicago Press, Chicago.

Kuhn, T.S., 1983: 'Commensurability, Comparability, Communicability' in P.D. Asquith and T. Nickles (eds.) *PSA 1982*, Vol. 2, Philosophy of Science Association, East Lansing, Michigan, pp. 669–688.

Lakatos, I. and Musgrave, A. (eds.), 1970: *Criticism and the Growth of Knowledge*, Cambridge University Press, Cambridge.

Laudan, L., 1990a: 'Normative Naturalism', *Philosophy of Science*, 57, 44–59.

Laudan, L., 1990b: *Science and Relativism*, University of Chicago Press, Chicago.

Lavoisier, A.L., 1776: *Essays Physical and Chemical* (English translation of the 1774 *Opuscules physiques et chimiques*), reprinted Frank Cass, London, 1970.

Lavoisier, A.L., 1862: *Oeuvres, Tome II: Mémoires de Chymie et de Physique*, Paris.

Lavoisier, A.L., 1790: *Elements of Chemistry in a new systematic order, containing all the modern discoveries*, translated by R. Kerr, reprinted with a new introduction by D. McKie, Dover Publications, New York, 1965.

Le Grand, H.E., 1972: 'Lavoisier's Oxygen Theory of Acidity', *Annals of Science*, 29, 1–18.

Le Grand, H.E., 1975: 'The 'Conversion' of C-L. Berthollet to Lavoisier's Chemistry', *Ambix* 22, 58–70.

Macquer, P.J., 1778: *Dictionnaire de Chimie*, 2nd edition, Paris.

Margolis, H., 1993: *Paradigms and Barriers: How Habits of Mind Govern Scientific Beliefs*, University of Chicago Press, Chicago.

McCann, H.G., 1978: *Chemistry Transformed: The Paradigmatic Shift from Phlogiston to Oxygen*, Ablex Publication Corporation, Norwood, New Jersey.

McEvoy, J.G., 1978: 'Joseph Priestley, "Aerial Philosopher": Metaphysics and Methodology in Priestley's Chemical Thought from 1772 to 1781, Part II', *Ambix* 25, 93–116.

McEvoy, J.G., 1997: 'Positivism, Whiggism, and the Chemical Revolution', *History of Science* 35, 1–33.

Melhado, E.M., 1983: 'Oxygen, Phlogiston, and Caloric: The Case of Guyton', *Historical Studies in the Physical Sciences* 13, 311–334.

Melhado, E.M., 1985: 'Chemistry, Physics, and the Chemical Revolution', *Isis*, 76, 195–211.

Metzger, H., 1926: 'La Philosophie de la Matiere chez Stahl et ses Disciples', *Isis* 8, 427–464.

Metzger, H., 1927: 'La Théorie de la Composition des sels et la Théorie de la Combustion d'apres Stahl et ses Disciples', *Isis* 9, 294–325.

Morris, R.J., 1969: 'Lavoisier on Fire and Air: the Memoir of July 1772', *Isis* 60, 374–380.

Morris, R.J., 1972–1973: 'Lavoisier and the Caloric Theory', *British Journal for the History of Science* 6, 1–38.

Morveau, G., 1772: *Digressions Académiques*, Paris.

Morveau, G. *et al.*, 1787: *Méthode de nomenclature chimique, proposée par MM de Morveau, Lavoisier, Berthollet & de Fourcroy*, Cuchet, Paris.

Newton-Smith, W.H., 1981: *The Rationality of Science*, Routledge and Kegan & Paul, London.

Nola, R., 1980: 'Fixing the Reference of Theoretical Terms', *Philosophy of Science* 47, 505–531.

Nordmann, A., 1986: 'Comparing Incommensurable Theories', *Studies in the History and Philosophy of Science* 17, 231–246.

Oldroyd, D., 1973: 'An Examination of G.E. Stahl's *Philosophical Principles of Universal Chemistry*', *Ambix* 20, 36–52.

Partington, J.R., 1962: A *History of Chemistry*, Vol. 3, Macmillan, London.

Partington, J.R. and McKie, D., 1937: 'Historical Studies in the Phlogiston Theory – I. The Levity of Phlogiston', *Annals of Science* 2, 361–404.

Partington, J.R. and McKie, D., 1938: 'Historical Studies in the Phlogiston Theory – II. The Negative Weight of Phlogiston', *Annals of Science* 3, 1–58.

Partington, J.R. and McKie, D., 1938: 'Historical Studies in the Phlogiston Theory – III. Light and Heat in Combustion', *Annals of Science* 3, 337–371.

Partington, J.R. and McKie, D., 1939–1940: 'Historical Studies in the Phlogiston Theory – IV. Last Phases of the Theory', *Annals of Science* 4, 113–149.

Perrin, C.E., 1969: 'Prelude to Lavoisier's Theory of Calcination – Some Observations on Mercurius calcinatus *per se*', *Ambix* 16, 140–151.

Perrin, C.E., 1970–1971: 'Early Opposition to the Phlogiston Theory: Two Anonymous Attacks', *British Journal for the History of Science*, 5, 128–144.

Perrin, C.E., 1973: 'Lavoisier's Table of the Elements: A Reappraisal', *Ambix* 20, 95–105

Perrin, C.E., 1982: 'A Reluctant Catalyst: Joseph Black and the Edinburgh Reception of Lavoisier's Chemistry', *Ambix* 29, 141–176.

Perrin, C.E., 1983: 'Joseph Black and the Absolute Levity of Phlogiston', *Annals of Science* 40, 109–137.

Perrin, C.E., 1986: 'Lavoisier's Thoughts on Calcination and Combustion, 1772–1773', *Isis* 77, 647–666.

Perrin, C.E., 1987: 'Revolution or Reform: The Chemical Revolution and Eighteenth-Century Concepts of Scientific Change', *History of Science* 25, 395–423.

Perrin, C.E., 1988a: 'Research Traditions, Lavoisier, and the Chemical Revolution', *Osiris* 4, 53–81.
Perrin, C.E., 1988b: 'The Chemical Revolution: Shifts in Guiding Assumptions' in A. Donovan, L. Laudan and R. Laudan (eds.), *Scrutinizing Science: Empirical Studies of Scientific Change*, Kluwer, Dordrecht.
Perrin, C.E., 1989: 'Document, Text, and Myth: Lavoisier's Crucial Year Revisited', *British Journal for the History of Science* 22, 3–25.
Perrin, C.E., 1990: 'Chemistry as Peer of Physics: A Response to Donovan and Melhado', *Isis* 81, 259–270.
Popper, K., 1959: *The Logic of Scientific Discovery*, Harper & Row, New York.
Priestley, J., 1774: *Experiments and Observations on Different Kinds of Air*, J. Johnson, London.
Priestley, J., 1796: *Considerations on the Doctrine of Phlogiston and the Decomposition of Water*, J. Johnson, London.
Sankey, H., 1991: 'Translation Failure between Theories', *Studies in the History and Philosophy of Science* 22, 223–236.
Sankey, H., 1994: *The Incommensurability Thesis*, Avebury, Aldershot.
Sankey, H., 1997: *Rationality, Relativism, and Incommensurability*, Ashgate, Aldershot.
Scheele, C.W., 1777: *Observations and Experiments on Air and Fire*, English translation J.R. Foster, London, 1780.
Scheele, C.W., 1786: *Chemical Essays*, trans. J. Beddoes, London and Edinburgh.
Scheffler, I., 1967: *Science and Subjectivity*, Bobbs-Merrill, Indianapolis.
Schofield, R.E., 1964: 'Joseph Priestley, the Theory of Oxidation and the Nature of Matter', *Journal for the History of Ideas*, 25, 285–294.
Schofield, R.E. (ed.), 1966: *A Scientific Autobiography of Joseph Priestley*, MIT Press, Cambridge, Massachusetts.
Siegel, H., 1985: 'What is the Question Concerning the Rationality of Science?', *Philosophy of Science* 52, 517–537.
Siegfried, R., 1972: 'Lavoisier's View of the Gaseous State and its Early Application to Pneumatic Chemistry', *Isis* 63, 59–78.
Siegfried, R., 1982: 'Lavoisier's Table of Simple Substances: its Origin and Interpretation', *Ambix* 29, 17–28.
Siegfried, R. and Dobbs, B.J., 1968: 'Composition, a Neglected Aspect of the Chemical Revolution', *Annals of Science* 24, 275–293.
Smeaton, W.A., 1957: 'L. B. Guyton de Morveau (1737–1816)', *Ambix* 6, 18–34.
Smeaton, W.A., 1962: *Fourcroy: Chemist and Revolutionary, 1755–1809*, Heffer and Sons, Cambridge.
Smeaton, W.A., 1963: 'New Light on Lavoisier', *History of Science* 2, 51–69.
Smeaton, W.A., 1974: 'Guyton de Morveau and the Phlogiston Theory', in *Mélanges Alexandre Koyré*, Hermann, Paris, Vol. 1, 522–540.
Snelders, H.A.M., 1979: 'The Amsterdam Experiment on the Analysis and Synthesis of Water (1789)', *Ambix* 26, 116–133.
Stahl, G.E., 1730: *Philosophical Principles of Universal Chemistry*, translated by Peter Shaw, London.
Thagard, P., 1990: 'The Conceptual Structure of the Chemical Revolution', *Philosophy of Science* 57, 183–209.
Toulmin, S.E., 1957: 'Critical Experiments: Priestley and Lavoisier', *Journal for the History of Ideas*, 18, 205–220.
Van Fraassen, B.C., 1980: *The Scientific Image*, Clarendon, Oxford.
White, J.H., 1932: *The History of the Phlogiston Theory*, Edward Arnold & Co., London.
Worrall, J., 1988: 'The Value of a Fixed Methodology', *British Journal for the Philosophy of Science* 40, 376–388.

JOHN WORRALL

KUHN, BAYES AND 'THEORY-CHOICE': HOW REVOLUTIONARY IS KUHN'S ACCOUNT OF THEORETICAL CHANGE?*

1. INTRODUCTION: KUHN AND THE HOLD-OUT

Book reviews supply a rich source of sharp, sardonic humour. Probably my favourite remark about reviewing was by the wonderfully droll Reverend Sidney Smith, who opined 'I never read a book before reviewing it – it prejudices a man so!' Another favourite – in similar (though strictly speaking contrary) vein – is from a psychologist friend of Wesley Salmon's, who, when asked for his opinion of the latest Dianetics tosh by Lafayette Ron Hubbard, apparently remarked 'I cannot condemn a book before reading it; but after reading it, I shall'. My favourite remark, though, from within a book review is probably: 'This book fills a much-needed gap in the literature.'

No one could, of course, seriously hold that this last remark applies to Thomas Kuhn's *The Structure of Scientific Revolutions* nor to his earlier wonderful book on *The Copernican Revolution*. Indeed the latter is an outstanding example of proper, largely internal history of science, while the former is one of the most influential and discussed, quoted and misquoted books of our time. But as for the whole secondary literature on what Kuhn did and did not *really* mean – a literature to which Kuhn himself contributed rather generously – I think that one could argue quite plausibly that *it* fills a much-needed gap.

Surely the sincerest tribute to an investigator is not endlessly and scholastically to interpret and reinterpret his or her writings, but rather to try to make progress towards solving the problems that he or she raised. At any rate, I shall try in this essay to arrive quickly at first-level concerns about the rationality of science – and especially the rationality of theory-change in science – using problems raised by Kuhn, and criticising claims that Kuhn seems to have made, without worrying too much about whether they express his 'real view', if indeed there is such a single unified entity.

The chief target of the critical fire from those who felt Kuhn challenged the whole idea of science as a rational process was always his apparent views about the process of paradigm *change*. (Lakatos, for example, notoriously claimed that Kuhn's views made theory-change in science a matter of 'mob psychology'.) Most of what his critics found objectionable in Kuhn's account of theory-change

Robert Nola and Howard Sankey (eds.), After Popper, Kuhn and Feyerabend, 125–151.
© 2000 *Kluwer Academic Publishers. Printed in Great Britain.*

is reflected in his remarks about 'hold-outs' to 'scientific revolutions'. He claimed that if we look back at any case of a change in fundamental theory in science we shall always find eminent scientists who resisted the switch to the new 'paradigm' long after most of their colleagues shifted. These 'hold-outs' – Priestley defending phlogiston against Lavoisierian chemistry is a celebrated example – are often (though by no means invariably) elderly scientists who have made significant contributions to the entrenched paradigm. Kuhn added to this interesting but relatively uncontroversial descriptive claim the challenging *normative* assertion that these 'elderly hold-outs' were no less justified than their more mobile contemporaries: not only did they, as a matter of fact, stick to the older paradigm, they were also, if not exactly right, then at least not wrong to do so. On Kuhn's view, 'neither proof nor error is at issue' in these cases, there being 'always some good reasons for every possible choice' – that is, both for switching to the revolutionary new paradigm and for sticking to the old. Hence the hold-outs cannot, on his view, be condemned as 'illogical or unscientific'. But neither of course can those who switch to the new paradigm be so condemned. In one sense, then, it is easy to see why Kuhn expressed mystification over the claim that he made the history of science an irrational affair: in Kuhn's cosy world, *everyone* is rational – revolutionary and reactionary alike. But a genuinely 'rationalist' account surely needs losers as well as winners: rationalists seek general rules of theory-appraisal which presumably will show that the hold-outs were, in some important sense, simply mistaken.

Discussions of this issue are likely to become overly-abstract and the significant questions missed unless real historical examples are investigated in some detail. In the next section, therefore, I outline the views of one hold-out (a not so elderly one as a matter of fact). This is a case I have discussed elsewhere,[1] so I shall be very brief. Having resketched the historical details, I extend – and I believe, improve on – my earlier attempt to use those details to illustrate some general methodological morals. In particular I shall draw on the case-study to provide what I believe is a much improved account of the relationship between Kuhn's views and those of contemporary personalist Bayesians. This improvement, which is partly inspired by a paper of John Earman's (Earman 1993), involves looking again at the issue of how far, and in which respects, Kuhn's views can be reconciled with personalist Bayesianism, and in particular investigating one point where the two positions seem radically at odds. Roughly speaking, I shall argue that Kuhn's account is inadequate *both* where it agrees with the Bayesians, *and* where it disagrees with them.

2. BREWSTER AND THE WAVE THEORY

The early nineteenth century 'revolution' in optics saw Fresnel's classical wave theory of light triumph over the material corpuscular theory of light, generally attributed to Newton. Although this episode's impact on man's whole worldview cannot match that of, say, the Copernican or Darwinian revolutions, it did involve a sharp change in accepted theory and is explicitly cited by Kuhn as counting as a revolution in his terms. Indeed, partly because it is a narrowly scientific affair, this particular theory-change provides, I believe, an especially clear-cut instance against which to test general methodological claims. The most significant hold-out to this revolution, from Britain at least, was Sir David Brewster.

Although perhaps chiefly remembered nowadays for his biography of the great Sir Isaac, Brewster was an important optical scientist in his own right. He was the discoverer of many of the properties of polarised light; he discovered 'Brewster's law' relating the polarising angle and refractive index of transparent substances; he discovered a whole new class of doubly refracting crystals – the biaxial crystals – which soon proved to have great theoretical significance; he discovered that ordinary unirefringent transparent media can be made birefringent by the application of mechanical pressure; and he discovered the hitherto unknown general phenomenon of selective absorption.

Brewster was certainly some sort of hold-out. In 1831 a fellow knight of the realm, Sir George Biddel Airy, published a *Mathematical Tract on the Undulatory Theory of Light* which begins:

> The Undulatory Theory of Optics is presented to the reader as having the same claims to his attention as the Theory of Gravitation, namely that it is certainly true . . .. (Airy 1831, p. vii)

This would, I think, have been regarded at the time as a rather extreme expression of what was, however, definitely the majority view among the (admittedly small) group of those qualified in optics. Certainly the great majority of that group felt that the corpuscular approach had been definitively superseded by Fresnel's ether-based approach. Brewster held some form of minority view. In the same year that Airy published his *Mathematical Tract*, Brewster presented a 'Report on the Present State of Physical Optics' to the British Association for the Advancement of Science in which he asserted that the undulatory theory was 'still burthened with difficulties and [so] cannot claim our implicit assent' (1883a, p. 318). Two years later he reported:

> I have not yet ventured to kneel at the new shrine [that is, the shrine of the wave theory] and I must acknowledge myself subject to the national weakness which urges me to venerate, and even to support the falling temple in which Newton once worshipped. (1833b, p. 361)

This rhetorical flourish notwithstanding, Brewster was no mere irrational, 'Newton-worshipping' reactionary. He produced some sensible and challenging arguments for the *ancien regime*. There are in fact, as I see it, three main elements in Brewster's views about the then current state of play between the wave and emissionist theories.

(i) Brewster accepted – fairly unambiguously – that as things stood the wave theory had proved to be empirically the more successful.

He frequently expressed great admiration for the wave theory and fully acknowledged that it had enjoyed outstanding explanatory, and especially *predictive* success. For example, he said:

> I have long been an admirer of the *singular* power of this theory to explain some of the most perplexing phenomena of optics; and the recent discoveries of Professor Airy, Mr. Hamilton and Mr. Lloyd afford the finest examples of its influence in predicting new phenomena. (1833b, p. 360; emphasis supplied)

(Here Brewster has primarily in mind the prediction drawn by Hamilton from Fresnel's theory of the existence of both internal and external conical refraction. Since they involve directing a narrow ray of light along a very precisely characterised path through crystals of a very particular sort, cut in a very particular way, these represent exactly the sort of phenomenon that could realistically only be discovered as the result of testing some precise predictions of some powerful theory. Humphrey Lloyd confirmed Hamilton's predictions experimentally in 1833.)

(ii) *Brewster believed that the wave theory – for 'all its power and all its beauty' – could not be true.*

He produced two main arguments for this belief. The first was of a general methodological kind, related to the recently fashionable thesis of underdetermination of theory by evidence. Brewster pointed out that the fact that a theory had enjoyed explanatory, and even predictive, success does not of course deductively entail that it is true. Instead:

> Twenty theories may all enjoy the merit of accounting for a certain class of facts, provided they have all contrived to interweave some common principle to which these facts are actually related. (1833b, p. 360)

He did allow that the wave theory's predictive success implies that 'it must contain among its assumptions some principle which is inherent in ... the real producing cause of the phenomena of light' (1838, p. 306). However, *first* other theories – notably the Newtonian one – might well be able to incorporate such a principle; and *secondly* there was no doubt in Brewster's mind that, despite itself incorporating such an assumption, the wave theory, considered as a fully realistically interpreted claim about the universe, had to be false. In particular, a fully realistically interpreted wave theory was committed to the existence of what Brewster himself described as 'an ether, invisible, intangible, imponderable, inseparable from all bodies and extending from our own eyes to the remotest verge of the starry heavens' (*ibid.*). This was always too much – or perhaps *too little* – for Brewster to swallow. So Brewster's view was that a fully acceptable theory would share many of the structural assumptions implicit in the current wave theory but would reject – at least – that theory's invocation of the luminiferous ether. And he had, moreover, not abandoned the hope that some (highly modified) version of the Newtonian corpuscular account might prove to be such a theory.

Alongside this general argument, Brewster produced a second argument relating to the details of the particular version of the wave theory then current. Brewster in effect pointed out that while the wave theory's predictive success might be impressive, it was by no means complete. The wave theory failed and failed badly, in the case of at least two empirical phenomena: those of dispersion and of selective absorption.

I concentrate here on the second, a phenomenon which Brewster actually discovered, though identical morals could be drawn from the first failure. Brewster found that if a beam of sunlight is passed through certain gases and then dispersed in a prism, the spectrum that emerges from the prism is marked by a series of dark lines – indicating that, speaking in wave-theoretical terms, the components of the sunlight of certain sharply defined wavelengths have been absorbed during passage through the gas. Brewster – entirely reasonably – pointed out that, rather than simply refuting some particular version of the wave theory, this phenomenon provided a general difficulty for the whole wave approach. Whatever the details, the general story the wave theory seemed forced to tell looked extremely far-fetched. Referring to one particular absorption line (in 'oxalate of chromium and potash'), the wave theory needed to claim that the ether within that gas 'freely undulates to a red ray whose index of refraction in flint glass is 1.6272, and also to another red ray whose index is 1.6274 while ... its ether will not undulate at all to a red ray of intermediate refrangibility whose index is 1.6273!'

In other words, an infinitesimal change in the length of a wave must be supposed to produce a discrete change from free passage through the ether within the gas to no passage at all. Brewster pointed out that:

> There is no fact analogous to this in the phenomenon of sound, and I can form no conception of a simple elastic medium so modified by the particles of the body which contains it, as to make such an extraordinary selection of the undulations which it stops or transmits .... (1833a, p. 321)[2]

(iii) Brewster disagreed with the wave theorists over the heuristic issue of the likeliest way forward.

The defenders of the wave theory in Britain, notably Airy and Baden Powell, had no problems in accepting Brewster's claims – so long as they were understood as simply about the *present state* of the wave theory. Each acknowledged (they could scarcely do otherwise) that, as it stood in the 1830s, the wave theory had, for example, no explanation for selective absorption. However they each went on to point out that the wave theory had earlier had no explanation for polarisation either and that in particular it had seemed to be refuted by Fresnel's and Arago's experiments on the interference of polarised beams. (If the famous two slit experiment is modified so that the light coming through the two slits is polarised in mutually orthogonal planes – by the interposition of suitably oriented quartz plates, for example – then the interference fringes visible in the original experiment disappear.) But, rather than give up the theory, Fresnel had taken the bold step of instead developing it – in fact by switching from the assumption that light waves are longitudinal to the assumption that they are transverse (and hence switching from the idea that the ether is an elastic fluid to the idea that it is an elastic *solid*). This step, they pointed out, had led to exactly the sort of predictive success that Brewster himself applauded. In particular, Fresnel's move had led to the prediction, mentioned earlier, of the hitherto unsuspected phenomena of internal and external conical refraction. Moreover, as Airy and Powell justly asserted, there was nothing to match this success in the whole track-record of the emission theory.

While Brewster seemed to have some relatively vague belief in the revivability of the emissionist/corpuscularian approach, the wave theory had recently, and more than once, shown its ability to change major problems into major predictive successes. In that situation, Baden Powell argued:

> No sound philosopher would for a moment think of abandoning so hopeful a track, and none but the most ignorant or perverse would find in the obstacles which beset the wave theory anything but the most powerful stimulus to pursue it. (1841, p. iii)

3. KUHN'S LATER ACCOUNT OF 'THEORY-CHOICE': 'OBJECTIVE' AND 'SUBJECTIVE' FACTORS

The idea of a scientist 'holding out' against a new theory seems hopelessly vague. We now have an altogether more detailed account of a hold-out-scientist's position. This will eventually enable us to ask more penetrating questions about Kuhn's general views on reason and theory-change. But first we need to have a clearer picture of those general views themselves.

Kuhn developed – initially in the 1970 *Postscript* to his *The Structure of Scientific Revolutions* and then, in rather more detail in chapter 13 of his (1977) book *The Essential Tension* – a fuller account of the factors underlying what he there called 'theory-choice', than anything found in the original book. The 'mob psychology' gibe, he argued in 1977, 'manifests total misunderstanding' because he had always allowed a crucial role to the 'objective factors' from the philosopher's 'traditional list' (and he mentions five such objective factors: empirical accuracy and scope, consistency, simplicity and 'fruitfulness'). Kuhn says:

> I agree entirely with the traditional view that [these objective factors] play a vital role when scientists must choose between an established theory and an upstart competitor ... [T]hey provide the *shared* basis for theory choice. (1977, p. 322)

His claim had simply been all along that, while important, these objective factors fail to supply an 'algorithm for theory-choice'. At any rate when the choice between rival theories is a live issue in science, the objective factors never *dictate* a choice. Amid a good deal of rather flabby talk about methodological rules operating as *values* that 'influence' choice rather than as *rules* that dictate it, Kuhn supplies two sharp reasons for this failure of objective factors to provide an 'algorithm' for 'theory-choice'. (The reason for the 'scare quotes' round 'theory-choice', which from henceforth will be taken as implicit, will be explained below.)

The *first* reason is that single objective factors often turn out to deliver no unambiguous preference when applied to the theories *as they stood at the time when the choice was being made* – 'Individually the criteria are imprecise: individuals may legitimately differ about their application to concrete cases' (1977, p. 322).

For example, it is often assumed that the Copernican heliostatic theory was empirically more accurate than the Ptolemaic theory. This *eventually* became true but only as a result of the work of Copernicus, Kepler, Galileo and others – who had clearly then 'chosen' the Copernican theory for other reasons (if, indeed, for any *reasons* at all).

The *second* source of the failure of the objective factors generally to deliver a definite choice of theory is that 'when deployed together, they repeatedly prove to conflict with one another'. That is, even where single objective factors do point clearly in the direction of one of the rival theories, different factors may – again at the time when the choice was actually being made – point in *opposite* directions: so, for example, while simplicity (in a certain sense) favoured Copernican theory, consistency (with other, then accepted, theories) undoubtedly favoured the Ptolemaic theory. Kuhn concluded that the objective factors, while supplying the shared criteria of choice,

> are not themselves sufficient to determine the decisions of individual scientists. For that purpose, the shared canons must be fleshed out in ways that differ from one individual to another. (1977, p. 325)

In other words,

> every individual choice between competing theories depends on a mixture of objective and subjective factors, or of shared and individual criteria. (1977, p. 325)

The intent of Kuhn's further explanation of his views was to show that, on the topic of theory-choice, they differed less from philosophical orthodoxy than had generally been believed: his critics' remarks about irrationality 'manifest total misunderstanding', and indeed he had chosen not to write on this topic earlier precisely because his real views on it diverge rather little from 'those currently received', as compared to other topics (1977, p. 321). Notice, however, that this more elaborate account is presented as explaining, and *endorsing* the earlier account in *Structure* and in particular as endorsing the claims about hold-outs. He also explicitly re-emphasised the specific entailment that, whenever a new theory is developed to challenge an older one, 'there are always at least some good reasons for each possible choice' (1977, p. 328) – that is, good reasons for sticking to the older theory as well as good reasons for switching to the new.

Much could be said about Kuhn's treatment of each of the 'objective' (or shared) factors, but the main point at issue in the present paper will be his general account of the distinction between, and necessity for, both 'objective' and 'subjective' factors.

4. KUHN'S ACCOUNT AND PERSONALIST BAYESIANISM

Kuhn himself seems, then, to have given here a direct answer to the question raised in my title: his account of theory-change in science, far from being revolutionary, is in close agreement with that given by 'the' philosophers of science. Several important issues can be clarified by pursuing this claim.

The first point to be made is of course that Kuhn's view of 'the' philosophers of science seems unjustifiably monolithic – it is difficult to think of a single issue on which philosophers of science speak with a single voice and certainly the issue of theory-change is not one of them. Much of Kuhn's account can indeed be interpreted as cohering quite well with one well-supported tradition within current philosophy of science – that of personalist Bayesianism. But that tradition, of course, as well as invoking fervent support, also invokes fierce resistance.

Bayesianism, as is well-known, makes the rationality of an 'agent' depend on two requirements, and – at least in the *pure* version (which, I would argue, is the only clear version so far articulated) – *only* two requirements. The first is that, at any given stage in the development of science, the agent distribute degrees of belief over the various statements available to her in such a way as to satisfy the probability calculus; and the second requirement is that, whenever new evidence *e* comes in and nothing else of epistemic significance occurs (that is, the agent's 'background knowledge', relative to which all probabilities are implicitly relativised, remains otherwise constant), then the agent's new degrees of belief be related to the old by the 'principle of conditionalisation'. This principle requires that the agent's new 'prior' degree of belief in any assertion *A* be, in those circumstances, her old degree of belief in *A*, *conditional on* the evidence e.[3]

The 'posterior probability' $p(A/e)$ is, *via* Bayes' theorem, dependent on, amongst other things, the prior probability $p(A)$. Personalist Bayesians think of the prior as measuring a purely subjective degree of belief. It is true that the agent may, and generally will, have arrived at those priors themselves by conditionalisation on *earlier* evidence – but that conditionalisation will itself have been dependent on an earlier prior and so on: the subjective element is ineliminable.

Given the subjectivity of the priors, then, as I suggested in my (1990), there seems to be no real problem in reconciling at least some aspects of Kuhn's account of 'theory-choice' with this particular Bayesian philosophy. (A similar point is made by Wesley Salmon in his (1990) article 'Tom Kuhn meets Tom Bayes'.) The 'subjective factors' are taken care of by the priors – the fact that two equally 'reasonable' investigators might disagree about the merits of two rival theories just means that they assign different priors to the same theory. Whether or not all the objective factors can be delivered as consequences of Bayes' theorem is another matter – but some have argued, with varying degrees of plausibility, that they can. If so, then, as Salmon pointed out, although Kuhn took himself to be denying the existence of an algorithm for ranking theories, he could be seen, at least in part, as *endorsing* an algorithm – namely Bayesian conditionalisation, while at the same time acknowledging, as personalist Bayesians anyway do, that subjective preferences need to be taken as inputs into this algorithm in order to produce a definite theory-choice. The fact (if it is one) that Airy and Brewster made different choices between the wave and corpuscular theories of light, even given all the evidence equally available to both, may be explained by the fact that they in effect assigned different *prior* probabilities to those theories ahead of the evidence.

But accepting that Kuhn's account is broadly coherent with that given by personalist Bayesians at best amounts, in the eyes of many philosophers (myself included), simply to a restatement of 'the problem with Kuhn': that he makes theory-change in a science a much more subjective affair than many of us believe it to be. (Remember that the outcome of Kuhn's analysis was still that there are always 'at least some good reasons for every choice'.) Conversely, to the extent that Kuhn's account can be reconciled with personalist Bayesianism, this simply underlines what critics of *Bayesianism* have always insisted on: that it allows much too large a role to subjective, personal factors to provide – without further augmentation – an adequate account of reasoning about theories in science.

Kuhn's claim that his views diverge relatively little from 'those currently received' is only in fact true of a proper subset of them; and is true of that proper subset because of a feature of their philosophy that many other philosophers think makes it ultimately indefensible.

The objections of the 'objectivists' stem first and foremost, of course, from the outright subjectivism of the priors. There are some much touted results concerning the 'swamping' or 'washing out' of priors – results which have led some commentators to hold that this source of subjectivism is, in the end, much less worrying than might initially be imagined. The relevant theorems prove that, under certain conditions, the posterior probabilities that two agents assign to some pair of rival theories will, given evidence of a certain sort, converge *whatever prior probabilities* (short of zero or one) they may have assigned to those theories. But, aside from detailed issues about whether or not the necessary conditions can plausibly be taken to hold in particular cases, the fact is that these results guarantee agreement – even agreement on the ranking of the rival theories – *only* in the limit, which of course is never achieved in practice. Given any actual theory-ordering in the light of the available evidence that a sensible person would regard as frankly ridiculous – such as, for example, a preference now for special creationism over Darwinism as an account of the present biological furniture of the earth – there must, quite trivially, be priors that the 'ridiculous' agent could have had, such that they conditionalised away fully and accurately in accord with Bayesianism on *all* the accumulating evidence, about the fossil record, homologies and the like and *still* arrived at their clearly unreasonable ranking.[4]

But, aside from the much-advertised problem of the priors, there are two further sources of subjectivism in the Bayesian approach that have received relatively little publicity yet which surely ought to be equally damning in the eyes of anyone who thinks of science as governed by strong objective principles of sound reasoning.

First the notion of evidence itself is subjective in this approach. When reconstructing episodes from the history of science in their terms, Bayesians invariably assume that all sensible agents come to take as evidence what we would all take as evidence. But this is an extra assumption for which there is no sanction in the 'pure' Bayesian account. According to that account, any synthetic assertion *e* is evidence for an agent (relative to epistemic situation *S*) if and only if she happens to assign, when in that situation, a subjective probability of one to *e*. (I here ignore wrinkles about Jeffrey conditionalisation which allows an agent to conditionalise on evidence about which she is less than certain – but again the, in that case non-extreme, probability the agent ascribes to *e* will be a subjective affair.) Anything that anyone comes to be subjectively entirely convinced of the truth of 'that they were abducted and raped by aliens', 'that god exists', 'that Jesus walked on water', 'that the needle in the meter points to "5" when all the rest of us see it as pointing to "10"', counts as new evidence for that 'agent'. At the point at which such an agent *becomes* convinced of the particular piece of 'evidence' at issue, she is required by Bayesian rationality to modify all her erstwhile degrees of belief by conditionalisation on it – however strange that 'evidence', and therefore those shifts in degrees of belief, look to you or me.

There is still more subjectivism. Let's assume that some Bayesian 'agent' in fact takes as evidence what any sensible scientist would take as evidence. There is still the issue of what counts for that agent as implicit 'background knowledge'. Bayesians are quite clear that all probabilities, at any given time, are to be thought of as relativised to the agent's background knowledge at that time. The distinction between evidence and background knowledge is blurred – perhaps inevitably so. Once an item of background knowledge is articulated, then, I suppose, it automatically counts as evidence for that agent (since she is bound to regard it as having probability one). Again in applications, Bayesians quietly assume that what counts as background knowledge at a given time in science is a more or less universal, intersubjective affair. But, *first*, this will be plausible even as an idealisation of the actual state of affairs in a particular science at a particular time only courtesy of a very selective attitude toward who counts as a competent scientist (as a *bona fide* member of the relevant 'scientific community'). And, *secondly*, there is, so far as I can tell, no official sanction within personalist Bayesianism for this assumption: on the contrary, according to the official 'pure' position, background knowledge is another entirely agent-specific factor – whatever the agent regards as 'given' or as delimiting the space of conceptual possibilities[5] is background knowledge for her, *whatever anyone else might think*.

This also means that there is another almighty slice of subjectivism lurking in the Bayesian's account of belief-*dynamics*. There is a crucial, but under-emphasised, clause in the principle of conditionalisation: your new prior probability on A must be your old probability on A, conditional on e, if in the meanwhile (that is, between the 'old' and 'new' times) e has turned from possible evidence to actual evidence, *and nothing else of epistemic significance has occurred* – that is, no other change has occurred between the two historical stages in your 'background knowledge'. There is again here a significant difference between the way Bayesian theory is standardly applied and the pure general theory. In applications, it is quietly assumed that the general epistemic situation, supplied by 'background knowledge', will be the same for all agents; and that, where the 'reasonable' assumption is that the only change to agreed background knowledge is the addition of some new evidence, this assumption too will be generally shared. However there are no such constraints in Bayesian theory itself – which, so far as I can tell, must leave this ingredient too as an agent-relative affair. A Bayesian agent is, apparently, officially allowed to assert at any point that her personal epistemic situation, her personal background knowledge, has suddenly changed, and hence call for an entirely new round of bets. Indeed she is *required* to do so in order to be Bayesian-rational, if her subjectively perceived epistemic framework somehow changes. Bayesian rationality imposes no need for any *argument* as to why the agent's 'background knowledge' has suddenly shifted and hence imposes no need for any particular argued connection between the new and old priors. Indeed this is not an area where normative considerations play any role – it will just be a descriptive matter whether or not the agent's background 'knowledge' (really, on this approach, set of background *beliefs*) remains unchanged.

Suppose, for example, a special creationist Bayesian sees her initially massive prior for creationism being steadily eroded by conditionalisation on evidence. Viewing her diminishing posterior with alarm, she (of course quite unrelatedly!) suddenly feels that her whole epistemic situation has undergone an abrupt change. Pointing out that all Bayesians agree that all degrees of belief are implicitly relative to general epistemic framework or background knowledge and that this general epistemic framework too is agent-dependent, she can simply 'call for a new round of bets', that is, insist on redistributing her priors against the, for her, new background. Suppose again that, against the new background, the special creationist theory has a massive prior.

This seems the antithesis of how to rank theories scientifically in the light of the evidence, yet the personalist Bayesian cannot but sanction it. Of course, the Bayesian does *not* sanction an agent's simply *pretending* to have undergone a shift in her general epistemic framework in order to 'defend' a theory favoured on non-evidential grounds (and we would all be suspicious of the veracity of any real scientific creationist who made such assertions about their subjective degrees of belief). Nonetheless, if there *were* to be such an agent who genuinely felt that the whole epistemic earth had moved and who ended up with a suddenly increased 'prior' for her scientific creationist view then the Bayesian could not regard her as in any sense irrational.

5. EARMAN AND SHIFTS IN THE 'SPACE OF CONCEPTUAL POSSIBILITIES'

Although such alleged shifts in general epistemic background may appear particularly suspect to the objectively inclined, John Earman has in effect argued (in his 1993) that there are cases where no such intuitions are elicited and where, on the contrary, the intuitively correct analysis does involve such a shift. Indeed Earman argues that the occurrence of such a shift is exactly what characterises a scientific *revolution*.

Earman, like Salmon and myself, investigates ways in which Kuhn's views on theory-change and those of the Bayesians can be reconciled. (His direct aim is to investigate relationships between Kuhn and Carnap, but the latter of course was or became a tempered personalist.) He arrives however at interestingly different conclusions: Earman in fact sees two ways in which Kuhn and the Bayesian must remain at odds and then a third issue on which fruitful cooperation is possible (indeed where the Bayesian must, in Earman's view, accept Kuhn's insights and hence radically augment her position).

On the first allegedly irreconcilable difference, Earman seems to me mistaken. He complains that:

> Kuhn's list of criteria for theory choice is conspicuous for its omission of any reference to the degrees of confirmation or probabilities of theories . . .. (1993, p. 21)

But surely the 'omitted' criteria should in fact be thought of as implicitly present on Kuhn's list – as either definable in terms of criteria that *are* explicitly mentioned (those of simplicity, perhaps consistency and especially empirical scope and adequacy), or, more strongly, as providing the means of defining those criteria. Simplicity, consistency (both internal consistency and consistency with other well-supported theories) and empirical scope and adequacy are criteria that, intuitively speaking, feed into judgements about degree of confirmation and the probabilities of theories in the light of evidence. Indeed some Bayesians would make the stronger claim that all these other criteria can themselves be defined in terms of probabilities. It seems hard to believe that Kuhn could consider that his views on theory-choice diverged comparatively little from 'those currently received' unless those views could accommodate, in one way or another, something very like degrees of confirmation.[6] This is how Salmon and I have reconstructed Kuhn.

Earman's second reason for seeing Kuhn and the Bayesian as at odds – essentially that Kuhn's notion of 'theory-choice' is fundamentally irreconcilable with a probabilistic approach – is altogether weightier and will be the subject of the next section. Here I want to concentrate on the issue on which Earman sees the Bayesian as needing to benefit from Kuhnian insights – basically in dealing with Kuhn's favourite topic: *scientific revolutions*.

Almost every postwar philosopher of science has implicitly recognised the importance of 'background knowledge' both in the generation and in the appraisal of particular scientific theories. But until recently, few have done much to turn this implicit recognition into a fruitful tool of analysis by looking more precisely at how science depends on 'background knowledge'. Even within the Bayesian approach, and despite the fact that that approach explicitly recognises the dependence of all probability assignments at a particular time on the background knowledge of the time, nothing is said about the exact nature of this dependence. On the contrary, and as we have seen, the way in which probabilities are 'relativised' to background knowledge is generally left as just another area in which subjective judgement inevitably intrudes. Earman – surely correctly – takes it that the Bayesian *needs* to say something more and (wearing his Monday-Wednesday-and-Friday clothes) starts to try to say it.

The chief role played by background knowledge for Earman is in specifying the background 'space of conceptual, theoretical possibilities'. If, in the light of background knowledge, this space is finite or can be finitely partitioned, then eliminative induction *via* observational evidence becomes possible, for example. The important feature for present purposes, however, concerns his view of *changes* in this background space of possibilities. According to Earman such changes are precisely what characterise scientific revolutions – of which he distinguishes two flavours: mild and strong.

A mild form of scientific revolution occurs with the introduction of a new theory that articulates possibilities that lie within the boundaries of the space of theories to be taken seriously but that, because of the failure of actual scientists to be logically omniscient, had previously been unrecognized as explicit possibilities. The more radical form of revolution occurs when the space of possibilities itself needs to be significantly altered to encompass the new theory. (1993, pp. 24, 25)

Earman claims that even the 'mild form' shows the inadequacy of the standard form of Bayesianism – even mild revolutions cannot be explained by

Bayesian conditionalisation:

> For conditionalizing (in any recognizable sense of the term) on the information that just now a
> heretofore unarticulated theory T has been introduced is literally nonsensical, because such a
> conditionalization presupposes that prior to this time there was a well-defined probability for
> this information and thus for T, which is exactly what the failure of logical omniscience rules out.
> (1993, p. 25)

And, of course, matters are still worse in the case of 'strong' scientific revolutions
when some genuinely new, as opposed to simply hitherto unrecognised, conceptual possibility is introduced. Even if an agent were logically omniscient she
could not assign at time t a well-defined probability to a theoretical
possibility that did not yet exist at t.

   Given this account, Earman sees Kuhn's subjective factors playing a role *not*
(as Salmon and I had) in the assignment of prior probabilities within a *given*
epistemic framework, but rather in *reassigning* probabilities after shifts in the
conceptual space:

> In typical cases [of either mild or strong revolutionary shifts] the scientific community will possess
> a vast store of relevant experimental and theoretical information. Using that information to inform
> the redistribution of probabilities over the competing theories on the occasion of the introduction
> of the new theory or theories is a process that, in the strict sense of the term, is *a*rational: it cannot be
> accomplished by some neat formal rules, or, to use Kuhn's term, by an algorithm. On the other
> hand, the process is far from being *ir*rational, since it is informed by reasons. But the reasons, as
> Kuhn has emphasised, come in the form of persuasions rather than proof. In Bayesian terms, the
> reasons are marshalled in the guise of plausibility arguments. The deployment of plausibility
> arguments is an art form for which there currently exists no taxonomy (1993, p. 26)

The first question that arises about Earman's account is whether or not the idea of
extensions of the conceptual space provides a satisfactory analysis of what
goes on in a scientific revolution.[7] This is clearly a big issue but there seem to me
several reasons for doubt.

   Notice two significant features of the account. *First*, Earman admits that the
distinction between the two flavours of revolution 'mild' (in which some hitherto
unrecognised but actually 'available' theoretical possibility begins to be taken
seriously) and 'strong' (in which the new theoretical possibility is genuinely new)
is 'blurred, perhaps hopelessly so' (1993, p. 25). *Secondly*, and relatedly, he admits
(indeed he insists) that it is generally possible *post hoc* to reconstruct even the
'strongest' revolutions (such as the transition from classical to relativistic physics) as having taken place within a common linguistic and conceptual framework (1993, p. 24).

   But, in view of these concessions, the Bayesian might seem well advised to
idealise and take it that the warring parties in a 'revolution' are working against a
common background of conceptual possibilities, while disagreeing only over
which possibility to prefer. After all, any account of the rationality of science is
bound to idealise in some ways. This is certainly true of personalist Bayesianism
in its standard form and even in the modified version towards which Earman is
working. Bayesian agents are assumed, for example, to be perfect deductive
logicians at least in the sense of assigning probability one to all logical truths.[8]
But suppose s, a complicated sentence of the propositional calculus involving

38 atomic sentences, is in fact a tautology. Even the most logically acute real Bayesian would have problems in immediately recognising – indeed she may live her life without *ever* recognising – that *s* is a tautology and hence that, whatever she may initially think (if indeed she thinks anything at all), her *real* degree of belief in *s* is one. Nonetheless the Bayesian supposes that this is what every agent's real degree of belief in *s* is – and that seems (to me at least) exactly right. So the complaint against taking warring parties in a 'scientific revolution' as working against a commonly agreed background cannot be merely that it idealises.

Earman himself, as we shall consider in more detail in the next section, reacts to the fact that, outright, some scientists *believe* certain theories (that is, in Bayesian terms, assign them probability one) by taking the line that they *ought not* to:

> One can cite any number of cases from the history of science where scientists seem to be saying for their pet theories that they set $p = 1$. Here I would urge the need to distinguish carefully between scientists as advocates of theories versus scientists as judges of theories. The latter role [alone] concerns us ... and in that role scientists know, *or should know*, that only in very exceptional cases does the evidence rationally support a full belief in a theory. (1993, p. 23 emphasis supplied)

But, if sensible idealisation is permitted, what is wrong with assuming, at least in the case of 'mild' revolutions, that, however it may appear psychologically to a given agent/scientist, she does in fact have some initial degree of belief in the 'revolutionary' possibility? The 'revolution' would then, for the Bayesian, simply consist in the (perhaps sudden) decrease, through conditionalisation, in the probability of the 'old' theory, and a corresponding increase in the probability of the 'new' theory.

And if this is the right way to treat 'mild' revolutions, and if the distinction between 'mild' and 'strong' is 'blurred, perhaps hopelessly so', and if it is always possible, as Earman claims it is, to reconstruct the theoretical dispute after the event as taking place against the background of a conceptual space held in common by the disputants, why not treat *all* scientific revolutions, admittedly somewhat idealistically, as taking place against the background of a fixed space of conceptual possibilities? So far as I can tell, Earman's main counter argument is, indeed, that it is psychologically unrealistic – the scientists involved in revolutions did not as a matter of fact themselves explicitly internalise the conceptual possibilities that make it possible to see the dispute as occurring against an agreed conceptual background. But, as I indicated, this seems to me, even if true, not necessarily either here or there.

But is it true? I find it difficult to see Earman's model as instantiated in a range of scientific revolutions. For example, neither the Copernican nor the Darwinian revolutions involved essentially new conceptual possibilities. The heliocentric model had, after all, been articulated as long ago as Aristarchus and was certainly not in any sense unthinkable for his Ptolemaic opponents – they conceded the possibility that the earth moved around the sun, and simply believed that this was (very likely to be?) false for a range of evidential and non-evidential reasons. As for Darwin, pretty well all of his contemporaries seem to have agreed that species have evolved, the only dispute was over the mechanisms, and their relative weights; and here Darwin spent much time stressing the analogy with artificial selection precisely in the attempt to make natural selection seem a natural, non-novel idea. Moreover, Wallace independently had arrived at essentially indistinguishable ideas – surely showing that they were 'in the air' at the time. (Indeed, the strikingly frequent phenomenon of simultaneous discovery in science, even of 'revolutionary' ideas, seems to indicate how much 'in the conceptual air' they *generally* are. Hooke, Wallis and Wren, for example, really did have the idea of universal gravitation and its inverse square relationship to distance – Newton really did hold that his genius was to have 'proved' the theory from Kepler's phenomena while the others 'merely' conjectured it.)

The Einsteinian revolution seems to provide the main stimulus for John Earman's general account. Although I would not dare cross swords with him concerning Einstein, it is a fact that there are other analyses that show the axioms of relativity as derived from new experimental results plus already generally accepted background principles.[9] (Also, remember that so far as *special* relativity goes at any rate, Poincaré is, with apparent justice, regarded as a simultaneous – or even pre- – discoverer with Einstein.)

If one were thinking – not too rigorously – about the history of science with Earman's intuitive distinction in mind, then probably it is the quantum revolution that would seem *prima facie* the 'strongest' of them all. But here too the quantisation of energy has been persuasively argued – in this case, by John Norton – to have been arrived at as a deduction from the phenomena (where this means, of course, deduction from the phenomena *plus* already existing background knowledge).[10] It took the genius of Bohr to show that energy-quantisation could be derived deductively from new experimental results plus already existing, and arguably, generally accepted background principles; but if there is such an argument, then it seems hard to deny that Bohr was showing that everyone, genius or not, *implicitly* recognised energy quantisation as in the space of conceptual possibilities ahead of his 'innovation'. New experimental results may be *surprising*, but it is hard to think of them ahead of their discovery as actually inconceivable; but if 'all' that it takes to arrive at a 'revolutionary' new theory is to plug some new experimental results into a general framework, that is not only conceivable but arguably part of generally accepted background knowledge, then it is hard to see how that new theory can itself have been outside the space of conceptual possibilities beforehand.

The situation so far as the historical episode considered in this paper goes is even clearer. Far from being an hitherto unrecognised conceptual possibility in the early nineteenth century, the wave theory of light had, of course, already been around (in altogether less impressive but still recognisably similar forms) for at least a century and a half. (Hooke, Huygens and, in the eighteenth century, Euler had all developed versions of it.) And again, the chief 'revolutionary' in this case, Fresnel, claimed (with perhaps surprising plausibility) that, given the premise that the corpuscular theory had proved so problematic in view of the experimental phenomena as to be out of the game,[11] his version of the wave theory (complete with the 'luminiferous aether') could be straightforwardly deductively inferred from experimental results plus uncontentious principles of background knowledge.[12]

In sum, it seems to me that Kuhn greatly exaggerated the revolutionary nature of 'revolutionary' change in science; and that John Earman is following suit. But now suppose that, for the sake of argument, we go along with Earman's analysis of scientific revolutions and therefore accept his claim that the Bayesian position needs to be augmented; and suppose further that we agree that that position needs to be augmented using Kuhn's account of the factors involved in 'theory-choice' as something like the right account of how prior probabilities get re-assigned after a revolutionary 'shake-up' of the space of conceptual possibilities. Such an account, in line with Kuhn, would admittedly be 'arational' since it does not conform to some 'neat set of formal rules' (there is no 'algorithm') but this does not mean, suggests Earman, that the process is actually *irrational*. The process is 'informed by reasons' – though Kuhn is right that these reasons take the form of 'persuasions rather than proof', or of 'plausibility arguments' – an 'art form for which there currently exists no taxonomy' (1993, p. 26).

This would then seem to amount to just another version of the earlier story of 'the problem with Kuhn' finding itself underlined by the partial agreement of his view with that of the personalist Bayesian. 'Objectivists' like myself want to insist that there is, at every stage of the development of science, such a thing as 'the intellectual argument' between two or more competing theories, and at each stage, there is an objectively correct view – no matter how complex it might be – about the state of that argument. No one expects such arguments to be purely deductive ones beginning from uncontentious premises and entailing one of the rival theories (if that is what Earman means by 'proof'). But if all that we have is 'persuasion' relying on 'plausibility' and if, as Kuhn's insistence on the idio-syncratic nature of the subjective factors seems to suggest, and as Earman's endorsement of his Bayesian version of Kuhn suggests he supports, one man's plausibility is the next woman's far-fetched implausibility, then all talk of 'rea-sons', 'persuasions' and 'art forms' is surely a smokescreen to cover the admis-sion of a sizeable chunk of relativism into the account of scientific theory-change. If, on the contrary, what counts as plausible is meant to be an objective matter, then the whole problem would seem to be to articulate and defend the objective principles that govern plausibility. If the 'neat formal rules' that John Earman recognises on behalf of the Bayesian are not up to this task, then we need to find other, stronger rules. To talk in Earman's way seems simply to surrender the game to the Kuhnian relativist.

And yet Earman must surely be right that we cannot plausibly expect to capture the whole of the complex and rich process of scientific theory-change in anything likely to count as a neat set of formal rules.

Not the *whole* of the complex and rich process of theory-change, certainly; but then, I shall argue, we should never have expected to. Philosophers of science, following Kuhn, have got themselves into a mess and have proved an easy target for *some* of the barbs of Kuhn-inspired, social constructivist-inclined critics by expecting too much. We need a proper (and at the same time more nuanced and yet more modest) identification of what features of this process of theory-change are, and what features *are not*, governed by considerations of 'rationality'. This identification can be made, as I shall indicate in the next section, by following through in some detail the second of the features of Kuhn's account that Earman sees as inconsistent with personalist Bayesianism: namely Kuhn's overly simple notion of 'theory-choice'.

6. WHY 'CHOOSE' A THEORY? RATIONALITY REGAINED

Both in the *Postscript* and in chapter 11 of *The Essential Tension*, Kuhn analyses theory-change in terms of scientists making theory 'choices'. As so often, Kuhn is less clear than one might like, and certainly he attempts no explicit definition of this notion. Implicitly however he *seems* to take choosing a theory to involve, not just the view that that theory is the best available so far in light of the evidence, not just the decision to work on it to see where it leads and how it can be developed, but also taking the theory fully to one's breast, 'accepting' it, believing it to be true.

Earman complains about this from a Bayesian perspective – arguing, that the only sense in which a scientist might 'choose' or 'accept' a theory consistently with the Bayesian approach is exactly 'the innocuous sense [of] choos[ing] to devote [her] time and energy to' that theory (1993, p. 22). To show how 'baffling' for the Bayesian is the idea of choosing or accepting a theory T in a sense that reflects a judgement about T's epistemic status, Earman considers a researcher who performs some introspection and decides that her subjective probability for T in the light of all evidence available to her is p. One Bayesian-kosher sense in which the researcher would surely be said to accept T is if $p = 1$ or is 'so near to 1 as makes no odds'. But

> Such cases ... are so rare as to constitute anomalies. Of course, one can cite any number of cases from the history of science where scientists seem to be saying for their pet theories that they set $p = 1$. Here I would urge the need to distinguish carefully between scientists as advocates of theories versus scientists as judges of theories. The latter role concerns us here, and in that role scientists know, or should know, that only in very exceptional circumstances does the evidence rationally support a full belief in a theory. (1993, p. 23)

While applauding, as indicated earlier, this willingness to override psychological facts (even about eminent scientists) in the name of good general sense, it is not at all clear to me that cases of full belief are either as rare or as unjustified as is here suggested. It surely depends how 'far down' the hierarchy of theories we go: the assertions that perpetual motion machines are impossible, that the heart pumps the blood round the body, that cells contain energy-providing mitochondria, that water consists of molecules consisting in turn of two atoms of hydrogen and one of oxygen, etc., all seem to me to be, given the present evidence, perfectly proper objects of outright or total belief (whatever that might precisely mean).

'Fundamental', 'explanatory' theories – precisely the sort of theory that has triumphed over others inconsistent with it as a result of a 'scientific revolution' – are, though, a different kettle of fish. And concerning them, John Earman is surely correct – although there are cases of scientists who seemed to assign them probability one (indeed I have been told by some scientists that they *need* to believe in the truth of their theories in order to work successfully on them), the

sensible view, precisely because of the historical record (a record that under-
writes the so-called pessimistic meta-induction), is that they *ought not* to.

Suppose, then, that T is such a fundamental theory (the general theory of
relativity, quantum theory, or whatever) and that some sensible research scientist
has a high but less than total degree of belief in T. What might the further fact that
she *accepts* (or 'chooses') T mean? One possibility, Earman points out, is that,
having decided initially that her degree of belief in T is, say, 0.75, she then, by
'accepting' T, converts that probability to one. Earman is again surely right that

> This is nothing short of folly, since she has already made a considered judgment about evidential
> support and no new relevant evidence occasioning a rejudgment has come in. (1993, p. 23)

But then the only other possibility is that she retains her initial degree of belief in T
($p = 0.75$) but 'acts *as if* all doubt were swept away in that she devotes
every waking hour to showing that [all relevant] observations can be explained by
the theory, she assigns her graduate students research projects that presuppose
the correctness of the theory' and so on. But this simply amounts to an alternative
expression of the view of 'acceptance' of a theory as a purely pragmatic decision not reflecting any judgement about the epistemic status of the theory.

I agree, then, that the Bayesian has good reason to be unhappy with Kuhn's
idea of theory-choice. But justified unhappiness on this score is not restricted to
Bayesians. Our historical case shows precisely why.

Did Brewster continue to 'choose' the Newtonian, corpuscular theory of light,
despite the availability of Fresnel's wave theory? This question, I suggest, with
its implicit commitment to measuring attitudes to theories along one dimension,
is inherently unsatisfactory. In so far as one can give an answer at all, it is 'yes-
and-no' (or perhaps, for reasons to be explained, 'yes-no-and-no').

Brewster had, remember, rather than a single view, three main, related but
independent views about the corpuscle/wave rivalry as it stood around 1830.
First he made various concessions about the empirical power and predictive
success of the wave theory, that can, I think, plausibly be interpreted as allowing that *as things stood* the wave theory had much the stronger empirical support. *Secondly*, he clearly held that, despite its empirical success, the wave
theory was not true (or at any rate, not at all likely to be true), and in particular
that the elastic solid ether it centrally postulated was not real (or at any rate, not
at all likely to be real). *Thirdly*, he seems to have disagreed with Airy, Baden
Powell and others about the way forward in optics – seeing grounds for optimism that developing the corpuscular theory further might turn the evidential
scales at present favouring its rival.

Kuhn's notion of theory-*choice*, as we saw, seems to involve not just preferring that theory as the best empirically supported theory, not just deciding to work
on it to see where it leads and how it can be developed, but also involves taking
the theory fully to one's breast, believing it to be true. Brewster 'chose' no theory
on this characterisation – I translate his view as entailing that he (a) regarded the
wave theory as presently best supported by the evidence, (b) believed in the truth
of no *available version* of *any* theory of light and (c) (roughly speaking) chose to
work on the corpuscular theory.

Earman, as we saw, complains from a Bayesian perspective about Kuhn's idea that choosing a theory involves 'accepting' it (that is, presumably, believing it to be true). But taking choice to involve commitment to truth was not Kuhn's only implicit mistake; his treatment also seems clearly to presuppose that choice is a single, all or nothing affair – you either choose a theory or you choose some rival. And here Bayesianism in a sense follows suit: it allows of course for *degrees* of belief, and suggests that the general case will be that several rival theories have non-zero probabilities, but it is still committed to the idea that brownie points for theories are, so to speak, scalars – an agent ranks theories simply according to the degree to which she believes the theory is likely to be true, given the evidence she has. In fact nearly all philosophers of science have been trapped into thinking about scientific rationality in terms of a single dimension: this theory is more probable than that, this research programme is progressive, that one is degenerating ... and therefore the reasonable guys 'prefer' the first.

But, as the case of Brewster illustrates, the truth is surely that what it is and is not reasonable to believe about a theory is a somewhat more complex matter – one with quite different aspects involving perhaps quite different considerations. So we should ask separately about the rationality of each of Brewster's three different views about the two rival theories he considered. Was any of the three views 'irrational' – or, perhaps better, in order to avoid the unnecessarily aggressive overtones of that word, was any of them contrary to sound scientific reasoning?

Well, clearly not the first view – that, as things stood, the evidence favoured the wave theory – since this was uncontroversial (and correct). According to Kuhn's much-discussed analysis, the claim invariably underlying the positions of the hold-outs is that the phenomena cited by the revolutionaries as telling evidence in favour of their new view can in fact be 'shoved into the box' provided by the older paradigm.[13] And one of the main reasons (perhaps the main reason) that hold-outs cannot justifiably be regarded as 'illogical or unscientific' is, he suggests, that this claim is not demonstrably incorrect. In fact something stronger can be said – there invariably are ideas around at the time of the revolution about which direction the proponents of the older paradigm should 'shove' in: that is, positive ideas about how the evidence that seems to tell in favour of the new theoretical framework might be accommodated by the old. In the case of the wave/corpuscle rivalry, for example, there were ideas around in the late eighteenth and early nineteenth centuries about how to give corpuscular explanations of the phenomena of interference and diffraction *either* as the result of very complicated diffracting forces emanating from 'gross matter' and, at different distances, either repelling or attracting the light-corpuscles, *or* as some sort of physiological effect.

Again Kuhn is not entirely clear, but *if* he is suggesting here that all it takes for the hold-outs to balance the evidential scales is for such 'shoving' to succeed, then he makes a major mistake about the nature of evidential support. As I and others have argued,[14] whatever one's precise account of evidential support, a general adequacy requirement is that such an account entail a big difference between the support lent by phenomena that are 'shoved' *ad hoc* into a theoretical framework and phenomena that are genuinely predicted by such

a framework. If Brewster, for example, stood ready to elaborate on his acceptance that the wave theory was, as it then stood, better supported by the evidence
by adding that all it would take to bring the evidential scales back into balance
would be *any sort* of *post hoc* accommodation within the corpuscular theory of
the phenomena predicted by the wave theory, then he too would be making a
significant mistake about the nature of empirical support in science. There is,
however, no historical evidence that this is the case: in particular Brewster is very
modest about his *suggestion* that interference may be a physiological phenomenon (of course this suggestion left him with a great deal to be modest about).[15]

What of the other two elements of Brewster's position?

Brewster could not bring himself to believe in the wave theory and in particular
in the ether, 'invisible, intangible, imponderable, inseparable from all bodies and
extending from our own eye to the remotest verge of the starry heavens.' He
predicted that the wave theory would eventually give way to a quite different one
'after it has hung around for another hundred years or so.' Was he being
'irrational' on this score? Well this would be a strange judgement to make in
view of the fact that Brewster was *right*! Indeed if anything he was overgenerous
to the wave theory and its elastic solid ether which was to last at best another
seventy or so years before being unambiguously rejected.

The history of theory-change in science in general surely requires a separation
of judgements about which of the available theories is currently picked out by
the evidence from judgements about which theory if any is true (or even likely to
be true). The fear, felt by many philosophers, is perhaps that the former sort of
judgement is weak to the point of vacuity if separated from the latter – what does
it mean for a theory to be 'favoured by the evidence', if not that the evidence
makes it more likely to be true than available rivals? This is a legitimate worry,
but it is nonetheless just true that Brewster's position – that the evidence favoured
the wave theory but that the wave theory was very likely to be false – was consistent, and indeed more than that: clearly reasonable. It seems to follow that we
had better make this separation. More on this after considering the third element
in Brewster's view of the then current state of play between the wave and corpuscular theories of light.

Brewster seems to have believed that the near-monopoly on talented advocacy
and development then enjoyed by the wave theory was bad for science. Let's
assume that this means that he believed that there was 'heuristic steam' left in
the corpuscular theory, so that development of *it* might eventually lead to a
version which was still better favoured by the evidence than the current version
of the wave theory. Was *this* view 'irrational'?

Well of course the dominant view in philosophy of science until two or three
decades ago was that the contexts of justification and of discovery are quite
separate and that rationality considerations come into play only in the former
context. Hence Brewster could think what he liked about the way forward in
optics without fear of contravening any rule of scientific logic. Nowadays we are
more sophisticated. But, however interconnected these two contexts might in
fact be, the connecting principle quite plainly cannot be the simple one that the
only reasonable course of action is to try to develop the theory that is presently

best empirically supported. The obvious point has often been made that such a principle would, apart from anything else, automatically condemn the great innovators of theoretical science – who, almost by definition, are those who start to work on a theory *before* it is the best empirically supported in its field and who, through their work, *turn it into* the best supported theory. There must therefore again be room for a separation between the theory one judges best on the available evidence and the theory one chooses to devote most effort to.

When Lakatos advocated the view that the *primary* domain of rationality is simply the area of empirical support – that is, judgements about which direction the evidence at present tends in, the almost universal reaction was that this was to weaken the notion of rationality to the extent of making it uncontroversial. If all that is needed for, say, a defender of classical physics in 1920 to count as a 'rational' is that she admit that relativity theory is ahead in terms of empirical support as things stand, but is then free to pursue any classical physics project she likes, then, Paul Feyerabend famously remarked, Lakatos' position is simply 'anarchism in disguise'.[16] In fact, though, such judgements of the present 'evidential score' and the fact that scientific rationality demands unanimity concerning them is surely not as trivial a matter as Feyerabend suggested. It is no easy matter, for example, to get a 'scientific' creationist to admit that her theory is presently massively behind evolutionary theory in terms of empirical support – even if you were to provide her with the comforting (though surely false) thought that there have been cases of theories that have started massively behind a rival in terms of the evidence and have eventually managed to turn the tables. But Feyerabend and of course others were right that there ought to be more to good reasoning in science than mere recognition of the present empirical score; and there is.

There is no straightforward connection between (i) the present evidential support enjoyed by some set of rival theories, (ii) the likely truth (or 'approximate truth') of those rivals and (iii) the reasonableness of various research strategies – in particular the strategy of concentrating all one's research efforts on the presently best supported theory. But no straightforward connection does not of course entail no connection at all; and the fact – if it is one – that the first thing to be straight about when it comes to good reasoning in science is the relative degrees of support enjoyed by the available rival theories does not entail that this is *all* one should be concerned about. *Of course* the fact that Darwinism is streets ahead of creationism on the evidence we have does not on its own entail that it is logically impossible for a creationist to produce a theory within her own approach that reverses the evidential tables. But if someone were to tell us that she intended to exploit this possibility it would be sensible to ask her exactly how she intended to proceed. It is difficult to see any sort of heuristic idea within the creationist programme the pursuit of which might turn the trick: indeed the whole *modus operandi* of that programme seems to be to come along after the (empirical) event and absorb evidence as it independently arises. God created the universe in 4004 BC roughly as it now is. How is that? Well, experiment and observe and whatever you find is how God made it! The programme's leading idea supplies an indefinite set of 'free parameters' that the creation scientist fills in as she goes along and this, I hold, is a recipe for creating specific theories that enjoy no real empirical support. The creationist who felt that there are unexploited heuristic possibilities within her programme would, I think, simply be making a mistake.

Returning to the more serious case of Brewster, just as in the case of the question of which theory he 'chose', the question as to whether or not Brewster was irrational or 'illogical or unscientific' (in Kuhn's terms) in holding out against the wave theory has no straightforward single answer. This does not mean, however, that it has no answer (as Kuhn suggests), but rather that it has a slightly more complicated answer. Brewster was right to concede that the wave theory was presently ahead in terms of empirical support. He was right that this does not entail that the wave theory is true (and of course right in particular that it could not be true unless it eventually gave an explanation of the phenomena of dispersion and of selective absorption).[17] As for his views about 'the way forward', we need to ask for more information.

How exactly, except by wishful thinking, did he think that developing the corpuscular theory in 1830 was going to lead to specific theories that might conceivably enjoy predictive and explanatory successes on par with, or perhaps surpassing, those enjoyed by Fresnel's wave theory? The corpuscular programme was by then as bereft of (unused) heuristic ideas as the scientific creationist approach always has been – the difference of course is that there had been significant heuristic ideas behind the corpuscular approach initially, it was just that by 1830 they had all been tried and failed.

In barest outline, the idea of the corpuscular programme was to reduce optics to the Newtonian mechanics of moving objects. Initially the idea had been to effect the reduction to *particle* mechanics – the particles of light being simple entities (though perhaps with different masses or different velocities according to the colour they produced) subject to forces emanating from 'gross matter' (at reflection, refraction and, in passing by the edge of ordinary matter, diffraction). Naturally, the theories that were thought of first gave these forces the forms of other already known forces, but it was clear right from the start that all such theories fail to yield the phenomena. There were special difficulties in the case of diffraction, where it became obvious that, if anything worked to accommodate the known phenomena of the diffraction fringes, it would have to be a highly complicated force law, one making the force switch from attractive to repulsive and back again as the distance from the 'diffracting object' (such as the slit-screen or straightedge) changed minutely. Polarisation phenomena (first discovered via double refraction through crystals such as calcite) clearly showed that light rays could be made to be 'sided' – that is, to exhibit different properties in different planes through the direction of propagation of the ray. This meant presumably that the light-'particles' themselves must be treated, not as Newtonian particles, but as extended bodies with different properties in different 'sides' – a suggestion made by Newton himself and investigated in gory detail by J.B. Biot in the early years of the nineteenth century. Biot succeeded, *partially*, in 'shoving' some of the phenomena predicted by the wave theory into the 'box' of the corpuscular approach, but without any hint of *independent* testability, without any hint of any testable prediction. Brewster faced a 'particle' theory that had already invoked the most complex of forces, had already endowed the light particles with 'poles' and complicated axial movements with respect to those poles and had still not produced anything resembling an empirical success.

If Brewster had some other view and believed that there was some unexhausted general idea behind the corpuscular approach that might yet yield a version of the theory that turned the evidential tables on its wave-based rival, then, so far as I can tell, he was just plain wrong.

Suppose he felt instead that by pursuing some already heavily pursued idea – perhaps if the expression for the diffracting force went to the 25th power of the distance, rather than the 24th – then everything would change: instead of lagging constantly behind the facts the corpuscular theory would suddenly become predictive. The right response then seems to me to be meta-inductive: of course it is logically possible that this might happen, but the evidence from the history of physics seems to be that no amount of flogging has ever revived a horse as dead as corpuscular optics was in 1830.

If, finally, he was simply relying on wishful thinking, serendipity, the idea that maybe by pondering the corpuscular approach some new idea would crop up that turned out to revolutionise the situation, then aside from making obvious remarks about flying pigs, one would need to ask whether the corpuscular approach with some essentially new idea would really be the corpuscular approach rather than some entirely new research programme (and one would need to ask whether even new research programmes arise 'out of thin air' rather than in some methodical way from old background knowledge and new phenomena).

Certainly Brewster's complaint that, in effect, the wave theory was ahead in terms of predictive success because it had more, smarter advocates is at 180° to the truth. Unlike the corpuscular approach, the wave approach had clear unexhausted heuristic resources in 1830. For example, dispersion – the fact that what the wave theory identifies as beams of light of different wavelengths travel at different velocities in the same transparent medium – was, as Brewster emphasised, an anomaly for the then current wave theory. But the wave theoretic prediction of the independence of velocity from wavelength followed only from the assumption that the ether within transparent bodies was the very simplest form of elastic medium: one that obeys Hooke's law exactly. This assumption was always too simple to be good – more complicated elastic media were known, there seemed every reason to think that by complicating the force law somewhat, dispersion would be dealt with.

This is precisely what Cauchy and others attempted. Moreover, and as Airy and Baden Powell pointed out, there were successful precedents to be cited in the wave approach – cases, such as Fresnel's shift from longitudinal to transverse waves, which had proved strikingly empirically (that is, predictively) successful. The wave theory (or rather wave programme) did not have more empirical success because it had more, smarter advocates; rather it had more, smarter advocates because they could see within the approach unexhausted theoretical opportunities for empirical success.

7. CONCLUSION

In this paper I have addressed, occasionally somewhat tangentially, the question of how revolutionary Kuhn's views – more especially, his views on theory or paradigm *change* – really are. I have argued in effect that, like the question of which theory Brewster 'chose' and the question of whether or not Brewster's hold-out views were 'rational', the answer is not straightforward.

Kuhn's general comments about hold-outs and their fundamental rationale are not revolutionary at all. His claim that these hold-outs are right (or, rather, not necessarily wrong) that the allegedly crucial phenomena can be 'shoved' into the older paradigm's 'box' amounts to no more than the Duhem problem with examples. And in so far as it implies (as it seems to) that shoving a phenomenon, predicted by a 'revolutionary' theory, into the box of the older theory means that that phenomenon can supply no reason to prefer the newer theory, it is plain wrong.

In so far as Kuhn's account can be reconciled with that of personalist Bayesianism it is not revolutionary enough – since this agreement simply underlines the insufficiency, the over-subjectivism of both accounts.

Finally I have argued that many of the problems, both with Kuhn and with Kuhn-influenced later studies, stem from another failure to be revolutionary enough: his talk of theory-choice repeats the mistake of taking scientists' attitudes toward the rival theories available to them as measured for rationality or reasonableness along only one dimension.

What is needed, then is a more elaborate and more revolutionary account of scientific 'rationality' – one that recognises the different elements of Brewster's view, explains more clearly what is involved in regarding a theory as the best supported by the evidence if this need *not* entail regarding that theory as the most likely to be true, and explains, more clearly than others have managed, the relationship between what are sometimes called 'acceptance' and 'pursuit'. I do not, of course, claim to have done any more than sketch some aspects of this more elaborate account here.

The right way to proceed, I think, is by concentrating in the first instance, not on individual scientists' choices in any sense of the term, but rather on reconstructing the *intellectual argument* between rival theories at different stages of science. The main objectivist claim is, or ought to be, that there is such a thing as the intellectual argument between competing theoretical views at any stage of science, and that there is such a thing as the objective state of that argument at each such stage. Once put in this way then it seems obvious that the 'state of the argument' may be a more complicated entity than can be reflected by a single set of numbers, in the way of the probabilists, or a single set of judgements – wave theory progressive, corpuscular theory degenerating, in the way of Lakatos, say.

Of course, nothing in the above account should be seen as denying that Kuhn was a major figure. Some aspects of his views will undoubtedly be recaptured in the promised, more sophisticated account. The account I see emerging from my current work will – by delimiting more carefully those attitudes towards rival theories where consensus amongst rational people really ought to be expected from those where different opinions are 'equally valid' – explain at least some of the motivation behind Kuhn's invocation of 'subjective factors', while preserving certain aspects of theory appraisal in the light of evidence as entirely objective (intersubjective). The progress of philosophy of science, like that of science itself (or so I have suggested), is really evolutionary rather than revolutionary.

London School of Economics

NOTES

* This paper is a modified and extended version of the Dyason Memorial Lecture given to the Australasian Association for the History, Philosophy and Social Studies of Science at Auckland in June 1997. I thank the Committee of the AAHPSSS, and especially Robert Nola, for honouring me with this invitation. And I thank Alan Chalmers both for comments on the lecture and for his kindness during my trip to Australasia. I am indebted for helpful research assistance to Antigone Nounou and James Ward.

1 See Worrall 1990.

2 The point about sound is, of course, that it was 'known' to consist of waves, transmitted in its case, through the air. Although by then Fresnel had shifted to the theory that light consists of *transverse* waves (sound waves are longitudinal, pressure waves) and hence it was known that the analogy was by no means complete, nonetheless given that they were, if the wave theory of light was correct, both wave phenomena provided a *prima facie* case for expecting any result found in light to have a counterpart in the case of sound.

3 The case of new evidence provides the most straightforward application of the principle of conditionalisation; but according to some versions of Bayesianism, at any rate (e.g., that advocated by Howson and Urbach, 1994), the idea of 'old' and 'new' probability assignments linked by conditionalisation may be applied whenever one is assessing the confirmatory weight of *any* piece of evidence, new *or* old. This requires some slick footwork concerning how to 'delete' known *e* from the operative 'background knowledge' relative to which all probabilities are assessed.

4 For elaboration of these arguments see my (1993).

5 The idea that the principal function of background knowledge is to delimit the 'space of conceptual possibilities' is one that John Earman has recently been developing in a number of ways, as we shall see below.

6 Earman in fact sees Kuhn's implicit rejection of probability and degrees of confirmation as intimately connected with his explicit rejection of a theory-neutral observation language and 'the largely tacit but pervasive anti-inductivism of *Structure*' (1993, p. 21). I see both of these views (especially the former) however, as confused and having no real influence on Kuhn's (1977) account of theory-choice. (I do though heartily endorse John Earman's remark that 'in the physical sciences there is in principle always available a neutral observation base in spatial coincidences, such as dots on photographic plates, pointer positions on dials and the like' (1993, p. 16). See for example my (1980) and (1985a).)

7 The second question, raised later, is whether, if we concede the accuracy of Earman's account of revolutions, his talk of plausibility arguments and art forms is anything more than a concession that relativism is correct.

8 Systems have been developed – for example in Garber (1983) and Niiniluoto (1983) – in an attempt to solve the old evidence problem, in which Bayesian agents *may* make purely logical discoveries, which may in turn affect their degrees of belief in substantive theories. (So the idea is that, although the facts about Mercury's perihelion may have been known ahead of Einstein's general theory of relativity hence those facts had probability one and so no confirmatory power, what was *not* known in 1914 was the *logical fact* that general relativity entails the precession of Mercury's perihelion.) However (i) this approach clearly requires a modification of the classical theory of probability; (ii) since this brings into the statement of the axioms themselves considerations of what the agent does or does not 'know', it involves replacing crisp mathematically precise axioms with vague ones; and (iii) the idea that it solves the old evidence problem is a non-starter (it is the substantive *evidence* of Mercury's perihelion, facts about Mercury's orbit that ought to confirm general relativity, not some logical truth).

9 See for example Zahar (1989).

150   JOHN WORRALL

[10] See especially Norton (1993).
[11] Needless to say, this 'premise' was itself not universally accepted.
[12] For details, see my (2000).
[13] See Kuhn (1970), pp. 151, 152.
[14] See, for example, my (1985b) and (1989a).
[15] The idea seems to have been that the light-particles in each of the two streams might arrive at the eye at distinctive intervals and that the two different intervals for the two streams might be such that the vibrations they each set up *within the eyeball* produce particular interference patterns at the retina. Of course there is nothing automatically unscientific in invoking physiology within optics (the wave theory, for example, *correctly* uses the limitations of our visual apparatus to explain the absence of observable interference patterns when two closely adjoining *but independent* point light-sources are trained on a screen). The problems with Brewster's suggestion on behalf of the corpuscular theory were that (a) no one ever succeeded in turning this explanation-sketch into anything like a full and adequate explanation and (b) there were never any independent tests of the idea. At best it showed how *one might* explain interference patterns on the corpuscular theory, but there was never any independent reason to take this possibility seriously.
[16] See, for example, Lakatos (1978, p. 110) and Feyerabend's (1975), dedicated, of course, to 'Imre Lakatos, fellow anarchist'.
[17] He even conceded, remember, that the wave theory's empirical success meant that 'it must contain among its assumptions some principle which is inherent in ... the real producing cause of the phenomena of light' (1838, p. 306). It might be argued from a structural realist perspective (see my 1989b) that, in the light of the history of scientific 'revolutions', this sounds like exactly the view it is reasonable to have concerning the truth claims of current theories.

REFERENCES

Airy, G.B., 1831: *A Mathematical Tract on the Undulatory Theory of Light*, London.
Brewster, D., 1833a: 'A Report on the Recent Progress of Optics', *British Association for the Advancement of Science, Report of the First and Second Meetings 1831 and 1832*, London.
Brewster, D., 1833b: 'Observations on the Absorption of Specific Rays, in reference to the Undulatory Theory of Light', *Philosophical Magazine*, 3rd Series, 2, 360–363.
Brewster, D., 1838: Review of Comte's *Cours de Philosophie Positive, Edinburgh Review* 67, 279–308.
Earman, J., 1993: 'Carnap, Kuhn and the Philosophy of Scientific Methodology' in P. Horwich (ed.), *World Changes: Thomas Kuhn and the Nature of Science*, MIT, Cambridge MA, pp. 9–36.
Feyerabend, P., 1975: *Against Method*, New Left Books, London.
Garber, D., 1983: 'Old Evidence and Logical Omniscience in Bayesian Confirmation Theory', in J. Earman (ed.), *Testing Scientific Theories*, University of Minnesota Press, Minneapolis, pp. 99–131.
Howson, C. and Urbach, P., 1994: *Scientific Reasoning: the Bayesian Approach* (2nd edn.), Open Court, Chicago and La Salle.
Kuhn, T.S., 1970: *The Structure of Scientific Revolutions*, second enlarged edition, University of Chicago Press, Chicago.
Kuhn, T.S., 1977: *The Essential Tension*, University of Chicago Press, Chicago.
Lakatos, I., 1978: *The Methodology of Scientific Research Programmes: Philosophical Papers*, Vol. 1, Cambridge University Press, Cambridge.
Niiniluoto, I., 1983: 'Novel Facts and Bayesianism', *British Journal for the Philosophy of Science* 34, 375–379.
Norton, J., 1993: 'The Determination of Theory by Evidence: the Case for Quantum Discontinuity, 1900–1915', *Synthese* 97, 1–31.
Powell, B., 1841: *A General and Elementary View of the Undulatory Theory of Light as Applied to the Dispersion of Light and Some Other Subjects*, London.
Salmon, W.C., 1990: 'Rationality and Objectivity in Science or Tom Kuhn meets Tom Bayes', in C. Wade Savage (ed.), *Scientific Theories*, University of Minnesota Press, Minneapolis, pp. 175–204.

Worrall, J., 1980: 'Feyerabend and the Facts' in H.-P. Duerr (ed.) *Versuchungen: Aufsatze zur Philosophie Paul Feyerabend*, Frankfurt (republished in G. Munévar (ed.), *Beyond Reason*, Kluwer Academic Press, Dordrecht, 1991, pp. 329–353).

Worrall, J., 1985a: 'The Background to the Forefront', in P.D. Asquith and P. Kitcher (eds.), *PSA 1984*, Vol. 2, Philosophy of Science Association, East Lansing, Michigan, pp. 672–682.

Worrall, J., 1985b: 'Scientific Discovery and Theory-Confirmation' in J. Pitt (ed.), *Change and Progress in Modern Science*, Reidel, Dordrecht, pp. 301–332.

Worrall, J., 1989a: 'Fresnel, Poisson and the White Spot: the Role of Successful Prediction in Theory-Acceptance', in D. Gooding *et al.* (eds.) *The Uses of Experiment – Studies of Experimentation in Natural Science*, Cambridge University Press, Cambridge, pp. 135–157.

Worrall, J., 1989b: 'Structural Realism: the Best of Both Worlds?' *Dialectica*, 43, 99–124 (reprinted in D. Papineau (ed.), *Philosophy of Science*, Oxford University Press, Oxford, 1996, pp. 139–165).

Worrall, J., 1990: 'Scientific Revolutions and Scientific Rationality: The Case of the "Elderly Hold-Out"', in C. Wade Savage (ed.), *Scientific Theories*, University of Minnesota Press, Minneapolis, pp. 319–354.

Worrall, J., 1993: 'Falsification, Rationality and the Duhem Problem: Grünbaum vs Bayes' in J. Earman, A.I. Janis, G.J. Massey and N. Rescher (eds): *Philosophical Problems of the Internal and External World*, University of Pittsburgh Press, Pittsburgh and Konstanz, pp. 329–372.

Worrall, J., 2000: '"Heuristic Power" and the "Logic of Scientific Discovery": Why the Methodology of Scientific Research Programmes is less than half the story', in G. Kampis, L. Kvasz and M. Stoeltzner (eds.), *Appraising Lakatos: Mathematics, Methodology and the Man*, forthcoming, Kluwer Academic Press, Dordrecht.

Zahar, E., 1989: *Einstein's Revolution: A Study in Heuristic*, Open Court, Chicago and La Salle.

JOHN F. FOX

WITH FRIENDS LIKE THESE ..., OR WHAT IS INDUCTIVISM AND WHY IS IT OFF THE AGENDA?*

1. INDUCTIVISM

There was a celebrated philosophical dispute. It was the very model of a philosophical dispute: protracted, much at cross purposes, confused, inconclusive. On one side, indisputably, was Karl Popper. The name he coined for his foes had even by the 1980s no entry in the *Oxford English Dictionary*, the *Encyclopaedia Britannica*, the *Encyclopaedia of Philosophy*, or even that repository of words too unrespectable to make such august lexica, Partridge's *Dictionary of Slang and Unconventional English*. He called them '*inductivists*'.

It was sometimes unclear who was in this camp. Sometimes they looked like Francis Bacon, Isaac Newton or Thomas Henry Huxley, at others more like Hans Reichenbach, Wesley Salmon or Yehoshua Bar-Hillel. Popper at times made it clear who some of his personal 'inductivist' targets were; Rudolf Carnap was always in there somewhere, though he often seemed to think the dispute was mainly misunderstanding. At times Popper made it clear that he considered certain doctrines 'inductivist'. Unfortunately, it was often clear that many of his personal targets did not hold his doctrinal targets. This is characteristic of polemicists – of Pius X attacking modernists, Ayn Rand attacking altruists, David Stove attacking anyone.

It is also conducive to unclarity. This unclarity is one reason I feel free to offer my own characterisation of inductivism.

In the past as now, philosophers like everyone else have usually taken many things for granted. Sixty years ago, and for quite a while afterwards, these were some of them.

Beliefs about the unobserved or about the future must be the result of inductive inference. This was considered almost self-evident. For such beliefs must be based on what we had experienced; which had of course been observed in the past or present. And to get from such a basis to a belief about the unobserved or about the future was precisely to make an inductive inference.

Much more was generally taken as obvious. For instance, that experience established for the experiencer the truth of those beliefs that were expressed by reports of the experience. After all, what else could? And laws and theories and generalisations and expectations about the future could be no part of the basis of

Robert Nola and Howard Sankey (eds.), After Popper, Kuhn and Feyerabend, 153–164.
© 2000 *Kluwer Academic Publishers. Printed in Great Britain.*

such beliefs; for they themselves were essentially later beliefs, that could only come to be believed on the basis of prior empirical beliefs.

Just how were such later beliefs grounded on prior ones? Clearly, by some kind of inference. The label 'inductive' came to characterise such inference. So since our beliefs in laws, predictions, etc. were obviously, at least to any sane person, at least often reasonable, at least some inductive inferences must sometimes be reasonable.

Some thought induction needed positive justification and some thought it did not. These latter usually considered that Hume had shown that it could not be justified, but that nevertheless it was obviously rational. For example, Strawson[1] and Goodman[2] scorned the demand to justify induction, but thought that the meanings of the words ensured the truth of 'induction is rational'. They also considered that the reflection that deduction could not be justified non-circularly was highly reassuring about induction, and undermined the case that induction needed justifying at all. In this many others joined with them. And after all, could not a defence of the claim that something was rational be taken, in a broad sense, as its justification?

Scientists had for a long while railed against premature, wild speculations that went way beyond the inductive evidence. That one should not do this was what not only every young scientist was taught, it was part of what educated layfolk knew about the method of science. This method, by contrast, was inductive, based on careful inductive reasoning from data well-established by observations. If they were very well educated, they also knew that it was a cardinal sin to prejudge matters by importing conjectures or hypotheses into one's readings of the empirical evidence; one was supposed to let this 'speak for itself'. And while – indeed, *because* – inductive inference might be incurably fallible, it was of the utmost importance at least to start from *premises* that were as certain as possible.[3]

The analogy between deduction and induction sounded reassuring, but one could point to books that spelt out systematically what were generally recognised as sound deductive patterns of inference. Was there any account that told us what inductive inferences were good? Alas, no. So a few philosophers saw a residual problem here, that of *inductive logic*: to formalise and systematise good inductive inference, much as deductive logic formalises and systematises good deductive inference.

Several factual and normative theses about knowledge and method were not only largely orthodox among philosophers but permeated the common sense of scientists and of educated layfolk. They were taken as so obvious that they were rarely spelt out. I will label them, collectively, *inductivism*. I have introduced all but one of them already. That one was also universally taken to be so obvious as to need no argument: that beliefs are justified only when they can be shown to be at least more probable than not. Of course, spelling things out is often a step towards their losing obviousness.

I_1: Our factual beliefs about the unobserved are the result of inference (perhaps implicit inference) from particular beliefs about the observed.

I_2: These particular beliefs are typically known to be true as a result of observation.

I_3: In particular, our knowing and our understanding such particular beliefs does not depend on our already accepting any laws or theories.

I_4: Beliefs about the unobserved are typically justified because of the soundness of such inference.

I_5: In particular, beliefs in generally accepted scientific laws and theories are typically justified.

I_6: Beliefs are justified only when they can be shown to be more probable than not.

I_7: An inductive inference to the generalisation is stronger, and its conclusion more probable, the more of its positive instances one has as premises.

I_8: An inductive inference is weaker, rasher, less justified, and more 'wildly speculative', the more its conclusion goes beyond its premises.

I_9: Scientific discoveries have been the result of following an inductive method, which involves starting from particular beliefs that observation certifies as true, understood as above.

I_{10}: This method next involves sound inductive inference, understood as above, from such beliefs to general laws and theories.

I_{11}: Sound inductive inferences are not rash; concluding to wild speculations is not sound.

I_{12}: Though inductive *inferences* may be less than conclusive, one must start from *premises* that are as certain as possible; it is crucial not to prejudge matters by importing conjectures or hypotheses into one's readings of the empirical evidence, but to let this 'speak for itself'.

When in the early 1930s Popper attacked almost all these tenets, each was still an almost universal orthodoxy.

2. THE FIRST BREACH

Nowadays, hardly any serious philosophers of science accept many of the tenets. Popper's arguments helped bring this about. But I argue that another crucial factor was a disguise. For those who crucially led the way in making anti-inductivism the new orthodoxy did not present themselves as anti-inductivist; on the contrary, they were waving the banner of induction, and seemed to be defending it against the Popperians. So I entitled an earlier version of this paper 'The Trojan Horse'. But I do not think that this disguise was cunning; I think that those who took the citadel for Popperism sincerely thought they were defending it. Hence the new title: 'With Friends Like These', who needed enemies?

This dismantling of inductivism by induction's defenders came in two phases. At first it was taken for granted that 'inductive logic' would provide inductive *rules of inference*, analogues of the deductive *Modus Ponens*. Reichenbach, for instance, had urged the 'Straight Rule'.[4] But Carnap[5] and Hempel[6] sought rather an analogue for inductive logic of deductive logic's *entailment*; and decided on a relation of *confirmation*. Most of its practitioners were aware that this new inductive logic licensed no inductive inferences at all, but few outsiders adverted to the fact. This lack was soon not only orthodox but presented as a merit; for instance, as the solution to the 'lottery paradox'. One avoids contradiction by never *making* particular predictions, but merely estimating their degree of confirmation relative to available evidence. With this shift I_1, I_4, I_7, etc. were not so much denied as shelved, put off the agenda.

In objecting that because science (and all of us) need predictions, inductive logic needs a rule of detachment, Salmon[7] and Kyburg[8] were voices in the wilderness. So this first major breach in the inductivist citadel was made by the inductive logicians.

Against his consensus of 'philosophers of very different persuasions' Salmon insisted that induction needed to be justified. What did he think needed doing that the majority did not? I suggest that at the very least, it was all these things: to justify some inductive inferences; to justify some particular claims about the future (the unobserved); to justify some generalisations which include instances about the future (the unobserved).

Inductive logic had abandoned all three tasks. It was when pointing this out, under the rubric 'the abdication of the inductive judge', that Lakatos coined the phrase 'degenerating problem-shift'.[9]

Carnap thought that inductive logic had to be a true logic; for him, this meant not so much dealing with inferences as being *a priori*. He took universal laws in an infinite universe to entail an infinite number of observations that were *a priori* less than certain. So on all of a variety of assumptions he considered plausible, e.g., that these observations were *a priori* independent, he had to allot universal laws an *a priori* probability of 0. But this meant that even after any amount of evidence they still had zero probability. So by I_6 they could not be rationally believed; so I_5 also had to go. So, too, the questions about inductive *inference* that had been put off the agenda were in effect refused readmission.[10]

Carnap knew what he was doing, and kept protesting that Popper was misrepresenting him with his stereotype of the inductivist.[11] But partly taken in by Popper, onlookers tended to see Carnap and Popper as the paradigms, respectively, of an inductivist and an anti-inductivist.

3. POPPER'S ANTI-INDUCTIVIST THESES

Popper worked out his ideas in conscious reaction to a package he took to be inductivist orthodoxy. His arguments were against various of the theses I distinguished; some were direct arguments and some merely pointed out conflicts among the theses. He tended to take anyone who held any of them to have committed inductivism, and argument against any to discredit inductivism. So it was understandable and often fair for his opponents to protest that they were being misrepresented. But his weakness at exegesis is of little interest compared with the boldness of his counter-theses. Here are some of them.

P_1: In trying to explain some phenomena, we do not seek out the least informative theory that will do the job; rather, we value more informative theories, and those with high explanatory content. But

P$_2$:   we should consider these highly *im*probable; for it is the least informative theories that are the most probable.

P$_3$:   We do not and should not consider the mere amassing of positive instances good evidence for theories, though

P$_4$:   on probabilist doctrines we should; in fact

P$_5$:   if it were reasonable to believe the most probable theories, and to try to gather evidence to render our beliefs as probable as possible, and if amassing positive instances increased probability, it would be a good scientific strategy to repeat the same experiment *ad nauseam*. But

P$_6$:   this is appalling scientific strategy.

P$_7$:   On the contrary, we should focus on testing theories as severely as we can, and

P$_8$:   we do and should (tentatively) judge theories to have shown their worth only when they have survived severe tests.

P$_9$:   Rationality pertains not to our accepting but to our rejecting beliefs; genesis is nothing, exodus everything.

P$_{10}$: Any supposed probabilistic confirmation of hypotheses by instances is due entirely to the deductive *component* of the relation of the instance to the hypothesis.[12]

In the mid-1960s, the massed Popperians of Aldwych hosted an international conference whose proceedings they edited in order to launch a concerted attack on rival tendencies in the philosophy of science. They saw these as inductivism, under Carnap's leadership,[13] and historicism, under Kuhn's.[14] Their tactics succeeded in getting Popper's ideas much more serious attention than ever before across the Atlantic, but also irritated many people severely. Several eminent philosophers produced counterattacks in the name of inductivism. In the course of this, they made heavy use of subjectivist or personalist Bayesian ideas. I claim that this defence in fact insinuated many more of Popper's central anti-inductivist ideas into orthodoxy, and that this was not realised largely because it was done under the banner of defending inductivism against Popper's attacks.

In abandoning inductive *inference* Carnap's 'objectivist' Bayesianism had already changed inductivism into something unrecognisable. The second, 'subjective' wave of Bayesianism, abandoned almost all Popper's remaining doctrinal targets.

4. BAYESIANISM VERSUS POPPER

Here is the gist of the Bayesian counterattack on Popper.[15]

B$_1$: P$_1$ is true; but

B$_2$: P$_2$ is false. Bayesianism allows any coherent distribution of prior probabilities, and in particular allows highly informative theories to be assigned very high probabilities. It is true that in the special case when one theory has strictly greater informative content than another, i.e., entails it without being entailed by it, it can be no more probable than that other and will in general be less probable; but when we are choosing between competing consistent theories this is never the case. For if theories compete, they differ on some point; so the more informative one, if consistent, does not entail its rival.

B_3: P_3 is true, but

B_{3a}: we *do* consider the mere amassing of positive instances good evidence for later *predictions* of theories, and

B_{4a}: on Bayesian grounds, so we should. However,

B_5: P_4 is false, as a mere fact of mathematics.[16]

B_6: P_6 is true, but

B_{6a}: Popper cannot explain why it would not be good strategy to repeat the same experiment *ad nauseam*,[17] while

B_{6b}: Bayesianism can.

B_{7a}: P_7 is correct, but

B_8:   Bayesianism can explain why it is reasonable to accept hypotheses that have survived severe testing,[18] while

B_{8b}: Popper cannot explain this.[19]

B_{9a}: P_9 is also false, and

B_{9b}: Popper and Miller's argument to it is fallacious.[20] However,

B_{10}: Rationality does pertain not to the origins but to the transformations of our beliefs.

I have not given the history of such changes, and my exclusive focus on Popper is in some ways distorting. For one instance: before Popper, Duhem[21] had attacked many of the tenets of inductivism; though he found few converts – as indeed Popper did for the first twenty years. For another, independently of Popper and the Bayesians, Quine promoted a holism about belief that undermined the inductivist picture, and his metaphors (of adding a drop of water of convention to the wine of fact, of beliefs facing the tribunal of experience as a body) did much to popularise P_9.

5. POPPER ON BAYESIANISM

A first step towards appraising the disputes is to recognise that of the nine theses I extracted from Popper, only P_{10} was aimed at Bayesianism, and it was developed much later than the others – in the 1980s, by when the state of the discussion had changed drastically.

In his early comments on Bayesianism Popper was respectful. This can astonish, till one realises that Popper did not see it as a form of inductivism at all. He carefully and accurately pointed out that it was not an inductive logic, in the sense in which he had been criticising inductive logic:

> Richard C. Jeffrey has recently developed a very interesting theory of the probability of universal laws. But this is a theory of purely subjective belief rather than a theory of 'logical probability'. Admittedly, being based upon the formal calculus of probability, it satisfies Ramsey's, de Finetti's and Shimony's so-called rationality criteria; yet the attribution of what I call absolute probabilities is purely subjective. It can therefore hardly be described as a logical theory, or an 'inductive logic'.[22]

A great deal can be said about Bayesianism, both for and against, that is not directly relevant to the controversy with Popper, and won't be said here.[23]

6. BAYESIANISM VERSUS INDUCTIVISM

The Bayesians focussed their attack on Popper. What was not noticed was how much, in the course of this, they were elaborating a position that was incompatible with inductivism in almost all the ways Popper's position was.[24] B_{10}, for instance, is startling agreement with Popperian anti-inductivism. Here are some things Bayesians say or imply about the theses by which I characterised inductivism.

I_1 is false. Definite degrees of belief in some claim U about the unobserved can be the result not of inference precisely, but of conditionalizing on the evidence. But this conditionalizing is only possible if there was already some degree of belief in U; such degrees cannot originate in but can only be modified by empirical evidence.

I_3 is in general false. If we have a high degree of belief in a theory, we can have no lower in any of its logical consequences, and this is often why we believe in such consequences.

I_4, I_9 and I_{10} are therefore false.

I_5 is true only in the sense that a high degree of belief in a theory so far not contradicted by observations is not irrational. But a high degree of belief *a priori* is not irrational. Nor is it irrational to have assigned it such a low *a priori* belief that all the observations so far have left it subjectively extremely improbable. Bayesianism tolerates, relative to however much or little confirming evidence, either extremely high or extremely low degrees of belief in theories, because of its tolerance of any prior probabilities.

I_{12} is therefore false.

I_6 is not true in the sense inductivists intended, for only logical truths can be shown to be more objectively probable than not. And in the subjective sense, the idea of 'showing' is not relevant. It is legitimate simply to *assign* something *a priori* a very high probability; as long as this is not incoherent with other probabilities one assigns.

7. GREY ON GREY

In this paper I have been mainly trying to redraw the map of the state of the controversies. The items of information on the map are not themselves new or particularly controversial. I have not been trying to adjudicate on the issues. I would like to conclude by making some very select and succinct criticisms, and by pointing out some ways in which the options mainly discussed do not exhaust the possibilities, and some alternatives I consider worth exploring.

On P_1 and B_1: We do and should seek to maximise explanatory power, which also tends to maximise content. Bayesianism does not forbid seeking high explanatory power; so this does not refute it as a partial theory of rationality. But nor does Bayesianism *recommend* such seeking. To this extent it is inadequate as a theory of rationality.

On P_2 and B_2: I think the Bayesians are right, both in what they concede to Popper and in the way they correct him. What Popper points out constitutes no problem for Bayesians.

It does tell against traditional inductivism. For this *does* urge prudence and risk-avoidance in forming beliefs about generalisations and about the future; and this does mean preferring more probable and so less informative theories. On this basis Whitehead urged assuming only that laws of nature hold for our short cosmic epoch; which would have vetoed their confident use in reconstructing early moments of the universe. On this basis van Fraassen still urges withholding assent from theories generally. However, Bayesianism gives a blanket permission to start out by arbitrarily assigning as high a probability as one wishes to any speculation whatever. This startling deviation from traditional inductivism is an 'opposite extreme' at least as dissonant with good scientific practice.

That P_4 is not accurate undermines the polemical point of P_3, P_5 and P_6. This illustrates Popper's frequent strawmanning in the enthusiasm of debate. Lakatos acknowledged this in his own polemic in favour of Popper's and against Carnap's methodology.[25]

On B_8: Bayesianism does not in fact imply that it is reasonable to accept hypotheses that have survived severe testing. I do not offer this as a criticism of Bayesianism, for I do not think that this is always reasonable. I offer it as a correction of a claim sometimes wrongly made for it.

If Bayesianism is correct, passing a severe test increases confidence in an hypothesis by a factor which can be taken as a measure of the severity of the test; an impressive enough result. But Bayesianism permits assigning such minuscule priors to unrestricted generalisations that no practically performable number of severe tests would raise confidence in any to as much as 1 in 10^{10}; and it permits assigning such high priors to some others that if they are never themselves very directly tested, they will still always be preferred to rivals that keep passing severe tests. Still, Bayesianism can provide a rationale for wanting severe tests, and for taking them more seriously than others; Popperism cannot, without abandoning strong inductive scepticism.

Still, there is an important way in which Popper's methodology is clearly better than the Bayesian. According to Popper, when Uranus was seen misbehaving, the appropriate response was to reject the conjunctive theory from which the wrong predictions had been derived, and devise another to replace it which explained both the successes of the old theory and the deviant phenomena, and which made new predictions; when these new predictions are confirmed, the new theory should be tentatively accepted. According to Bayesianism, the appropriate response was to inspect all one's priors and conditionalize on the new observations of Uranus. Here Popper fits the world's best practice of Leverrier and Adams like a glove, and the Bayesian prescription is wildly askew.

P_8 is not quite correct. Many theories are accepted confidently on the basis of some piece of confirming evidence, before they have been subjected to further severe tests.[26] For instance, the chlorine theory of the hole in the ozone layer was established on the basis of one observation of precise anti-correlation of ClO and O_3 levels across the hole.[27] If they fail their subsequent tests this acceptance may be short-lived, but that is another issue; this is true even of theories that have passed several severe tests, for all acceptance is tentative.

P_{10} is vague enough to be defensible; but to be defensible, it must not be understood as denying that there are cases where E confirms H even though neither of E and H entails the other, for there demonstrably are such cases.[28]

I have not here presented any position systematically or coherently in its own terms. I have concocted 'inductivism' from the intersection of what was orthodox in say the 1920s with what Popper attacked under that rubric; I have excerpted from Popper merely some positions that were meant to discredit these, from Bayesianism merely some polemic against Popper. So while I have tried to be accurate, I have not given here enough to provide an overall fair picture of any of the three traditions.

8. A MERIT OF INDUCTIVISM

That caveat made, I will end by suggesting a crucial inadequacy of both Popperism and Bayesianism; a respect in which Popperism is even worse off, but in which Bayesians are so badly off that this comparison provides them little rational consolation.

Since Newton hammered it out in his controversy with Hooke, the distinction between what is reasonably well established and what is mere conjecture has been basic to the self-understanding of the scientific tradition.

It is at least as crucial for the daily tasks of living in the world. We always assume that certain things can in effect be known, and assumed to be known, to hold as a rule; e.g., that turning a steering-wheel clockwise is not a good way of turning left, that pulling the trigger of a loaded gun aimed at someone is likely to harm them, that humans are the biological offspring of other humans. It is on such assumptions that we test for competence in licensing drivers of school buses, convict people of murder or criminal negligence, assess people for sanity. Yet even such elementary judgements of rationality fail to be vindicated by Bayesian resources, as Earman rightly 'fears'.[29]

Indeed, Bayesianism seems to provide no rationale even for making any confident predictions on the basis of past experience, and can explain the rationality neither of predictions, nor of choice among generalisations that conjoin any speculation at all about the future course of events with the same description of observations so far.

For if the H_n are the as yet unrefuted hypotheses that agree about their first n instances but disagree about some later predictions, the ratios of their probabilities after their first n instances have been observed are as the ratios of their priors. So at every stage, if one of the H_n is more probable than another, it is purely a function of their prior probabilities. So any Bayesian explanation of theory-preference between two such as yet unrefuted theories is purely a function of the allocation of the two priors; and Bayesianism allows the opposite allocation.[30]

So if what were rational were only what Bayesian resources vindicate, strong scepticism about such predictions and generalisations (i.e., deeming them not rational) would stand. It is no accident that half-lapsed Popperians become Bayesians; or that some of the brightest not-lapsed-enough Bayesians come close to Popperian scepticism about theories. Thus van Fraassen counsels against believing theories, there being no reason to think them probable. Popper wrote 'we tentatively "accept" this theory – but only in the sense that we select it as worthy to be subjected to further criticism, and to the severest tests we can design'. This was anti-Bayesian: no rational beliefs or betting quotients here. Earman offers a transatlantic echo, but hopes to make it acceptable by offering it 'on behalf of the Bayesian':

> Scientists do choose theories, but on behalf of the Bayesian I would claim that they choose them only in the innocuous sense that they choose to devote their time and energy to them: to articulating them, to improving them, to drawing out their consequences, to confronting them with the results of observation and experiment.[31]

Against Popper and Earman, I consider this factually indefensible. We often consider worthy of our time and our testing theories that we do not believe and would not rely on. In this sense, Newton 'chose' the vortex theory even more than Descartes did, and Duesberg 'chose' the theory that HIV causes aids even more than Gallo did. We also constantly accept theories in a much stronger sense than this. In research we typically take many laws for granted in the course of confronting others with the result of experiment; and in practice we constantly stake our lives on our law-like beliefs.

Philosophies of science that cannot allow for the distinction between what is reasonably well established and what is mere conjecture are patently inadequate to practice. For all the inadequacies of the inductivist tradition, in insisting on this distinction it remains in one crucial respect superior to the Popperian and Bayesian traditions.

LaTrobe University

NOTES

*   An earlier version of this paper was delivered at the AAHPSSS conference in July 1997 at Auckland University, under the title 'The Trojan Horse: or, How Popperism Took the Inductivist Citadel'.
[1]   Strawson 1952, pp. 248, 249, pp. 261–263.
[2]   Goodman 1955, pp. 59–64.
[3]   This was a central motivation for empiricism.
[4]   Reichenbach 1938.
[5]   Carnap 1945.
[6]   Hempel 1945.
[7]   Cf. e.g. Salmon 1968.
[8]   Kyburg 1968.
[9]   Lakatos 1968, p. 273.
[10]   It is sometimes thought that use of infinitesimals solves these difficulties. Not so; if universal laws are assigned infinitesimal prior values, after any amount of evidence they still have only infinitesimal probability.
[11]   Cf., for example, Carnap 1968a.
[12]   Let us accept, as stipulative, the common Bayesian definition of confirmation: E confirms H just if $P(H/E) > P(H)$. Popper and Miller, 1983 published a simple theorem of probability theory: that for any H and E, H is equivalent to the conjunction of something E *entails* and something E *does not confirm*. For $H \leftrightarrow (H \vee E)$ & $(H \vee \neg E)$; E entails $(H \vee E)$; and E does not confirm $(H \vee \neg E)$.
[13]   The relevant proceedings were published as Lakatos (ed.) 1968.
[14]   The relevant proceedings were published as Lakatos and Musgrave (eds.) 1970.

[15] I am here using 'Bayesianism' for its commonest, 'purely personal' or 'subjective' variant.
[16] Howson and Urbach 1989, pp. 259, 260.
[17] Musgrave 1975.
[18] For it is one of the most elementary theorems of probability that when $P(H)$ and $P(E)$ are not 0, $P(H/E)/P(H) = P(E/H)/P(E)$. What would count as a severe test of H? Surely testing whether E, where E would be expected given H, but would otherwise be extremely unlikely; that is, where the ratio on the right of the equation is high. But the ratio on the left is the factor by which E confirms H!
[19] Jeffrey 1975, Rosenkrantz 1977, Grünbaum 1976a.
[20] Howson and Urbach 1989, pp. 395–398; Earman 1992, pp. 95–98.
[21] Duhem 1905.
[22] Popper, in discussion of Carnap 1968, in Lakatos (ed.) 1968, p. 290.
[23] For some that can be said for and against it, see Earman 1992; cf. also Fox 1996a. For much more for it, Howson and Urbach 1989.
[24] Rosenkrantz 1977, though an objectivist Bayesian and so in some ways closer to Carnap's position than to the subjectivist position for which I here use the term 'Bayesianism', was unusual in recognising that Bayesians and Popperians were allies in attacking an inductivist orthodoxy.
[25] Lakatos 1968.
[26] For example, statistical thermodynamics was widely accepted on the basis of Einstein's analysis of Perrin's experiments on Brownian motion; the chlorine theory of the hole in the ozone layer was accepted on the basis of an initial measurement that showed a beautiful anti-correlation between ozone and chlorine monoxide concentrations across the hole.
[27] Cf. Christie 1996.
[28] As long as there are at least 3 mutually exclusive and jointly exhaustive propositions (A, B and C), of which two (A and C) have probability > 0, there are cases where H does not entail E or vice versa, yet $P(H/E) > P(H)$; e.g., where H is $B \vee C$ and E is $A \vee B$.
[29] Earman 1992, p. 160.
[30] For elaboration of this argument, see Fox 1996a, and for its use in rebutting Earman's use of 'convergence' theorems to provide a Bayesian vindication of predictions, see Fox 1996b.
[31] Earman 1992, p. 193.

REFERENCES

Carnap, R., 1945: 'On Inductive Logic', *Philosophy of Science* 12, 72–97.
Carnap, R., 1968a: 'Reply', in Lakatos (ed.) 1968, pp. 307–314.
Carnap, R., 1968: 'Inductive Logic and Inductive Intuition', in Lakatos (ed.) 1968, pp. 258–271.
Christie, M., 1996: 'Scientific Theories are Richer than Popper Imagined', *Victorian Centre for the History and Philosophy of Science Preprint #6/96*.
Duhem, P., 1905: *La Theorie Physique: Son Objet et Sa Structure*, translated as *The Aim and Structure of Physical Theory* 1974, Princeton University Press, Princeton.
Earman, J., 1992: *Bayes or Bust*, MIT Press, Cambridge MA.
Fox, J., 1996a: 'The Limits of Bayesianism', *Victorian Centre for the History and Philosophy of Science Preprint #3/96*.
Fox, J., 1996b: Review of Earman 1992, *Australasian Journal of Philosophy* 74, 214–218.
Goodman, N., 1955: 'The New Riddle of Induction', in *Fact, Fiction and Forecast*, Bobbs-Merrill, New York.
Grünbaum, A., 1976a: 'Is the Method of Bold Conjectures and Attempted Refutations *Justifiably* the Method of Science?', *British Journal for the Philosophy of Science* 27, 105–136.
Grünbaum, A., 1976b: '*Ad Hoc* Auxiliary Hypotheses and Falsificationism', *British Journal for the Philosophy of Science* 27, 329–362.
Grunbaum, A., 1976c: 'Is Falsifiability the Touchstone of Scientific Rationality? Karl Popper versus Inductivism' in Cohen R., Feyerabend P. K. and Wartofsky M. (eds.), *Essays in Memory of Imre Lakatos*, Reidel, Dordrecht, pp. 213–252.
Hempel, C.G., 1945: 'Studies in the Logic of Confirmation', *Mind* 54. 1–26, 97–121; reprinted in Hempel, C.G., 1965: *Aspects of Scientific Explanation and Other Essays in the Philosophy of Science*, pp. 3–46, Free Press, New York.
Howson, C. and Urbach, P., 1989: *Scientific Reasoning: the Bayesian Approach*, Open Court, Chicago.

Jeffrey, R., 1975: 'Probability and Falsification: Critique of the Popper Program', *Synthese* 30, 95–117.

Kyburg, H.E., 1968: 'The Rule of Detachment in Inductive Logic', in Lakatos (ed.) 1968, pp. 98–119.

Lakatos, I., 1968: 'Changes in the Problem of Inductive Logic', in Lakatos (ed.) 1968, pp. 315–417.

Lakatos, I. (ed.), 1968: *The Problem of Inductive Logic*, North-Holland, Amsterdam.

Lakatos, I. and Musgrave, A. (eds.)., 1970: *Criticism and the Growth of Knowledge*, Cambridge, Cambridge University Press.

Musgrave, A., 1975: 'Popper and "Diminishing Returns from Repeated Tests"', *Australasian Journal of Philosophy* 53, 248–253.

Popper, K.R., 1963: *Conjectures and Refutations*, London, Routledge and Kegan Paul.

Popper K.R. and Miller D., 1983: 'A Proof of the Impossibility of Inductive Probability', *Nature* 302, 687–688.

Popper K.R. and Miller, D., 1987: 'Why Probabilistic Support is not Inductive', *Philosophical Transactions of the Royal Society of London*, A321, 596–591.

Reichenbach, H., 1938: *Experience and Prediction*, University of Chicago Press, Chicago.

Rosenkrantz, R., 1977: *Inference, Method and Decision*, Reidel, Dordrecht.

Salmon, W.C., 1968: 'The Justification of Inductive Rules of Inference' in Lakatos (ed.) 1968.

Strawson, P.F., 1952: *Introduction to Logical Theory*, Methuen, London.

LARRY LAUDAN

IS EPISTEMOLOGY ADEQUATE TO THE TASK OF RATIONAL THEORY EVALUATION?

1. INTRODUCTION

The philosophy of science is generally understood to have two broad branches, one dealing with the conceptual foundations of the sciences and the other with the certification of the knowledge claims of the sciences. The first corresponds to what we might, if we were feeling pretentious, call the metaphysical foundations of science. The second is generally seen as applied epistemology. I shall have nothing here to say about the former. My focus, rather, will be on the initially plausible, and broadly held, view that theory testing and theory evaluation are – at root – *epistemological* activities. The appraisal of theories is seen as a special case of the epistemic evaluation of particular statements or beliefs. Whilst everyone concedes that a theory such as the special theory of relativity poses difficulties of appraisal over and above those associated with (say) 'The sun rose this morning', both activities are thought to be – at least in the ideal case – epistemic in essence. I believe this view to be fundamentally mistaken and systematically misleading. This essay will attempt to explain why that is so.

My strategy here will be simple and straightforward. I shall argue that analytic epistemology invites us to think of the testing and evaluation of any statement (including such complex statements as scientific theories) as a matter of exploring whether we have grounds for believing all the consequences of the statement to be true or truth-like. 'Tests', within this way of thinking, involve determinations as to whether *some* of those consequences – especially the ones likely to be false if the theory is false – stand up to empirical scrutiny. Epistemology thereby defines a very strict sense of evidential relevance. *A fact is epistemically relevant to the appraisal of a theory just in case that fact, or its negation, is among the consequences of the theory.*[1] I shall show that this conception of relevance is wholly at odds with the ways in which scientists use empirical information to appraise theories. Given that contrast, we shall have the options of either: (a) granting that theory appraisal is not entirely or even principally an epistemic activity; or – sure to appeal to the less naturalistic among us – (b) holding that scientists understand very little about theory appraisal and urging scientists to mend their epistemically errant ways.

165

Robert Nola and Howard Sankey (eds.), After Popper, Kuhn and Feyerabend, 165–175.
© 2000 *Kluwer Academic Publishers. Printed in Great Britain.*

2. SAVING THE PHENOMENA

The argument here is a very simple one. On the one hand, a key thesis of epis-temology, indeed *the* key thesis of analytic epistemology, is that a theory – like any other statement – determines the relevant phenomena to be consulted in appraising it. On the other hand, the best scientific practice makes it clear that theories do not fully define what they are responsible for. Let me put it less opaquely. When someone proposes a new theory for some domain of scientific inquiry, we ask, of course, whether that theory is a good or acceptable one. Being a good theory is a devilishly tricky notion to explicate in its details but in very general terms, it is easy to describe what we mean. First, we want to know how the theory fares with respect to the already acknowledged problems or phenomena in the domain. Would accepting this theory require us to give up vast amounts of problem-solving or explanation-giving capacity which existing theories already enjoy? Secondly, we typically ask whether the theory in question has predictive or explanatory resources that enable us to use it to anticipate phenomena in the domain that are unknown and unexpected, given existing theories. Finally, we ask whether the extensions to the domain which the theory's surprising predic-tions (if any) promise are correct. In other words, good theories account for what we already know, they make surprising predictions and those predictions suc-cessfully stand up to serious tests. The account I have just offered would, I sus-pect, be accepted as pretty close to the mark by most philosophers of science and by most (natural) scientists. The story will not come to any reader as a shocking revelation about the nature of science.

Nonetheless, we already have in the account I have rehearsed a picture of science that raises fundamental challenges to the epistemological enterprise overall and to specific epistemologies like scientific realism or Bayesianism in particular. To see what that challenge is, we must focus on the first of the three elements I mentioned. When assessing a theory, I wrote, we ask whether it can account for most of the facts already acknowledged as phenomena in the domain or field of investigation. This is *not*, let me emphasise straightaway, a demand for anything like full cumulativity from one theory to another. I have long stressed that, on this issue, I am with Kuhn and Feyerabend and against those positivists and realists who once demanded that rational theory choice required the reten-tion by later theories of all the successes of their predecessors.[2]

But even though full cumulativity is too strong a requirement, something close to such cumulativity or retention must be insisted on. If a theory lacks this ability, that is, if there are many already explained phenomena in the field which a new theory cannot account for, we properly regard this as a liability, that is, as a reason – more or less weighty – to reject the theory. This liability emerges *even when the new theory is logically consistent with the results in question*, that is, even if the theory is not refuted by any of the phenomena it cannot account for. If the theory fails to account for such phenomena, it is likely to be judged inadequate. In *Progress and Its Problems*, I invented a technical term to describe such situations. I called them *non-refuting anomalies*. They were a species of anomalies, I thought, because – like more traditional sorts of anomalies – they are legitimate *prima facie* grounds for rejecting a theory. Yet what was interesting about this class of cases was that, unlike the usual anomaly, they did not arise from a situation where a theory made a prediction that turned out to be false. So they were non-refuting. What they indicated was not the falsity of a theory but its *incompleteness*, its inability to solve problems that such theories *should* solve.

How do scientists know which problems a new theory should offer a solution for? After all, phenomena do not typically come with labels attached indicating the domain to which they belong. Characteristically, this issue is resolved by asking what problems have already been solved by the rivals to the theory in question. For instance, when Tycho Brahe sat down in the sixteenth century to produce a new system of planetary astronomy, he knew perfectly well the sorts of things his theory would have to explain. It would have to account for diurnal motion of the heavens, for planetary retrogressions, for solar and lunar eclipses, for periodic changes in apparent size and brightness of the planets. When early modern astronomers assumed the burden of developing a theory which would 'save the phenomena,' as they called it, they had in mind something quite specific and precise. The phenomena to be saved were those contained in the 2,000 year record of the positions and brightness of the wandering stars (along with the Sun and the Moon), along with the regularities which one could educe from those tables of star positions. Failure to grapple with *any* of these problems would have been fatal to the Tychonic project since rival theories had already exhibited their capacity to solve or resolve many such problems. Thus, when we say that a theory is incomplete, we make this determination, not necessarily by having some theory-independent access to what the phenomena in a particular domain are, but by comparative reference to the successes of rival theories.[3]

Similar, familiar examples abound, where theories are criticised, and often rejected, not because they made false predictions but because they failed to address relevant phenomena. For instance, stable-continent theories of geology offered no explanation as to why the continents fit together so neatly. Uniformitarian geology gave no account of how the Earth evolved to its habitable state. Steady state cosmology offered no explanation for residual background radiation. Newton's physics (prior to Laplace) offered no account of why the planets all move in the same plane and in the same direction. Ptolemaic astronomy did not explain why – even within Ptolemy's own theory – all the planets have a solar component to their motion. Phlogistic chemistry is wholly silent about why gaseous elements combine only in integral multiples by volume. Theories of the terrestrial causes of dinosaur extinction leave unexplained the worldwide iridium spike that occurred towards the end of the Cretaceous. Geostatic models of the Earth cannot explain the Coriolis effects associated with large bodies of wind and water.

In every one of these cases, and dozens like them, we have a situation in which a theory is found wanting not because it made a prediction that was false but because it was silent where it should have spoken and where its rivals did speak. Where one theory gives an account of a phenomenon, the other attributes it to mere coincidence.

168   LARRY LAUDAN

Holding in mind examples of the sort I have just mentioned, I hope that no one will regard as problematic or controversial my claim that scientific theories are, and should be, judged against this yardstick, among others. Nor is it just scientific theories that must satisfy this demand. In philosophy as much as in science, it is commonplace to criticise a point of view, not because it says something false, but because it fails to address key problems or issues in its domain. (Indeed, the structure of this essay is itself of this sort, for I am alleging that epistemology is badly flawed precisely because it says nothing about an important class of judgements that routinely enters into the process of rational theory appraisal.)

What *is* controversial here is not the claim that non-refuting anomalies can be genuine grounds for faulting and even rejecting a theory but rather what *follows* from that acknowledgement. For *if* it is true that such anomalies are important and genuine, then most of epistemology – including the epistemology of science, so-called – is badly flawed, perhaps fatally so. My brief here is to spell out why there is a tension between the existence of non-refuting anomalies, on the one hand, and conventional theory of knowledge, on the other. Properly understood, the pervasiveness of non-refuting anomalies in the appraisal of theoretical beliefs throws into sharp relief the poverty of the epistemological project overall and, thereby, of all specific epistemologies, whether empiricist, realist, Bayesian or Neyman-Pearsonian.

But before I proceed to develop that case, a word or two is in order about this requirement. I chose as the heading for this section the classic phrase 'saving the phenomena'. This may strike some readers as a misnomer for the problem in question. After all, some contemporary philosophers (most prominently, van Fraassen) hold that a theory 'saves the phenomena' precisely when all its observable consequences are true. (This is his usage throughout *The Scientific Image*.) This notion is, of course, straightforwardly semantic and epistemic. If we accept this gloss on the traditional phrase, then it is no part of the obligation of a theory to give an account of things which go beyond that theory's empirical consequences. But, as I have argued in other contexts, van Fraassen's usage here represents a major break with the tradition. From the time of Plato forward, the demand that a theory should save the phenomena involved the insistence that it should account for *all* the salient and known facts of a given domain of investigation. This idea is quite different from van Fraassen's that a theory saves the phenomena whenever all its observational consequences are true. Van Fraassen transforms what had traditionally been a demand for explanatory completeness (of a sort) into a demand for observational correctness. One can, of course, use terms in whatever fashion one likes and I am not faulting van Fraassen's redefinition of this hoary old notion. But precisely because his formulation of the concept of saving the phenomena has been widely quoted, and perhaps even widely accepted, I think it important to make quite clear that I am not using this term in van Fraassen's sense.

3. THE EPISTEMIC AS TRUTH-RELEVANT

If the full implications of this problem are to be correctly perceived, we must begin by asking ourselves briefly about what epistemology is and what it is not.

By epistemology, I do not mean the foundationalist/sceptical sort that flourished from Plato and Aristotle to Hume and Kant. I mean contemporary epistemology, especially that associated with the analytic tradition in philosophy of science and the general theory of knowledge. The concern of pre-nineteenth-century theories of knowledge, and of much ordinary language epistemology even in the twentieth century, has been to identify the conditions under which we can say that someone 'knows' something. Everyone acknowledges that this is not going to take us very far if we are talking about science, since every right thinking person concedes that scientific theories are not the sorts of things that can be known in any philosophically robust sense of the term. If we would focus on theories of knowledge that are to be even conceivably germane to scientific activity, we obviously must look at those which concern themselves with identifying the conditions under which *rational belief* is possible, i.e., with the circumstances under which it would be reasonable to accept a theory as true or – supposing these notions to have any content – close to the truth or truth-like or highly probable. The point I am making, which I trust is not particularly controversial, is that epistemic concerns are always necessarily directed at specifying the circumstances under which we can make a warranted judgement about the truth of a belief, or its likelihood of being true; if the beliefs in question are scientific theories, then we are addressing the truth or likelihood of truth of a particular scientific theory.

Now, if that is so, it obviously follows that factors which have nothing to do with the truth or the probability of a theory inevitably fall outside the range of epistemically relevant considerations.[4] You can call them pragmatic factors or aesthetic factors or whatever you like; for purposes of this paper I will simply call them *non-epistemic factors*. To be explicit, a trait or property of a theory is non-epistemic just in case it fails to be indicative of, or germane to, or to co-vary with, the truth status of the theory in question. Everyone grants that, as a contingent matter of fact, non-epistemic factors often play a role in theory appraisal. Nevertheless, this concession is generally regarded as non-threatening to the epistemic enterprise since, in the *rational reconstruction* of the act of theory appraisal, such factors – it is supposed – can be systematically eliminated and the choice can ultimately be represented as driven entirely or at least primarily by epistemic factors.[5] That anyway is the ideal, and it is that ideal which I aim to challenge in this essay. For if that ideal is wrong – if, as I will argue, non-epistemic factors *must* play a role in the rational evaluation of every theory – then we will have grounds for claiming that epistemology is intrinsically inadequate to the task of representing rational theory choice, even as a limiting case.

Let us return now to the question of non-refuting anomalies and see what, if anything, the epistemologist – of whatever persuasion – might have to say about them. Someone offers, let us suppose, a new theory in some arena of investigation, perhaps a new theory about the structure of the universe. Let us suppose that this theory, T_1, says nothing whatever about conditions of the cosmos when its structure was drastically different from its present one. However, let us suppose that T_1 does a good job of surviving whatever empirical tests we subject it to. That is, every time we check its predictions, they are correct. Perhaps it even makes some surprising predictions that turn out to be true. In short, it looks good *epistemically*. A scientific realist would perhaps even say it is reasonable to suppose T_1 is true. A Bayesian would tell us that T_1 had earned a high degree of rational belief. A constructive empiricist would tell us that it was reasonable to suppose that all T_1's empirical consequences were true.

Under such circumstances, would or should scientists accept the judgement of their philosophical brethren and accept T_1? Almost certainly not, for they would be very suspicious – and rightly so – of T_1's refusal to address the question of how the universe got from its initial to its present state, especially if rival cosmologies can account for a great deal of the evolutionary history of the universe before its having reached its current state. T_1, we are supposing, cannot do that. Recall it is not that T_1 gets the evolutionary history wrong; it simply does not get it at all. Under these circumstances, scientists will properly reject T_1.[6] Still, that rejection cannot possibly be epistemically motivated, since the grounds for the rejection have nothing to do with the truth of T_1. Indeed, the scientists in question have every reason to believe that T_1 is true, as far as it goes. What troubles them, of course, is that it does not go far enough; it leaves too many important phenomena unaccounted for. It is drastically incomplete, relative to its rivals.

But incompleteness is *not* an epistemic attribute since the fact that a statement is incomplete is wholly independent of the question whether it is true. Obviously, there are plenty of true statements and probable statements that are incomplete. If we believe that holding our well-tested T_1 responsible for its failure to address phenomena solved by its rivals is a reasonable thing to do, then the basis for such a belief must be *non-epistemic*. It will involve cognitive values and cognitive virtues that are irreducible to epistemic values and virtues.

Yet I have already suggested to you that it is plausible to believe that every new theory in every scientific domain is held up to this yardstick: can it account for most of the well-known phenomena in the domain? If, as I have tried to suggest, that demand cannot be parsed or justified epistemologically, then we have clear evidence of the inadequacy of epistemological tools to make sense of the theory appraisal process in the sciences.

But are we perhaps moving too quickly here? Might there not be some way for an epistemologist to justify such factors as truth conducive? Bill Lycan, for one, has tried to mount a defence of principles similar to this requirement that a theory is responsible for accounting for the successes of its rivals. In chapter 7 of his *Judgement and Justification*, he explores the possibility that such requirements as this (he would likewise include the demand for simplicity, for generality and for explanatory power) have emerged from the processes of our mental evolution. In his scenario, a benign Mother Nature equips us with various desires which prove to have high utility in enabling us to cope with the world we find ourselves in. Now, I do not doubt for a moment that this requirement that later theories should capture the successes of their predecessors is a useful requirement. Among other things, it insures that we do not hastily abandon previous cognitive successes in the name of currently fashionable theories. It is even possible, although I doubt it, that – as Lycan claims – we hold this value because evolutionary pressures may have found it adaptive.

But what I firmly deny is that any of these gestures in the Darwinian or the pragmatic direction do anything to establish the epistemological or truth-related features of this principle. At the risk of repeating what I already said, true statements need not be complete. Complete statements need not be true. Hence the incompleteness of a statement is no indication of its falsity any more than its ability to save the known phenomena is a necessary condition of its truth. Towards the end of his discussion of such principles, Lycan concedes their non-truth related character:

> I am not claiming that our basic methods' adaptive utility justifies them in the epistemological sense or that it *per se* provides any guarantee of true beliefs as output. (Lycan 1988, p. 143)

But this concession, which Lycan makes *en passant*, should be deeply troubling to those keen to pursue the epistemological enterprise *per se*. The fact that key tools of theory appraisal are not epistemic and are not amenable to epistemic analysis implies that epistemology lacks the tools to rationally reconstruct the scientific enterprise. Values other than truth – what I called *cognitive* values in *Progress and Its Problems* – enter *essentially* into the appraisal of scientific theories. I note parenthetically that many readers of that book wondered then, and perhaps still wonder to this day, why I was so perverse as to focus on what I called 'cognitive' values rather than 'epistemic' ones, and 'cognitive' decisions rather than 'epistemic' ones. I hope what I have just said will make that choice of terminology a bit more intelligible.

4. SOME SPECIFICS

I have claimed thus far that scientific rationality cannot, not even in principle, be reduced to the theory of knowledge. But that claim is still at a pretty abstract level. What I should like to do now is to fill it out by turning to look at some of the specific attempts that have been made to provide epistemic moorings for scientific practice by some of the major epistemological camps within the philosophy of science. If what I have said thus far is correct, we should expect none of them to be able to account for practices such as the one I have described.

Let me turn first to my perennial epistemological target, scientific realism. I hope that in previous writings about realism, I have shown some appreciation of the fact that realism is a highly nuanced position, with a number of importantly different variants. For the purposes of this essay, however, I am going to gloss over those variations and talk about the features that virtually all forms of scientific realism have in common. I think that Wilfrid Sellars caught much of the essence of contemporary epistemic realism when he said that what the scientific realist believes is that the world is pretty much the way that our best confirmed scientific theories say it is. The task of realist epistemology is to show that belief to be justified. All card-carrying realists believe that there are certain tests of theories (they differ among themselves about exactly what sort of test is required) which, if a theory passes them, provide us reasonable grounds to hold that the theory in question is true or nearly true or verisimilar. Realist epistemology, as one might expect, consists in providing arguments to show that a theory which passes said tests can be reasonably held to be true.

If this were another occasion, I might with profit explore the question whether the tests that realists offer up are indeed truth conducive. But I shall leave this point to one side for my aim here is to show that, insofar as scientists have values other than truth, those can find no place in realist epistemology. Indeed, they can find no place there because realist epistemology – if it be epistemology pure and simple – is concerned exclusively with exploring the circumstances under which a statement can be held to be true.

There is an interesting irony lurking just beneath the surface here that will already have occurred to many readers. Realist epistemologists like Sellars, Boyd, Putnam and Popper have been in the vanguard of those insisting – as I do – that acceptable theories must be able to account for many of the phenomena saved by their rivals. Indeed, realists have sometimes treated the cumulativity thesis as if it were their own creation or at least integral to their epistemic designs. But we must ask ourselves: on what grounds *realist* epistemology can insist on retention of explanatory or predictive success? Why must later theories explain what their predecessors could? Is this retention requirement a part of realism because successful retention of the successes of a rival makes a theory more likely to be true? Of course not, because – to repeat myself one last time – completeness has nothing to do with truth or high likelihood. Indeed, the statements about whose truth we are most confident (e.g., the Sun rose this morning, the Earth is older than 6,000 years, heavy bodies fall downwards near the surface of the Earth) are all either low-level generalisations or statements of particular facts. Grand, general theories of the type that realists lust after (a lust I share by the way, although for rather different reasons) come up very short on the epistemic virtues scale. More importantly, the demand that a new theory must not only pass demanding tests of its consequences but also be such that its successes include the successes of all its rivals is a demand that has nothing to do with whether the theory is true.

The epistemology of scientific realism – by linking itself so strongly to the retentionist idea – has unwittingly undermined its status qua epistemology, that is, as a theory of justified true belief by virtue of insisting on a requirement for the acceptability of theories that has nothing to do with truth. Now, at one level, that is all to the good since I hold epistemology to be inadequate to the task of rationally reconstructing scientific choices. But let us be clear and explicit about what we are saying. Science may or may not be after the truth. But, as the earlier arguments already show, it is certainly after something more, or other, than the truth. Call it generality or saving the known phenomena, or cumulativity. Truth, if it figures at all among the cognitive values of the scientific community, is but one among several values that scientists seek to promote. Nor is truth *primus inter pares* among these cognitive values. Retaining the successes of a predecessor is, I have claimed, at least as central a requirement as that a theory pass whatever specific tests of its truth that we can devise. Accordingly, the idea that the core of rational scientific behaviour can be subsumed under, or reduced to, the axiological rubric, 'scientists aim to discover true theories,' simply will not do.

There is another way of framing the critique of epistemology that I am trying to articulate. In the early 1990s Jarrett Leplin and I undertook a series of investigations into the thesis of underdetermination. One conclusion we came to was that most advocates of the thesis of underdetermination were (what we called) consequentialists. This included philosophers such as Quine, Goodman, Popper and most of those concerned to develop theories of qualitative confirmation. We sought in that research to discredit the idea that the evidence for or against a theory may be found only among the things which that theory entailed or forbade. We tried to make the case that many important forms of information relevant to theory appraisal are non-consequential. In my view, non-refuting anomalies constitute one important class of such examples.

Although Leplin will probably not follow me this far, I now want to claim that it is no accident that most theories of evidence have turned out to be consequentialist. It is because epistemology itself, especially in the twentieth century, has been through-and-through consequentialist in character. The received semantics of statements construes the meaning of a statement as nothing but the set of things it entails. A theory is true, on this view, just in case, all its empirical consequences are true. Given this semantics, this construal of what makes a statement true, given as well the view that epistemology is concerned to identify the circumstances under which we are warranted in believing that a statement is true – that is, in believing all its consequences to be true – it is no wonder that theorists of knowledge have imagined that the only relevant evidence for a statement is evidence of the truth or falsity of one or another of that statement's consequences.

If realist epistemology is in trouble here, so too are the other familiar epistemologies of science. Consider briefly Bayesianism. The Bayesians purport to tell us how to determine the posterior probability of a scientific theory, given its initial or prior probability, on the one hand, and a piece of relevant evidence, on the other. I have always been suspicious about Bayesianism because it seemed able to concoct a story to explain virtually any feature of scientific behaviour, rational or otherwise. But here, I suspect, the Bayesians are reduced to silence. To see why, let us imagine that a new theory, T, is introduced into a field of investigation. T's prior probability is some value, r. Already well established in that field, let us suppose, are certain phenomena, p. Now suppose that T is neutral with respect to these phenomena, i.e., it neither entails them nor their negations. More strongly, suppose that the conditional probability of those phenomena, given T, is the same as their prior probability – which is technically what a Bayesian means when he says that T does not address them. Under such circumstances, the likelihood of T – in the face of these phenomena that it does not account for – is precisely the same as its prior probability. That is to say, for the Baycsians, *the failure of T to address phenomena in its domain of application need do nothing to alter our initial confidence or prior degree of belief in T.* That is precisely how it should be, *if* we are doing epistemology, for we have already emphasised the point that the truth status (or probability) of a theory is unaffected by its incompleteness. But if, as I have also argued, completeness is a cognitively relevant, indeed, cognitively essential, component of the evaluation of the acceptability of a theory, then we see once again the limitations of epistemology, in this case Bayesian epistemology, in making sense of normative theory evaluation in the sciences.[7]

Likewise realist epistemologists must implicitly countenance non-epistemic objectives if they are to preserve the spirit of their project. If their exclusive concern were with finding theories most likely to be true, realists would all become van Fraassen-ites, willing, even eager, to formulate ontologically ema-ciated versions of theories rather than the theories themselves, since – as van Fraassen never tires of reminding us – the former are always likely to have more truth and less error in them than the latter. The fact that realists resist such Ockhamization of theories can be explained only, I submit, by noting that they are not doing realist epistemology but something grander and more interesting than that. Similarly, when Bayesians allow that scientists can be rational even when they do not maximise the probabilities of their beliefs, they too have implicitly abandoned the epistemological project for maximising true belief.

Some years ago, van Fraassen correctly observed that a theory does not have to be true to be good, meaning that false theories can be of value. Van Fraassen still believed, however, that a theory, if true, would *eo ipso* be good. What I have been trying to argue is that a true theory is not necessarily good any more than a good theory is not necessarily true. Whether it is good depends on how it fares with respect to certain relevant non-epistemic criteria such as its ability to handle non-refuting anomalies. Transposing van Fraassen's aphorism, we can say that 'a theory doesn't have to be false to be bad.' Indeed, *most true statements are bad theories in just this sense*; to wit, their scope is insufficient to qualify them as viable scientific theories.

Thus far, I have been trying to suggest that familiar theories of knowledge make no allowance for the importance of non-refuting anomalies in the appraisal of theories. Before I conclude, I want to suggest, albeit sketchily, that precisely the same criticism applies to familiar accounts of the *testing* of theories. From Bayes, Whewell and Peirce to Popper and Mayo, most of the plausible philo-sophical theories of testing have laid central stress on the ability of a theory to make surprising, successful predictions. I share the view that the ability of a theory to achieve such feats is relevant and important to its evaluation. Theories should be faulted when they fail such tests. But are such tests more important than the failure of a theory even to address central phenomena in its domain? The theory of testing, like epistemology in general, has been driven by a con-sequentialism which maintains that a theory is tested only by exploring whether what it says is true or false. Where a theory has nothing to say about a range of phenomena, it cannot be tested in the technical sense of that term against those phenomena. Statisticians routinely acknowledge two types of testing errors: accepting a false theory and rejecting a true one. Failing to address a phenom-enon is not acknowledged as an error or as a failed test on any account of testing of which I am aware. Yet, as we have seen, such failures can be just as significant for judgements about the acceptability of a theory as any failed prediction is.

5. CONCLUSION

Philosophers of science learned in the nineteenth century that science is not knowledge, as least not as epistemologists understood that idea. I am suggesting that a lesson of this century is that neither is science identifiable with what it is rational to believe to be true. Scientific theories can never be shown to be justified true beliefs nor would justified true beliefs, if we could find such, necessarily be theories. The appraisal of a scientific theory is not chiefly an appraisal of whether it is true nor yet is the acceptance of a theory acceptance as true. It is doubtless plausible that there are epistemic elements in the appraisal of scientific theories. That is not in dispute. What is at issue is whether an epistemically driven appraisal of theories could possibly capture the range of key considerations that scientists properly regard as relevant to the acceptability of a theory. The thesis of this essay has been that such epistemically based tools of evaluation cannot in principle render reasonable the demand for saving the phenomena. If I am right, perhaps it is finally time to recognise that the tools of epistemology are inadequate to the task of understanding scientific progress and rationality. It is methodology, not just epistemology, that we should be concerned about.

Guanajuato, Mexico

NOTES

[1] I shall below reformulate this thesis in ways that will make it relevant to probabilistic epistemologies.

[2] My objections to full cumulativity were developed at length in Laudan 1977.

[3] Even if one supposes that we have access to the phenomena in a domain independent of rival theories (as certainly seems plausible in my earlier example of planetary astronomy), the only phenomena which a theory is liable for failing to save are those which at least one of its rivals has already saved. For if there are phenomena which no extant theory has saved, those would count equally against all the theories in the domain and their cognitive effect would thus be neutralised.

[4] Popper, I think, was quite right when he observed: 'The proper epistemological question ... is whether the assertion made is true ... And we try to find this out, as well as we can, by examining or testing the assertion itself; either in a direct way, or by examining or testing its consequences' (Popper 1963, p. 27). Little did Popper realise that this made his demand that we want *interesting* truths extra-epistemic.

[5] Indeed, the *raison d'être* of rational reconstruction for writers like Carnap and Reichenbach was to redescribe situations of theory and belief appraisal in a way that separated the epistemic from the non-epistemic factors.

[6] If this example seems perverse and improbable, it should not. After all, the geological cum cosmological project of Hutton, Playfair and Lyell was precisely one of developing a theory of the Earth which eschewed any account of its origins or its history prior to its becoming habitable; this, in the face of dozens of previously cosmological theories (more especially Buffon's) which accounted for many of the surviving traces of an earlier, pre-habitable state of the Earth.

[7] There are other ways in which Bayesian practice departs from the epistemic straight and narrow. For instance, a strict Bayesian epistemological acceptance rule for theories has to be one which seeks to maximise probabilities. That in turn implies that the only acceptable hypotheses are those entailed by the evidence, since it is trivially provable that only such hypotheses maximally promote the objective of seeking the truth and nothing but the truth (or, the correlative objective of minimising error). When Bayesians countenance the acceptance of hypotheses which are not entailed by the evidence, it is because they implicitly concede that scientific inquiry ineliminably involves non-epistemic objectives.

REFERENCES

Laudan, L., 1977: *Progress and Its Problems*, University of California Press, Berkeley.

Lycan, W., 1988: *Judgement and Justification*, Cambridge University Press, Cambridge.

Popper, K., 1963: *Conjectures and Refutations*, Routledge and Kegan Paul, London.

KEVIN T. KELLY

NATURALISM LOGICIZED

1. INTRODUCTION

The approach to scientific methodology developed in my recent book *The Logic of Reliable Inquiry* (*LRI*) shares many general features with that summarized in Larry Laudan's concurrently published collection of papers *Beyond Positivism and Relativism* (*BPR*). Nonetheless, this fact might not be apparent, as my own work emphasizes mathematical theorems, whereas Laudan's draws primarily upon historiography. It is, therefore, of some interest to discuss the extent of the agreement and the significance of the differences. More generally, the discussion will (i) provide a logical analysis of the instrumental significance of empirical meta-methodology and (ii) redefine the role of logic in a post-positivistic, naturalized approach to epistemology and scientific method.

2. NORMATIVE NATURALISM

First, some important points of agreement. (1) We both view methodological principles as hypothetical imperatives (i.e., methods are recommended as means to an end) (*BPR*, pp. 132, 133, *LRI*, p. 3). (2) We both identify an empirical component in these hypothetical imperatives (*BPR*, p. 133, *LRI*, p. 5). (3) We agree that hypothetically normative epistemology is consistent with naturalized epistemology (*BPR*, p. 133). (4) We agree that aims can be criticized for being unachievable (*BPR*, p. 77, *LRI*, pp. 158–160, 190). (5) We agree that methodological norms should in some sense explain scientific progress (*BPR*, pp. 138, 139). (6) We agree that contemporary norms need not be satisfied by exemplary historical practice. Laudan's apt term for the position just sketched is *normative naturalism*. So far as this description goes, I am also a normative naturalist.

Our agreement does not end there. (7) We agree that the historicist attack on normative epistemology is founded, to some extent, on persistent positivistic dogmas; (8) we both question the normative force of methodological intuitions (*BPR*, pp. 137, 138); and (9) we agree that progress is not necessarily a matter of accumulating information.

Broad agreement on normative naturalism leaves considerable room for fundamental differences in emphasis, however. Whether a rule advances or inhibits our interests depends on such substantive matters of fact as the circumstances in

177

Robert Nola and Howard Sankey (eds.), After Popper, Kuhn and Feyerabend, 177–210.
© 2000 *Kluwer Academic Publishers. Printed in Great Britain.*

which it is applied, our ability to follow it correctly, the quality of the input, and so forth. But there is evidently a structural dimension as well, for the form of a methodological rule, like that of a computer program, has a great deal to do with what it does and, hence, with its success or failure in promoting our ends.

Laudan emphasizes the empirical dimension of means–ends claims. Given this emphasis, Laudan's guiding metaphor for naturalized epistemology is Baconian empirical science. Instead of deductively unpacking the formal structures of methodological rules prior to consulting experience, he treats the rules like black boxes and recommends that we empirically estimate the chance of success of the rule by consulting the results of historical practice.

I prefer to emphasize the analogy between methodological rules and computational procedures. My guiding metaphor is not Baconian inquiry, but theoretical computer science. Computer scientists are, after all, in the business of recommending rules and procedures based on their ability to achieve desirable goals. Of course, the means–ends relations investigated in algorithm analysis are to some extent empirical: the algorithm may fail outside its appropriate domain of application, the software has to be installed correctly, it has to be free of mistakes in its code, and so forth. But the explanatory core of such a recommendation is, nonetheless, an *a priori* analysis of what a rule with a given formal structure *would* do in various possible circumstances if it were correctly followed. In fact, this approach better reflects genuine practice in mature empirical sciences. Newton's genius was to fully unpack the geometry of orbital motion prior to consulting experience, so that, for example, null precession over the centuries provided an extremely accurate estimation of inverse square centripetal attraction. I propose that the theory of computability and computational complexity can serve to focus and to organize naturalistic methodology in much the way that geometry organized mechanics.

The *a priori* version of normative naturalism that I have just described is not new. It has been developed over the past forty years to a level of some sophistication by computation theorists under the heading of 'formal learning theory'.[1] The name of the subject is perhaps misleading, until one realizes that for computer scientists, 'learning' is a matter of reliable convergence to a correct answer to an empirical question, so that a theory of learning is actually a general theory of the existence of feasible, reliable, empirical methods.

3. LAUDAN'S PROGRAM

In this section, I review Laudan's position in some detail, marking the points at which we differ. The discussion follows the outline of Laudan's programmatic paper 'Progress or Rationality? The Prospects for Normative Naturalism' (*BPR*, pp. 125–141).

Laudan introduces normative naturalism as a response to recently popular nihilistic views about scientific method. According to Laudan, this nihilism arises from two assumptions. (1) Most great scientists have chosen rationally among alternative theories and (2) a methodology of science is an account of unconditional or categorical rationality. It follows that an account of scientific method must be satisfied by the practice of most great scientists.

Laudan rejects (2), responding that scientific rationality depends on the scientist's methodological aims and on her current beliefs about which acts are likely to further those aims. Our methodology should reflect our own aims and beliefs rather than those of historical figures. Laudan then distinguishes methodological 'soundness' from 'rationality'. Presumably, a 'sound' method really promotes our goals, whereas rationality reflects an individual's beliefs about what would further her own goals.

While I agree with Laudan in rejecting (2), I do not think this maneuver responds effectively to methodological nihilism. For example, Feyerabend's nihilism requires neither (1) nor (2). Rather, it is based on a straightforward means–ends argument with respect to an aim that seems plausible in the present day, namely, progress.

We find ... that there is not a single rule, however plausible, and however firmly grounded in epistemology, that is not violated at some time or other. It becomes evident that such violations are not accidental events . . . .. On the contrary, we see that they are necessary for progress (Feyerabend 1975, p. 23).

Since this argument is offered in the spirit of Laudan's empirical normative naturalism, it is hard to see how Laudan's position could respond to it, except by reinterpreting the verdicts of history. I prefer to criticize Feyerabend for claiming to have proved that no general methodological directives exist after discrediting a few proposed examples. In computer science, where impossibility results are routinely proved, pessimism based on the failure of a few, particularly simple, programming attempts is not taken seriously, and properly so. I recommend that naturalized epistemology reform itself in a similar direction.

Moreover, I am not as eager as Laudan and other historicists to trace methodological variation to divergent ends and beliefs. Even for scientists who share goals and beliefs (e.g., finding a correct answer to an empirical question), different scientific problems require very different means for their solution. For example, Bacon's methods of similarity and difference demonstrably lead to the truth when it is assumed in advance that the truth is a conjunction of monadic predicates. When disjunctions of such predicates are relevantly possible, more powerful methods are required (*LRI*, chapter 12). These strategies are very different from strategies for estimating limiting relative frequencies. Inferring conservation laws in particle physics suggests still other strategies exploiting the richer structure of linear spaces (cf. Schulte 1998, 1999a,b). This is analogous to the situation in computer science. Some formal problems seem to require search, others succumb to recursive 'divide and conquer' techniques, and still others are unsolvable unless we weaken our notions of success. If one's aim is to get the right answer as soon as possible, it is hard to see what sorts of interesting algorithmic principles would be suitable independently of the specific type of empirical problem one faces. It would be more plausible to discuss relational methodological principles that depend on the structure of the problem at hand. That is precisely the approach of formal learning theory.

Laudan then sketches normative naturalism, as described above. He first observes that methodological rules like 'avoid *ad hoc hypotheses*' are really disguised hypothetical imperatives of the form 'if you want to develop theories which are very risky, then you ought to avoid ad hoc hypotheses.'[2] Such a condition is 'warranted', according to Laudan, if we 'find' that following the recommendation is the best way we have yet thought of to promote the intended aim. Thus, hypothetical imperatives are subject to empirical investigation.

Laudan next addresses the obvious, skeptical charge that empirical justifications of empirical methods are circular. Faced with this problem, other epistemologists have advocated genuinely circular, coherentist epistemologies. Laudan opts for a methodological version of foundationalism in which a single, unobjectionable method is used to justify more sophisticated rules, which are in turn used to justify still more sophisticated rules, and so forth. The rule he chooses is something like maximization of expected (methodological) utility with respect to objective chances of success estimated using the straight rule of induction.[3]

> (R1) If actions of a particular sort, m, have consistently promoted certain cognitive ends, e, in the past, and rival actions, n, have failed to do so, then assume that future actions following the rule 'if your aim is e, you ought to do m' are more likely to promote those ends than actions based on the rule 'if your aim is e, you ought to do n' (*BPR*, p. 135).

Laudan's proposal bears some resemblance to Hilbert's foundational program in mathematics, for both approaches propose the use of more elementary, uncontroversial means (finitist arithmetic, the straight rule of induction) to vindicate the soundness of more controversial means (the infinitary methods of analysis, sophisticated scientific practice).

> I hasten to add that (R1) is neither a very sophisticated, nor a very interesting, rule for choosing between rival strategies of research. But then, we would be well advised to keep what we are taking for granted to be as rudimentary as possible. After all, the object of a formal theory of methodology is to develop and warrant more complex and more subtle criteria of evidential support (*BPR*, p. 135).

The pivotal notion of 'consistently promoting' in the definition of (R1) is vague in a manner that masks difficult questions. What if one method succeeded the only time it was tried, while the other was tried thousands of times with a few failures? Also, what if the current application has a rare feature on which the most successful method always failed and on which an infrequently applied competitor always succeeded? Or even worse, what if the current application has a feature that one can see by computational analysis to defeat the rule even though the method has never been used in such circumstances in the past? So although (R1) is simple, its recommendations are hardly as uncontroversial as Laudan suggests.

Another objection, due to Robert Nola (1999), concerns Laudan's requirement that goal achievement be an observable variable in the historical record. Laudan's proposal to use method (R1) to determine the instrumentality of a method M may work for observable goals such as maintaining consistency with the current data. But it cannot work for such aims as truth, empirical adequacy, or even future problem solving effectiveness because they are not observable in the historical record, and hence cannot generate instances of the kind (R1) requires as input.[4] One might use another inductive method M' to determine whether such an (unobservable) goal G is, in fact, achieved, but then (R1) would not be able to vindicate the instrumentality of M' with respect to the goal of determining whether unobservable goal G is satisfied, and so forth, for chains of any finite length. So there is no way in which to 'bootstrap' up from (R1) alone to methods vindicated with respect to aims like truth, empirical adequacy, or problem-solving effectiveness.

Perhaps the most serious objection to Laudan's proposed, meta-methodological program is that for all its emphasis on means and ends, it does not explain what would be *achieved* by a chain of meta-methods, each of which oversees the performance of its predecessor. Although his program holds out the hope of replacing intuition mongery with objective means–ends relations, this standard of intelligibility is not applied reflexively to his own program.

In spite of these objections, the idea of using one inductive method to empirically justify another as a means to a goal raises interesting logical and epistemological issues when it is presented with sufficient generality and without the encumbrance of Laudan's empiricistic and foundational commitments. In section 9 below, I employ learning theoretic techniques to establish *a priori* when reliable meta-methodological chains of various kinds are possible and what can be accomplished by them.

Laudan next observes that his naturalistic approach eliminates the need to base methodology on 'methodological intuitions'. I agree with that, but this feature of naturalism is independent of Laudan's strongly empirical approach to naturalistic methodology. The computationally informed naturalism I advocate is both instrumental and largely *a priori*, appealing not to historical data but to the respective formal structures of the particular empirical problem addressed and of the various methods that might be employed to solve it.

Although Laudan understands the primary aim of methodology to be the empirical justification of methodological rules as means for local, observable ends, he is also interested in explaining scientific progress as the result of repeatedly achieving such ends through time. I prefer a more direct approach, in which progress is viewed as an aim in its own right. Learning theory is directly concerned with such hypothetical imperatives as 'if you want to converge to the truth (in a given sense) then use method M.'[5] Laudan's conceptual detour through more proximate aims is thereby eliminated.

A key feature of Laudan's position is its emphasis on axiology, or the appropriateness of goals. This emphasis stems from Laudan's desire to rout methodological relativism, for he realizes that viewing methodological norms as hypothetical imperatives opens the door to relativism with respect to goals. Laudan's response is to claim that the appropriateness of scientific ends is itself an objective fact, since (a) appropriate ends must be feasible and (b) appropriate ends must have been reflected in the history of science (*BPR*, pp. 157, 158). This gives rise to a 'reticulated' account of justification in which changing theories of feasibility lead to changes in aims which lead to changes in methods, which lead to changes in theories, etc. (Laudan 1984, pp. 79, 80).

I agree strongly with Laudan's emphasis on feasibility of aims. Feasibility is a matter of problem solvability by agents of a given kind. Some empirical problems are unsolvable even by logically omniscient agents. Others are solvable by logically omniscient agents, but not by computable agents. Still others are solvable by computable agents, but not by any agent with a finite memory store, and so forth. Learning theorists are keenly interested in discerning the general features of empirical problems that make them solvable in one sense rather than another.

I also agree, to some extent, with Laudan's requirement that aims share some continuity with the past. Such sensitivity to practice is essential if the theory of computability is to yield relevant results. When computer scientists face such ill-defined problems as 'planning' or 'learning', they cannot begin to apply computability theory until they associate the informal problem with a spectrum of mathematically precise models of what 'planning' or 'learning' require. It is understood that this extra-theoretical process of explication must reflect, to some degree, actual planning and learning behavior. Actual behavior need not turn out to be an optimal solution, but it should at least appear to have been directed toward a solution to some mathematically precise problem in the spectrum. It is always open, in a computation theoretic analysis that yields highly counter-intuitive results, to question whether the formal problem addressed reflects what people actually want to accomplish.

But practice is not supreme. Computability analysis, by its very nature, forces one to turn a logical microscope on the problem under study, to an extent that intuitive, philosophical, or historical discussions rarely achieve. When practice and analysis disagree, it is possible that theory has unearthed structural possibilities that never would have come to light in the historical record because historical figures did not notice them either. That is why I oppose Laudan's particular emphasis on history in the philosophy of science, an emphasis which has been the received view in the field for some decades. If the philosophy of science is to earn its keep, it should do more than report back to scientists what they actually do. It should, like science itself, open new and exciting possibilities. History may suggest plausible goals and methods, but these suggestions are merely suggestions.

4. ELEMENTS OF LEARNING THEORY

Although formal learning theory is sometimes thought to be rather forbidding in detail, it is refreshingly simple in outline.[6] For all the scientist knows (or cares) the actual world may be one of many relevantly possible worlds. Each relevantly possible world responds to the scientist's acts with inputs through time. The scientist is capable of responding to these inputs in different ways. If the scientist's task is to determine whether a given hypothesis is empirically adequate, she may respond to the inputs with successive test outcomes (accept, reject) or with successive assignments of degrees of belief or of confirmation to the hypothesis in question. Any task involving such responses about a given hypothesis will be referred to as a hypothesis *assessment* problem.

In other circumstances, the scientist starts not with a hypothesis but with a *question* to be answered. Some hypotheses will be *relevant* to this question and a question may for our purposes be identified with its potentially relevant answers. An answer to the question is a potentially relevant answer that is also correct (e.g., true or empirically adequate). Such tasks are called hypothesis *generation* or *discovery* problems.

In either case, the scientist hopes to converge, in some sense, to a correct output; whether it be a correct assessment of a given hypothesis or a correct answer to a given question. Many scientific discoveries have resulted from happy accidents, but methodology is about guaranteed or *reliable* success, meaning success over a 'broad' range of relevant possibilities. To summarize, learning theory concerns the ability of a method or strategy to converge to a correct output (test result or relevant hypothesis) over a specified range of relevant possibilities. An *empirical problem* is a specified range of possibilities, together with a hypothesis to assess or a question to answer. Thus, learning theory concerns solutions to and the solvability of empirical problems.

5. STRATEGIC GOALS FOR HYPOTHESIS ASSESSMENT

Much variation is possible within the vague framework just described. The Socratic spirit demands that such vague terms as 'relevant possibility', 'success', and 'convergence' be provided with precise explications at the outset. Learning theory follows a different approach, providing a scale or spectrum of clear interpretations rather than a single one. This leads to a range of different types of scientific goals, each of which has a unique, epistemological character.

For example, consider the case of hypothesis assessment. Very ambitiously, one might hope for a method guaranteed to produce the truth value of the hypothesis by some time established in advance. But such ambitions usually cannot be achieved in science. More leniently, one might hope for a method that eventually halts with the truth value of the hypothesis. This is called *decision with certainty*. Decision with certainty is an empirical analogue of the computational concept of 'decidability'. But whereas many interesting formal problems are computationally decidable with certainty, the point of the classical problem of inductive generalization is that most general empirical hypotheses are not. At this point, the axiology of feasibility recommends moving to a weaker goal. Popper's original idea was that universal generalizations can nonetheless be refuted with certainty even though they cannot be verified with certainty. Similarly, purely existential hypotheses can be verified with certainty but not refuted with certainty.

Unfortunately for Popper's original idea, most scientific hypotheses are not really refutable with certainty either. Notoriously, a hypothesis can be saved from refutation by tinkering with the rest of the theory. And even in an idealized, empirical setting in which experimental outcomes are unproblematically theory-independent, probability estimates are logically consistent with any data in the short run, even if such an estimate is understood to imply a limiting relative frequency of outcomes in the future data. The same is true of the hypothesis that there are only finitely many types of elementary particles to be discovered, the hypothesis that a system is chaotic as opposed to orderly, and the hypothesis that a given sequence is produced by a Turing machine rather than by some uncomputable process.

Popper's response (1968) was to reconceive falsificationism as an injunction against coddling pet views rather than as a criterion of success. An alternative option is to weaken the criterion of success again, so that certainty is never required, whether the hypothesis is true or false. For example, one might require only that a method stabilize to the state of correctly rejecting or accepting the hypothesis under assessment without necessarily halting or providing a sign that it has done so. It is natural to call this standard *decision in the limit*. As Peirce and James emphasized, limiting convergence, unlike convergence with certainty, allows for an appealingly fallible sense of methodological success, according to which following the method is guaranteed, eventually, to reach a correct answer, but certainty is never forthcoming because there is never any guarantee that the method will not change its conjecture after seeing the next datum. Within the comfortable confines of a viable research paradigm, the possibility of a major crisis is a mere, philosophical curiosity. But from the outside looking in, the history of science is a history of broken certainties and no amount of 'inductive support' or other holy incantation can ensure that the same will not happen again. At best, we can hope that inquiry is organized so as to eliminate surprises after some future time that will not be recognized as such.[7]

Hypotheses about limiting relative frequency, computability, or the finite divisibility of matter are not decidable in the limit either. This remains true of limiting relative frequencies even if we assume *a priori* that the limit of the observed frequencies exists. Feasibility demands yet weaker aims. It turns out that a hypothesis specifying a particular value for a limiting relative frequency is *refutable in the limit* given that the limit exists, where *limiting refutation* requires convergence to rejection just in case the hypothesis is true and *limiting verification* requires convergence to acceptance just in case the hypothesis is false. Computability and finite divisibility are verifiable in the limit. Also, if chances are understood to entail limiting relative frequencies, the existence of an 'unbiased' statistical test is equivalent to verifiability or refutability in the limit, depending on whether the 'rejection' zone is defined with a strict or a non-strict inequality (*LRI*, chapter 4).

Limiting verification and refutation are very weak, in the sense that the vacillations witnessing nonconvergence may come with arbitrary rarity. Surely, we would like to do better. But if there were an *a priori* bound on how long one must wait to see the next vacillation, if it occurs at all, this bound would allow one to construct a limiting decision procedure, which is impossible in the examples mentioned. So again, I agree with Laudan's feasibility condition: if limiting verification is possible and limiting decision is not, then don't demand an upper bound on the frequency of a limiting verifier's rejections when the hypothesis under test is false.

If it is not assumed that a limiting relative frequency exists, a hypothesis asserting that it exists with a given value is not even verifiable in the limit.[8] This leads, by the axiology of feasibility, to even more attenuated notions of success. *Gradual decision* requires that the real values assigned to the hypothesis by the method approach the truth value of the hypothesis, possibly without ever actually reaching it. Gradual decidability is in fact equivalent to limiting decidability, because a gradual decision procedure can be converted into a limiting decision procedure by means of accepting or rejecting according to whether the gradual method's output exceeds or fails to exceed a cutoff (e.g., 0.5). The one-sided versions of gradual decision are strictly more lenient than their limiting analogues, however.[9] *Gradual verification* requires that the outputs approach unity just in case the hypothesis is true and *gradual refutation* requires that the outputs approach zero just in case the hypothesis is false. In fact, limiting relative frequency is gradually verifiable but not gradually refutable.

6. THE LONG RUN IN THE SHORT RUN

Limiting success occasions the natural objections (a) that the limit is too long to wait for and (b) that limiting correctness provides insufficient constraints on what to believe in the short run. These objections can be met, to some extent, by requiring that no reliable method converge *as fast* as our method in each relevant possibility and *faster* in some relevant possibility, in which case our method may be said to be *data minimal*. To demand the truth faster than a data-minimal method can provide it is to demand the impossible.

As Kuhn emphasized, it is both practically and cognitively costly to retool when a theory is retracted. Taking this concern seriously, we would prefer reliable methods that not only minimize convergence time, but retractions as well. Note that a single retraction could occur arbitrarily late, so convergence time and number of retractions are two different considerations.

It turns out to be too strict to require that no reliable method performs as few retractions in any relevant possibility and fewer in some relevant possibility (Schulte 1999a), for this is only possible when the potential answers to an empirical question are all decidable with certainty, and hence there is no genuine problem of induction. Suppose, however, that there is an *a priori* bound on the number of retractions required prior to convergence. In the case of hypotheses that are refutable with certainty, at most one vacillation is required: start out accepting and then reject when the hypothesis is refuted. The hypothesis that exactly one star of a given mass exists is decidable in the limit with at most two retractions: reject until a star of that mass is encountered and then accept until another one is encountered. When such a bound exists, it is natural to insist on methods that decide the hypothesis in question in the limit, that minimax retractions and that are data minimal with respect to all limiting decision procedures.

In *The Will to Believe*, William James (1948) remarked that finding the truth is different from avoiding error and that the two aims are usually in tension. Data-minimality suggests the aim of finding the truth, since a method that refuses to conjecture a potential answer to the question at hand could not possibly have converged to the right answer yet, whereas a method that produces a potential answer consistent with the data *might* have already succeeded. Minimizing retractions suggests the aim of avoiding error, since a method that withholds judgment until the evidence is conclusive performs no retractions at all.

Reliability, data-minimality, and minimaxing mind changes can jointly impose strong requirements on methodology in the short run. To illustrate this point, suppose we know either that each stage will be green, or that at some finite stage n, green will give rise to blue forever after, in which case we may say that each stage is 'grue(n)'. If these are the only relevant possibilities, then the *unique* data-minimal, mind-change-minimaxing, limiting decision procedure is the one that conjectures that all stages are green until seeing a blue outcome (say at stage n), after which the method conjectures forever after that each stage is grue(n).[10] The same result obtains if we consider as relevant possibilities all hyper-grue predicates of the form grue($n_0, n_1, \ldots, n_k$), where $k \leq m$, for some fixed m (Schulte 1999b). The predicate grue($n_0, n_1, \ldots, n_k$) means green through stage n_0, blue from then through stage n_1, green from then through stage n_3, etc. If the fixed bound m is dropped, then success with bounded retractions is impossible.

One may think of decision with a bounded number of retractions as a criterion of success in its own right (Case and Smith 1983, *LRI*), where *decision with n retractions* requires that the method decide the hypothesis in the limit, vacillating between acceptance and rejection (or vice versa) at most n times. When retractions are being counted, it turns out to matter what one's leading conjecture is. For example, refutation with certainty is equivalent to decision with at most a single retraction, starting with acceptance, for a method that refutes with certainty starts out accepting the hypothesis and then, when trouble is encountered, retracts its former conjecture and replaces it with rejection. Similarly, verification with certainty is equivalent to decision with at most a single retraction, starting with rejection. What about decision with certainty? Since decision implies both verification and refutation, decision with certainty is equivalent to decidability with at most one retraction starting with an *arbitrary* conjecture (either acceptance or rejection). As we allow more retractions, we therefore arrive at generalized notions of verification, refutation, and decision. For example, decidability with at most two retractions starting with acceptance allows the method to begin with acceptance, change its mind thereafter to rejection, and finally switch back to acceptance. Once all the allowed retractions are used up, the method's output is certain.

7. LEARNING THEORETIC QUESTIONS

Laudan's normative naturalism focuses on hypothetical imperatives for particular methodological principles and on the feasibility of particular aims. From a learning theoretic viewpoint, the former question concerns the relation 'M solves problem P'. The latter concerns the property 'problem P is solvable', which is definable as the existence of a method M solving problem P.

Hypothetical imperatives and feasibility axiology work in tandem. Methodological understanding is obtained by formally solving for the strongest aim in the hierarchy of convergent goals that can be satisfied for a given empirical problem. Thus, the hierarchy of goals can be viewed as a kind of classification system for empirical problems. All of the problems within a given classification are in a precise sense 'methodologically equivalent', giving rise to intuitively similar methodological difficulties and calling out for similar sorts of solutions.

The ability to formally isolate the strongest aim achievable for a given empirical problem addresses a weakness in Laudan's empirical version of normative naturalism. It is very difficult to show by means of empirical data that no stronger aim *could* have been realized for a given problem. Thus, Laudan adopts an empiricistic stance and asks history only if a *known* method was *observed* to do better. Learning theoretic negative results cover all possible methods, and hence allow one to show that no possible method could have done better. This is precisely the role that computability theory plays in computer science.

Once one has seen a good number of solvability and non-solvability results, one wishes to know if there is an elegant structural characterization of solvability. That is, one desires a purely structural property Φ (making no explicit reference to methods or to success) such that an arbitrary problem *P* is solvable just in case it has Φ. Such results are called *characterization theorems*. Since they provide necessary and sufficient conditions for the possibility of reliable inquiry, they might be viewed as logically valid *transcendental deductions*.

Characterizations of the concepts of assessment introduced above are easily presented (cf. *LRI*, chapter 4). Assume a given set of relevant possibilities to be specified. Assume, also, that the hypothesis is not *globally underdetermined*, in the sense that the same infinite input stream arises from worlds in which it is respectively true and false (else no possible method could find the truth value of the hypothesis in each relevant possibility). Such a hypothesis is verifiable with certainty just in case each relevant possibility satisfying the hypothesis eventually presents inputs whose occurrence entails that the hypothesis is true. A hypothesis is refutable with certainty just in case its complement is verifiable with certainty, and is decidable with certainty just in case it is both verifiable and refutable with certainty. At the next level, a hypothesis is verifiable in the limit just in case it is a countable disjunction of hypotheses that are refutable with certainty. A hypothesis is refutable in the limit just in case its complement is verifiable in the limit, and is decidable in the limit just in case it is both verifiable and refutable in the limit. A hypothesis is gradually verifiable just in case it is a countable conjunction of hypotheses that are verifiable in the limit. It is gradually refutable just in case its complement is gradually verifiable and is gradually decidable just in case it is decidable in the limit. Thus, each notion of reliable success corresponds to a structural recipe for building up all the hypotheses for which that sense of success is achievable.

These results illustrate the grain of truth in the positivists' attempt to relate 'cognitive significance' to logical form. *If* hypotheses are expressed in a first-order language and *if* the input stream presents all the true, quantifier-free sentences in the language, and *if* each object is named by some constant in the language, *then* the quantifier prefix of the hypothesis determines the senses in which it can be reliably assessed. Specifically, quantifier-free hypotheses are decidable with certainty, existential hypotheses are verifiable with certainty, universal hypotheses are refutable with certainty, hypotheses with quantifier prefix ∃∀ are verifiable in the limit, hypotheses with quantifier prefix ∀∃ are refutable in the limit, finite, boolean combinations of existential and universal hypotheses are decidable with bounded retractions and hence in the limit, hypotheses with quantifier prefixes of form ∃∀∃ are gradually refutable and hypotheses with prefixes of form $\forall\exists\forall$ are gradually verifiable. But none of this is a function of logical form *per se*; nor is it a characterization of meaning. It is a *contingent* relationship between logical form and levels of achievable reliability, where the contingency relating the two is an assumption about the kind of data that would arise in a given relevantly possible world.

A simple structural characterization of decision with bounded retractions can also be given (*LRI*, chapter 4). Hypotheses that are verifiable with certainty are decidable with one retraction starting with rejection and refutability with certainty characterizes one retraction starting with acceptance. Decision with $n + 1$ retractions starting with rejection is possible exactly when the hypothesis under test can be expressed as the disjunction of a verifiable hypothesis with a hypothesis that is decidable with n retractions starting with acceptance. Dually, decision with $n + 1$ retractions starting with acceptance is possible when the denial of the hypothesis can be decided with the same number of retractions, starting with rejection.

8. HISTORICISM RECONSIDERED

Verification and refutation with certainty can be understood in two very different ways. Refutation with certainty requires that the data logically contradict the hypothesis and verification with certainty requires that the data logically entail the hypothesis. Thus, refutation and verification are logical entailment relations. So when it is discovered that a hypothesis is neither verifiable nor refutable with certainty, one response is to look for a 'generalized' entailment relation (degree of confirmation or inductive support) that does hold between the data and the hypothesis.

But as suggested above, refutation and verification with certainty may *also* be viewed as success criteria for empirical methods, just as they are viewed as success criteria for formal methods in mathematical logic and computability theory. The shift in type is important. Success criteria are goals (ends) rather than methods (means). So on this perspective, intuitive or historical arguments for the propriety of a fixed method (generalized entailment relation) give way to objective, computation theoretical arguments about achievability of the various goals. When it is discovered that a hypothesis is neither verifiable nor refutable, it is natural to move to weaker criteria of success that are achievable (e.g., limiting and gradual success).

Limiting goals, such as decision in the limit, are different from verification and refutation with certainty because it is no longer an option to view limiting success criteria as fixed logical relations that determine *when* to reject or to accept a given hypothesis. For example, one limiting method may converge to the truth on a given data stream faster than another such method does, and thereby converge more slowly on some other data stream, so the two methods disagree for some arbitrary length of time on both data streams. The logic of efficient, limiting convergence does not favor one solution over the other, leaving ample room for hunches, predilections, and scientific 'bon sense', so long as they do not inhibit the strategic goal of finding a correct answer as soon as possible.

Viewing verification and refutation with certainty as success criteria, rather than as generalized logical relations, leads to a reconception of the debate between historicism and logic. First of all, one argues for a generalized notion of entailment by a process of explication or reflective equilibrium, which is a kind of spiral process of correcting the explication with practice and correcting practice with the explication. This leaves methodology open to the plausible charge that it is merely armchair sociology in logical dress. Since learning theory focuses on objective, computational questions about the solvability of empirical problems, it does not invite this objection.

Portraying scientific method as a fixed, generalized entailment relation also occasions the objection that following such recommendations would have precluded scientific progress when the social character and costs of inquiry are considered (Feyerabend 1975). Such arguments cannot be directed against a logical approach to scientific method based on learning theory, because they *are* learning theoretic arguments. In fact, many of the results of learning theory can be viewed as *formally* grounded Feyerabendian critiques of particular methodological proposals (*LRI*, Osherson *et al.* 1986, Martin and Osherson 1998). Such critiques have the form that some empirical problem would have been solvable had the recommendation not been insisted upon. A computational critique of the rule of rejecting a theory when it is logically contradicted by the data will be discussed in detail below.

Finally, the logical relation conception of methodology invited Kuhn's nihilism concerning the logic of scientific change. Kuhn's basic argument in *The Structure of Scientific Revolutions* (1962) is that in major episodes of scientific change, no logical relation or generalization thereof holding between theory and evidence *rationally compels* one to drop the theory *when it is dropped*, so the change is arbitrary. According to Kuhn, the momentous empirical question facing a scientist is not the correctness of a given hypothesis, but the viability of her paradigm. Viability is a vague matter, but it has something to do with the potential of the paradigm to generate puzzles and solutions to them. *Piecemeal* viability means something like: *for each* new anomaly that a competitor can handle at the time, *there exists* an articulation of the paradigm that resolves it. *Uniform* viability is more ambitious, requiring that the paradigm possess some as-yet unknown articution that will once for all absorb all new anomalies handled by competing theories (e.g., the 'end of science' foretold by some advocates of the fundamental particle paradigm in physics).[11] Piecemeal viability is of $\forall\exists$ form, and is therefore refutable in the limit, whereas uniform viability is of $\exists\forall$ form and is therefore verifiable in the limit. Barring *a priori* bounds on how long it would take to find such an articulation, neither question is decidable in the limit (cf. Kelly *et al.* 1997). Recall that in order to verify a hypothesis H in the limit, a method must reject H infinitely often if H is false. So whereas it is not arbitrary that a limiting verifier perform these rejections at some times or other, it is up to the method rather than to logic when, exactly, they occur. Nonetheless, there are still normative recommendations to be made on a logical basis, for some methods will fail even to verify the hypothesis in the limit and others will converge more slowly than necessary. Thus, the absence of an objective compulsion to drop the paradigm at a particular time is explained by, rather than raising a difficulty for, the learning theoretic logic of the paradigm selection problem.

The ultimate, historicist argument is relativism. Relativism is a danger to the generalized logical relation conception of methodology for the obvious reason that others may reject the proposed relation in light of different, culturally informed intuitions. Since there is no further reason for following the relation, the discussion ends there. On the instrumental approach, there can at least be agreement about the possible circumstances in which various methods would work, even when there are differences in aim and in beliefs about what the actual circumstances are.

It might be thought that relativism poses a serious problem for learning theory nonetheless, for how can we converge to the truth, even if we want to, if meaning and truth change in incommensurable ways through time? But there are still intelligible, strategic goals that do not presuppose translatability across scientific revolutions, so long as we do not measure progress in terms of increasing *content* (which requires content comparisons across incommensurable languages) or *verisimilitude* (which requires a fixed metric defined across incommensurable languages).[12] We might require, for example, that science eventually stabilize the truth value of the hypothesis under investigation and then learn what it is. Or we might require, more weakly, that whatever the truth value is in the future, we eventually always know what it is. This raises another Feyerabendian sort of question: would it injure the power of inquiry to require that the truth value eventually stabilize? The answer is affirmative (*LRI*, chapter 14, Kelly and Glymour 1992), so learning theory provides a *logical* argument in *favor* of inducing incommensurable changes.

The historicist quarrel with logic is actually a quarrel with the 'generalized logical relation' approach to methodology.[13] Reconceiving refutation as the first success criterion in a sequence of ever weaker criteria leads to a logical perspective on methodology that embraces the central premises of the historicist position without drawing the nihilistic conclusions.

9. WHAT EMPIRICAL NATURALISM CAN AND CANNOT DO

The core of Laudan's epistemological program is the idea of using one empirical method to investigate the conditions under which another method will succeed in achieving a given goal. I criticized Laudan for depicting such inquiry as a search for empirical correlations between means and ends, since such ends as achieving empirical adequacy are not directly observable in the historical record. Furthermore, I objected that Laudan's correlational approach leaves no room for *a priori* computational analysis of the conditions under which a method would succeed. Finally, I objected that for all the emphasis on means and ends, Laudan did not say what could be accomplished by methods assessing methods assessing methods. But the general idea is of sufficient interest to warrant a fresh approach.

Suppose we are interested in finding out whether a given method will succeed. This question has several empirical dimensions. If we ignore the structure of the method and treat it as a black box, as Laudan seems to suggest, then it is an empirical question even to determine what the method would direct us to do in a given situation. But if we look at what the method is, and if a precise sense of convergent success is specified, then in principle[14] it is an *a priori* matter to determine the set of seriously possible future trajectories along which the method succeeds (in the specified sense). Call this set the *presupposition* of the method (relative to the intended sense of success).[15] Once a method's presupposition has been determined *a priori*, the problem of empirical meta-methodology reduces to determining whether the presupposition of the method under investigation is actually true. In what follows, I will assume that methods are transparent to the meta-methods investigating them, so the meta-methods merely assess the pre-suppostions of the methods they investigate.

The pressing means–ends question raised by empirical, normative naturalism is, then, what one could *do* with methods that check the presuppositions of methods that check the presuppositions of methods. ... Could one, for example, by looking only at what the meta-methods do, converge to a correct conjecture about the original hypothesis H? If not, then it is hard to see what the point of all the assessing is supposed to be. In such a case, one might say that the sequence represents a *vicious* empirical regress of the sort condemned by skeptics like Sextus and Hume. But if there is a strategy for assembling the conjectures of the meta-methods in the chain into a single conjecture that converges to the truth, then the chain can be used to achieve a cognitive goal and the regress may be exempted from the charge of pointlessness.

If the converse is also true (i.e., a single method that succeeds in a given sense can be turned into a given kind of meta-methodological chain of methods), then we may say that the chain is informationally or methodologically *equivalent* to the single method. Methodological equivalence imposes some discipline on our epistemological hopes in much the way that the concept of energy imposed dis-cipline on our hopes for perpetual motion machines. There is no question of a meta-methodological chain allowing us to do the impossible (i.e., to construct a single method that succeeds in an unachievable sense). But a meta-methodo-logical chain could do far *worse* than to be methodologically equivalent to the best sort of solution that a given problem admits, just as a heat engine may fall far short of being perfectly efficient in terms of energy transfer. The degree of viciousness of an epistemic regress may be viewed as the extent to which the sense of success to which the chain is equivalent falls short of the best achievable sense of success. For example, if there exists a certain refutation procedure, then a chain equivalent to a limiting decision procedure is inefficient, but less so than a chain equivalent to a limiting refutation procedure.

For a simple example of a methodological equivalence, consider the following situation. We throw method M_1 at the problem of trying to refute H with cer-tainty given background knowledge K. But M_1 works only when an empirical presupposition P_1 is satisfied. Meta-method M_2 is supposed to refute with cer-tainty whether this presupposition is, indeed, satisfied. Meta-method M_2 actually does refute the presupposition P_1 of M_1 with certainty given K. Now suppose that we only observe what the two methods conjecture through time, without looking at the data they receive. What could we tell about H?

Without saying more, not much. For suppose M_1 is a crazy method that alternates forever between acceptance and rejection without ever looking at the data, and suppose that M_2 rejects no matter what, without looking at the data. Then M_1 fails on every data stream in K so P_1 is unsatisfiable. And M_2 is correct on every data stream in concluding that P_1 is false, so M_2 refutes P_1 with certainty given K, as required. But one could conclude nothing about H from watching these two methods, since they both say the same thing no matter what they observe, and therefore erase all of the information in the data. According to the above criterion, such a pair represents a vicious empirical regress.

The situation changes, however, if both methods *aspire* to refute with certainty, in the sense that they outwardly appear to be refuting their respective hypotheses with certainty even if they really are not. More precisely, say that M_1 aspires to refute H with certainty given K just in case K entails that M_1 starts out accepting and retracts at most once. We may also speak of aspirations to verify with certainty given K, verify in the limit given K, etc. For example, M aspires to decide in the limit given K just in case on each data stream satisfying K, M converges either to acceptance or to rejection.

Now suppose that M_1 aspires to refute H with certainty given K and that meta-method M_2 refutes with certainty given K whether the aspirations of M_1 will actually be realized. Then we can construct a method M that decides H given K with two retractions starting with acceptance that succeeds just by watching what M_1 and M_2 do. Method M may be defined as follows. Make M start out accepting. Thereafter:

1. M accepts when M_1 agrees with M_2.
2. M rejects when M_1 disagrees with M_2.

To see that M works as claimed, suppose that K is satisfied and let P_1 be the presupposition of M_1. There are four easy cases to consider:

1. P_1 and H are satisfied: then M_1, M_2 always accept so M always does so as well, with no retractions.
2. P_1 is satisfied but H is not: then M accepts until M_1 rejects and continues to reject thereafter, using one retraction.
3. H is satisfied but P_1 is not: since M_1 is an aspiring refuter, M_1 starts with acceptance and can retract at most once, so since P_1 is not satisfied, it must be that M_1 eventually reverses its initial acceptance to a rejection. Meta-method M_2 correctly reverses its initial acceptance to a rejection as well. So M converges to acceptance after at most two retractions, starting with acceptance.
4. Neither H nor P_1 is satisfied: again, M_1 can only have failed by never reversing its initial acceptance to a rejection. Meta-method M_2 eventually reverses its initial acceptance to a rejection. So M converges to rejection after at most one retraction.

In each case, M converges to the right conjecture about H using at most two retractions, starting with acceptance.

What if M_1 aspires to verify H given K and M_2 verifies the presupposition of M_1 given K? Then exactly the same construction again implies that H is decidable with two retractions starting with acceptance. The situation is similar if we have a refuter of verification or a verifier of refutation. In either of these cases, the result is the same except that M starts with rejection rather than acceptance.

So for aspiring methods, refutation of refutation and verification of verification imply two retraction decidability starting with acceptance and refutation of verification and verification of refutation imply two retraction decidability starting with rejection. Methodological equivalence requires the converse implications as well. Let us consider whether they hold in the present example. Suppose that M decides H given K with at most two retractions starting with acceptance. Can we construct M_1, M_2, P_1, such that M_1 aspires to refute H given K and does so under presupposition P_1 and M_2 refutes P_1 given K? It is up to us to choose the both the presupposition P_1 and the methods M_1 and M_2.

Here is how to do it. Choose P_1 as the (naturalistic, methodological) proposition that M retracts at most once. Let M_1 be the aspiring refuter (given K) that watches M and accepts until M retracts once, rejecting thereafter whatever else M does. Let M_2 start with acceptance and then reject, with certainty, when M retracts for the second time. Evidently, M_2 refutes P_1 with certainty given K. Moreover, M_1 refutes H with certainty under presupposition P_1, because when P_1 is satisfied, M_1 converges correctly to whatever M converges to. If P_1 is not satisfied, then M uses its second retraction and converges to acceptance, but M_1 incorrectly converges to rejection. Thus, M_1 refutes H with certainty if and only if P_1 is satisfied. The converses of the claims for verification of verification, refutation of verification, and verification of refutation are similar. So we have arrived at a simple example of a meta-methodological equivalence theorem:

Proposition 1 *The following situations are methodologically equivalent:*

1. *There are two methods M_1, M_2 such that*
   (a) *M_1 aspires to verify [refute] H with certainty given K and does so under presupposition P_1, and*
   (b) *M_2 refutes [verifies] P_1 with certainty given K.*
2. *H is decidable given K with at most two retractions, starting with rejection.*

The analogous proposition in which M starts with rejection and the aspirations of M_1 and M_2 mismatch is also true.

Let us now generalize the preceding analysis along two dimensions at once. We will move from a single meta-method to an arbitrary, finite chain of meta-methods, each of which second-guesses the presuppositions of its predecessor. And we may as well also allow each method in the chain to succeed under a fixed bound on retractions. When is such an attenuated meta-methodological situation possible? Just when there is a single method that uses the sum of the retractions of all the methods in the chain and whose first conjecture depends in a systematic manner on what the methods in the chain achieve. The exact statement of the equivalence is as follows. For convenience of notation in dealing with chains, let P_0 henceforth denote the original hypothesis H under investigation.

Proposition 2   *The following situations are methodologically equivalent:*

1. *There exists a finite chain $M_1, \ldots, M_k$ of methods such that*
   (a) *for each $i < k$, method M_{i+1} aspires to decide P_i given K with n_{i+1} retractions starting with c_{i+1}, and does so under presupposition P_{i+1}, and*
   (b) *K entails P_k, the presupposition of the final method in the chain.*
2. *There exists a single method M that decides P_0 with $n_1 + \cdots + n_k$ retractions given K, starting with conjecture c, where c is 'reject' if an odd number of the c_i are 'reject', and c is 'accept' otherwise.*

The general proof of this proposition, and of all those that follow, is given in the Appendix. By way of illustration, consider a situation in which M_1 decides H with two retractions starting with acceptance, M_2 decides the presupposition of M_1 with three retractions starting with rejection and M_3 decides the presupposition of M_2 with one retraction starting with acceptance without presuppositions. The result tells us that this is possible exactly when there is a single, presupposition-less method M that uses $2 + 3 + 1 = 6$ retractions. Method M uses at most six retractions because it retracts once each time one of the component methods retracts. Since an odd number of the three methods start out rejecting, so does M.

The preceding analysis provides a clear motivation for empirical meta-methodology. It does not give us something for nothing (nothing could). Rather, adding more empirical meta-methods to the chain amounts to an even episte-mological trade in which the sense of success is weakened (more retractions are countenanced) in exchange for weaker methodological presuppositions. Although it does not show up in the statement of the proposition, another such trade-off concerns time to convergence, for it will typically take longer for the single method constructed from the chain to converge than it would have taken M_1 to converge when its narrower presupposition is satisfied. Whether this trade is rational will depend upon the plausibility of the presuppositions and on the costs of retractions and delayed convergence. This is where history, individual psychology, and external cirumstances figure in. Logic presents the possible options and the systematic trade-offs among them.

The result just presented assumes that each meta-method in the chain succeeds with bounded retractions. What could we do with a finite, meta-methodological sequence of limiting decision procedures for which no such bounds exist? Assuming that the methods in the chain are all guaranteed to converge to acceptance or to rejection, we could turn the whole sequence into a single method that also decides the original hypothesis in the limit. One may describe the situation by saying that limiting decidability is preserved or closed under finite meta-methodological regresses.

Proposition 3   *The following situations are methodologically equivalent:*

1. *There exists a finite chain $M_1, \ldots, M_k$ of methods such that*
   (a) *for each $i < k$, method M_{i+1} aspires to decide P_i in the limit given K and does so under presupposition P_{i+1} and*
   (b) *K entails P_k, the presupposition of the final method in the chain.*
2. *There exists a single method M that decides P_0 in the limit given K.*

Before, we saw that bounded retraction meta-methodology trades retractions for weaker presuppositions and delayed convergence. The same is true here, except that the increase cannot be measured by a uniform bound as in the bounded retraction case (Proposition 2).

10. EMPIRICAL NATURALISM WITHOUT FOUNDATIONS

In the finite meta-methodological chains considered in the preceding section, the method at the end of the chain serves as an anchor or foundation for the entire chain, since it is required to succeed in every serious possibility. This is reminiscent of Laudan's idea of picking a single method to anchor the process of empirically investigating what other empirical methods can do. But what if there is no foundation for the chain? What if every method in the chain has empirical presuppositions and more methods can always be added, on demand, to assess them? Then there is nothing to science but assessments of assessements of assessments, without end. That is not to say that scientists ever use infinitely many meta-methods all at once. The relevant infinity is potential rather than actual: in the face of yet another challenge to her reliability, the scientist is disposed to respond with yet another meta-method to test the presuppositions of the method challenged.

The instrumental question, once again, is what one could *do* with a potentially infinite chain of methods, each of which investigates the presupposition of its predecessor. Suppose, then, that there is a (potentially) infinite sequence of meta-methods, each of which, say, refutes the presupposition of its predecessor with certainty under some presupposition. Moreover, suppose that none of the methods in the chain works without empirical presuppositions. Inquiry floats on an infinite abyss of presuppositions.

It is natural to assume that although no method works without presuppositions, the presuppositions tend to get weaker, so that for each $i \geq 1$, P_i entails P_{i+1}. Call such a sequence *increasingly reliable*. In other words, even though there is no 'foundational' method, the infinite sequence is nonetheless *directed* in the sense that each successive meta-method is at least as reliable as the method it assesses.

When is an infinite, foundationless, increasingly reliable chain of aspiring refutation meta-methods possible? Whenever H is refutable with certainty by a single, 'super-method' that succeeds over the disjunction of all the nested presuppositions. Thus, we can say that refutation with certainty is closed under infinite, increasingly reliable regresses.

Proposition 4   *The following situations are methodologically equivalent:*

1. *There exists an infinite chain $M_1, \ldots, M_k, \ldots$ of methods such that*
    (a) *for each $i \geq 0$, method M_{i+1} aspires to refute P_i with certainty given K and does so under presupposition P_{i+1},*
    (b) *for each $i \geq 0$, P_i entails P_{i+1}, and*
    (c) *K entails $(P_1 \vee \cdots \vee P_n \vee \cdots)$.*
2. *H is refutable with certainty given K.*

A result of this kind is double-edged. On the one hand, an infinite, meta-methodological chain of refuting methods is not pointless, since it is equivalent to a single refutation method that has weaker presuppositions than any method in the chain. But in another sense it may seem pointless, since we could have used the equivalent, single method to begin with! There is, however, increasing interest in the philosophy of science these days in 'local' or 'piecemeal' methodology (e.g., Mayo 1996). The preceding result says that adding more and more 'local' refuting methods when challenged can add up to performance methodologically equivalent to having a single refuting method, without committing one's self to a single method handling all possible contingencies from the outset. Since scientists do not really commit themselves to fixed methods for eternity, the applicability of learning theoretic analysis is thereby greatly enhanced.

What could we do with an infinite, nested, sequence of verification methods? One might well expect a similar closure result, to the effect that an infinite, increasingly reliable chain of verifiers adds up to a verifier. But this is far from being the case, for an infinite, meta-methodological chain of certain verifiers is equivalent to an infinite chain of *limiting refutation* methods! It also turns out that an infinite chain of limiting refutation methods is equivalent to a single limiting refutation method, so limiting refutation, like certain refutation, is closed under infinite, increasingly reliable meta-methodological chains. Since the power of limiting decision lies between that of certain verification and of limiting refutation, it should come as little surprise that infinite chains of limiting deciders are also equivalent to having a single limiting refuter. Thus, limiting refutability is a fairly robust necessary condition for the existence of increasingly reliable meta-methodological regresses. These results cannot be improved to equivalence with an infinite, increasingly reliable chain of limiting verifiers, since even a single limiting verifier can succeed on hypotheses that are not refutable in the limit (LRI).

Proposition 5   *The following situations are methodologically equivalent:*

1. *There exists an infinite chain $M_1, \ldots, M_k, \ldots$ of methods such that*
    (a) *for each $i \geq 0$, method M_{i+1} aspires to verify P_i with certainty given K and does so under presupposition P_{i+1},*
    (b) *for each $i \geq 0$, P_i entails P_{i+1}, and*
    (c) *K entails $(P_1 \vee \cdots \vee P_n \vee \cdots)$.*
2. *Situation 1, with decision in the limit replacing refutation in the limit.*
3. *Situation 1, with certain verification replacing refutation in the limit.*
4. *P_0 is refutable in the limit given K.*

Corollary   *In conditions (1–3), P_i can be chosen to be of form $H \vee R_i$, where R_i is refutable with certainty given K, so all of the M_i can be chosen to converge to the truth given that H is true.*

The preceding result assumes that $P_0 = H$ entails P_1, so that M_1 converges to acceptance when H is true. If this condition is dropped, (1) and (2) become equivalent both to the existence of a limiting refuter of a certain verifier and to the existence of a certain verifier of a limiting refutation procedure. These 'mixed' chains do not collapse into anything more elementary, and may be viewed as criteria of success in their own right.[16]

Infinite, nested chains of certain refuters are equivalent to certain refutability and infinite, nested chains of certain verifiers are equivalent to limiting refutation. Is there some kind of infinite, meta-methodological chain that characterizes limiting decision? Here is one example of such a constraint. Say that M *converges as fast as M'* given K just in case for each data stream e satisfying K, for each stage k of inquiry, if M' has converged by k, so has M.

Proposition 6   *The following situations are methodologically equivalent:*

1. *There exists an infinite chain $M_1, \ldots, M_k, \ldots$ of methods such that*
   (a) *for each $i \geq 0$, method M_{i+1} aspires to decide P_i in the limit and does so under presuppositon P_{i+1},*
   (b) *For each $i \geq 1$, P_i entails P_{i+1},*
   (c) *K entails $(P_2 \vee \cdots \vee P_n \cdots)$, and*
   (d) *for each $i \geq 1$, M_{i+1} converges as fast as M_i given K.*
2. *H is decidable in the limit given K.*

This result may be understood, intuitively, as follows. Increased reliability requires a method to cope with more possibilities, which delays convergence. So requiring the successive meta-methods in the sequence to have non-decreasing reliability and *also* non-increasing convergence time implies that reliability eventually stops increasing after some point in the sequence. The tail of the sequence provides no further essential information after that point leaving us with what is essentially a finite sequence of limiting decision methods (Proposition 3).

11. EMPIRICAL NATURALISM WITHOUT FOUNDATIONS OR DIRECTION

The preceding section provided an analysis of unfounded epistemic regresses. But it still assumed that the infinite, empirical regress is at least *directed*, in the sense that the meta-methods are increasingly reliable. What if we drop that assumption as well? Is foundationless, directionless meta-methodology necessarily pointless, in the sense that one could not turn the conjectures of the methods into a recognizable notion of convergence to the truth?

If no further conditions are added, then the answer is affirmative, since every hypothesis whatsoever possesses such a chain; so such a chain cannot be equivalent to methods succeeding in any of the convergent senses defined above.

Proposition 7   *Every H has an infinite chain $M_1, \ldots, M_k, \ldots$ of methods like the one described in Proposition 4, except that the nesting condition is dropped.*

The argument is simple: every method succeeds over some set (possibly empty) of relevant possibilities. So every infinite sequence of methods starting with acceptance and using at most one retraction witnesses the preceding proposition. Such chains are, therefore, extreme examples of vicious or pointless empirical regresses so far as reliable convergence to the right answer is concerned.

Are there further conditions we can impose on the methods in the undirected chain in order to end up with a condition equivalent to limiting decidability? Consider the following two properties. (1) A meta-method is *positively* [*negatively*] *reliable* given K just in case it never converges to acceptance [rejection] incorrectly on any data stream satisfying K. Let A_{i+1} denote the proposition that M_{i+1} eventually stabilizes to acceptance and let R_{i+1} denote the proposition that M_{i+1} eventually stabilizes to rejection. Then positive [negative] reliability requires that A_{i+1} [R_{i+1}] entail $P_i[\neg P_i]$. A meta-methodological chain is positively [negatively] reliable just in case each meta-method occurring in it is. (2) Another, possible property of infinite meta-methodological chains is *positive* [*negative*] *covering*. A chain positively [negatively] covers K just in case K entails $(A_2 \vee A_n \vee \cdots) [(R_2 \vee R_n \vee \cdots)]$.

The positive [negative] covering and reliability conditions do not imply objective directedness in the sense that methods get more reliable farther out in the chain. Convergence to acceptance [rejection] implies truth, and every data stream is accepted [rejected] in the limit by some meta-method in the sequence, but that implies neither that later meta-methods accept more than earlier ones nor that later methods commit fewer errors than earlier ones. Indeed, a later method might reject every data stream. So these properties steer clear both of foundations and of directedness (i.e., increasing reliability).

Nonetheless, such a non-directed, foundationless chain exists precisely when the hypothesis is decidable in the limit by a single method! And the same is true even if the methods in the chain merely decide in the limit rather than refuting with certainty. So under suitable conditions, even unfounded, undirected meta-methodology can have an appealing point.[17]

Proposition 8   *The following situations are methodologically equivalent:*

1. *There exists an infinite chain $M_1, \ldots, M_k, \ldots$ of methods such that*
    (a) *for each $i \geq 0$, method M_{i+1} aspires to refute [verify] P_i with certainty given K and does so under presuppositon P_{i+1},*
    (b) *the chain is positively [negatively] reliable, and*
    (c) *the chain positively [negatively] covers K.*
2. *There exists an infinite chain $M_1, \ldots, M_k, \ldots$ of methods such that*
    (a) *for each $i \geq 0$, method M_{i+1} aspires to decide P_i in the limit given K and does so under the presuppositon P_{i+1},*
    (b) *the chain is positively [negatively] reliable, and*
    (c) *the chain positively [negatively] covers K.*
3. *P_0 is decidable in the limit given K.*

12. REFUTATIONS OF REFUTATIONS AND THE LOGIC OF DISCOVERY

The preceding, meta-methodological characterizations of limiting decidability relate in an interesting way to the problem of discovery. Recall that a 'discovery method' outputs propositions in response to new data and that an empirical question specifies a range of possible answers. Say that a method *answers* such a question *in the limit* just in case after some finite time it always produces a true conjecture that entails a correct possible answer to the question.

When the potential answers partition the relevant possibilities, it is well known that the question is answerable in the limit if and only if each potential answer is decidable in the limit (*LRI*, chapter 9). Combining this fact with the preceding results yields

Proposition 9 *A question is answerable in the limit given K just in case each potential answer has a meta-methodological sequence S satisfying one of the following conditions:*

1. *S is a finite sequence of limiting decision procedures, the last of which has a presupposition covering K.*
2. *S is an infinite, nested sequence of limiting decision procedures whose presuppositions cover K, such that each later method is guaranteed to converge at least as fast as any preceding method, given K.*
3. *S is an infinite sequence of positively [negatively] reliable refuting [verifying] methods whose acceptance [rejection] sets jointly cover K.*
4. *S is an infinite sequence of positively [negatively] reliable limiting decision procedures whose acceptance [rejection] sets jointly cover K.*

Thus, even foundationless, undirected meta-methodology suffices for (and indeed is equivalent to) the existence of a method that answers the question in the limit. This result provides some logical vindication of Popper's otherwise perplexing faith that refutations of refutations of refutations without end ultimately add up to convergence to the truth in the limit.

13. COMPUTABLE METHODOLOGY

The constructions occurring in the proofs of the above results are all computable (cf. the Appendix). So if the infinite meta-methodological sequences they operate upon are effectively presented, the result of composing the construction with the effectively presented meta-methodological sequence is a single, computable method that succeeds in the required sense. More precisely, say that meta-methodological sequence $M_1, \ldots, M_n \ldots$ is computable just in case there exists a computable function C such that for each i and for each finite data sequence ϵ,

$$C(i, \epsilon) = M_i(\epsilon).$$

Thus, all of the above propositions continue to hold when the meta-methodological sequences and the methods equivalent to them are required to be computable.

This situation is not untypical. Since both computability theory and learning theory study similar criteria of problem solution, it often happens that results holding for 'ideal' or 'logically omniscient' methods can easily be transformed into closely analogous results concerning computable inquiry.

The same cannot be said for the alternative tradition in methodology, which models scientific method as a generalized entailment relation reflecting 'confirmation' or 'empirical support'. According to that view, methodological norms are not based on computationally achievable aims, but on the maintenance of logical relations that computational agents cannot maintain, so that computational strategies are judged normatively deficient.

This tendency to model methodological norms using uncomputable logical relations is one of the most persistent features of the positivistic legacy. The *sine qua non* of logical positivism was a sharp distinction between questions depending on matters of fact and on mere relations of ideas. In methodology, this translates into a sharp distinction between empirical methods, which face inductive skeptical arguments, and formal methods, which do not. It has therefore seemed acceptable to deal with the problem of induction in its own right, reserving formal considerations like computability as an afterthought.

But once again, learning theory invites a very different viewpoint, despite its strongly logical character. If Hume had an excuse for thinking that all formal problems should be decidable *a priori* (since relations of ideas fall under the gaze of the mind's eye), we, as heirs to Gödel's legacy, do not. The message of Gödel's logical revolution is that from the viewpoint of a computational agent, formal problems are for all intents and purposes empirical, since a computer can no more see to the end of its computational process than a scientist can gaze at the indefinite future of her discipline. Indeed, when the formal problem appears to the computational agent to pose the problem of induction, the result is uncomputability! The easiest example of an undecidable formal problem is the *halting problem*, which requires one to determine whether a given Turing machine halts on a given input. The epistemic dimension of the problem is obvious: no matter how long the program has refused to return an output, it may nonetheless do so at the very next moment. Although the unsolvability of the halting problem is not usually proved by means of this skeptical argument, it can be (cf. *LRI* and Kelly and Schulte 1995a), and such a proof does much to clarify the structural analogy between the problem of induction and uncomputability.

Moreover, one may entertain limiting notions of success when a formal problem is not computably decidable. For example, the halting problem is computably verifiable with certainty and its complement is computably refutable with certainty. Similarly, non-halting is refutable with certainty even though it is not verifiable with certainty. The limiting concepts of success are represented as well. For example, determining whether a given Turing program computes a total function is computably refutable in the limit but is not computably verifiable in the limit. Moreover, solvability in each of these attenuated senses can be characterized in terms of alternating quantifiers, in just the manner indicated above for empirical problems.[18]

Since learning theory's treatment of induction is parallel to the approach to formal problems in the theory of computability, it should come as little surprise that learning theory leads to a precise account of the power of computable inquiry. On this approach, computable inquiry may be viewed as posing a twofold problem of induction, an external one, reflecting the degree of interleaving of the data streams for and against the hypothesis through time, and an internal one, corresponding to the interleaving of epistemically possible future trajectories of one's own internal computations (i.e., to uncomputability). Accordingly, the respective characterizations of ideal learnability and of computability in terms of quantifier alternations can be neatly assembled into a characterization of computable learnability (*LRI*, chapter 7). Methodological approaches based on Bayesian updating, on the other hand, assume an ideal account of probabilistic coherence that is uncomputable over sufficiently rich formal languages. It is impossible to integrate computability considerations into such an account without compromising the required sense of coherence.

I claimed, above, that learning theory's uniformly instrumental perspective on formal and empirical methodology yields logical arguments for Feyerabendian conclusions. One such critique concerns the proposal, shared by Bayesians and Popperians alike, that a hypothesis should be rejected when it becomes inconsistent with the data. There are familiar historicist objections to this principle based on Duhem's thesis that no hypothesis is ever really refuted. But let us suppose for the sake of argument that the data are perfectly reliable and that the hypothesis really is ideally refutable with certainty: if it is false, the data will eventually say so. Would not the consistency principle be rationally mandated in this case? After all, hanging onto a refuted hypothesis delays convergence to the truth, so a method obeying the norm would weakly dominate in convergence time a method that did not.

But what if the consistency problem is uncomputable, so that it is impossible for a computer to verify whether the current data are consistent with it? One might suppose that we are rescued from such cases by feasibility: we cannot be required to do the impossible. But escape is not that easy. The rule requires that we never hold onto a hypothesis that is refuted, not that we decide consistency. Even if the full consistency problem is effectively unverifiable, we can still effectively satisfy the rule by erring on the side of caution and rejecting unrefuted hypotheses. But another goal is finding the truth. The question is whether there are problems in which the two goals clash for computable methods in the following sense: either can be computably achieved by itself, but no computable method achieves both. In that case, the consistency condition would no longer accelerate convergence to the truth: it would *prevent* convergence to the truth. So insisting on the rule would have effects entirely contrary to its intended function.

The answer is resoundingly affirmative: one can construct an empirical hypothesis that a computer can refute with certainty, but such that no method that always maintains consistency with the data succeeds even gradually; even if the method is, in a precise sense (i.e., hyperarithmetically definable) infinitely more powerful than a computable method (*LRI*, chapter 7, Kelly and Schulte 1995b).

An interesting corollary of this result is that any such method whose (hyperarithmetically definable) subroutine for detecting logical consistency is insulated from the empirical data as a separate 'subroutine' fails to achieve anything close to what a mere Turing machine can do, namely, refute the hypothesis with certainty. So the very idea that theorem proving can be functionally isolated from empirical information radically restricts the potential power of computable science! This last vestige of the traditional, methodological dichotomy between matters of fact and relations is mistaken.

This is the kind of result I had in mind when I distinguished methodological discoveries from reflections on existing practice. The proof is based on a construction involving mathematical concepts that historical scientists had no idea about, since the theory of computability had not been invented yet. Combing through the history of science will never yield such an insight (unless scientists do the logical work themselves so that historians can read about it in the historical record). Whether such logical possibilities will be realized in future scientific inquiry is hard to say. But here I agree entirely with Laudan: it is the objective, means–ends relations that matter. Whether or not the relationship has arisen in practice has to do with evidence for the relation (which in this case is established *a priori*) rather than with its normative force.

14. CONCLUSION

This paper provides some idea of the similarities and differences between two divergent images of normative naturalism. The first emphasizes structural analysis of the conditions under which a method *would* succeed, whereas the second focuses on historical surveys of apparently successful applications of a method in the past. I explained how the *a priori* approach provides a compelling role for logic in post-positivistic, naturalized methodology that embraces, rather than resists, much of the historicist critique of positivism and that avoids the inherent conservatism of historical surveys. I also presented a new, logical framework in which to distinguish useful empirical regresses from 'vicious' ones. An important feature of this analysis is that useful empirical regresses can be unfounded and undirected, in the sense that later methods may fail to be more reliable than earlier ones. Finally, I illustrated how logic can be used to provide computational critiques of methodological principles whose instrumentality is obvious when computability is ignored.

Logic is not all there is to normative naturalism. But neither is history. As history, itself, suggests, scientific success demands detailed attention to the mathematical implications of the structures under investigation. Learning theory is just normative naturalism informed by this lesson.

ACKNOWLEDGEMENT

Oliver Schulte's careful and detailed corrections led to major improvements in a late draft of the paper. Clark Glymour also provided prompt comments on an early draft.

Carnegie Mellon University

APPENDIX

Proof of Proposition 2 $(1) \Rightarrow (2)$: Suppose we are given a finite meta-methodological sequence $((M_1, P_1), \ldots, (M_k, P_k))$ such that each M_i is an aspiring n_i retraction decision procedure given K, M_1 decides H with n_1 retractions starting with c_1 under presupposition P_1 and for each i from 1 to $k-1$, M_{i+1} decides P_i with n_{i+1} retractions starting with c_{i+1} under presupposition P_{i+1}. Moreover, let P_k include all serious possibilities in K.

We must construct a single method M that decides H with $n_1 + \cdots + n_k$ retractions over K starting with c, where $c = $ 'accept' if an even number of the c_i are 'reject', and $c = $ 'reject' otherwise.

The construction of M is as follows. M simulates all the methods in the sequence on the finite sequence of data input so far. Then M calculates its current conjecture by setting $b := $ the number of methods among $M_1, \ldots, M_k$ that currently reject. If b is even, M accepts, and M rejects otherwise.

Let data stream e satisfying K be given. Since each method in the sequence uses at most a finite number of retractions and there are only finitely many such methods, there is a stage m after which each method M_i has stabilized to its ultimate conjecture u_i. We may now reason by 'backward induction' as follows. Since M_k's presupposition is trivially satisfied, u_k is correct. So if u_k is rejection, and M_{k-1} is an aspiring n_{k-1} retraction decision procedure, M_{k-1} converges to the wrong answer, so we may reverse u_{k-1} and we agree with u_{k-1} otherwise. Call this corrected conjecture v_{k-1}. Now v_{k-1} is correct, so if it is reject we reverse u_{k-2} and otherwise agree with u_{k-2} to obtain the corrected conjecture v_{k-2}. Proceeding in this way, we ultimately obtain the corrected v_1, which correctly indicates whether e satisfies H. Since one reversal occurs for each 'rejection' occurring in $(u_1, \ldots, u_k)$ and since two reversals cancel, the correct conjecture u_1 is 'accept' just in case an even number of 'reject' conjectures occur in $(u_1, \ldots, u_k)$. So M converges to the correct answer.

Observe that M retracts only when the number of rejecting methods in the sequence changes from even to odd. In the worst case, each method in the sequence retracts at a different time (if two retract simultaneously, M does not retract). So at worst, M retracts $n_1 + \cdots + n_k$ times.

M starts with the leading conjecture c_1 of M_1 if there are an even number of retractions among the initial conjectures $c_2, \ldots, c_n$ of the meta-methods and starts out with the reversal of c_1 otherwise. But by the backward induction argument, this is the correct conjecture about H if all of the conjectures $c_1, \ldots, c_k$ are correct for data stream e.

$(2) \Rightarrow (1)$: Induction on the binary examples provided in proposition 1. To see how to generalize the construction, consult the proof of proposition 8.

Proof of Proposition 3 $(1) \Rightarrow (2)$: Given the chain, the backwards induction construction used to prove proposition 2 works here as well.

$(2) \Rightarrow (1)$: In the other direction, suppose M decides H in the limit given K. Then extend M with k meta-methods, all of which accept no matter what.

Proof of Proposition 4 $(1) \Rightarrow (2)$: Suppose we are given an infinite sequence $((M_1, P_1), \ldots, (M_i, P_i), \ldots)$ of methods, such that each M_i is an aspiring refuter given K and for each $i \geq 0$, M_{i+1} refutes P_i with certainty under presupposition P_{i+1} (where $H = P_0$). Also, suppose that for each $i \geq 0$, P_i entails P_{i+1} and that the P_i cover K. We must construct a single M that refutes H with certainty given K. The construction is as follows. M starts out accepting and rejects if any method M_i in the chain ever rejects. Evidently, M aspires to refute given K. So it suffices to show that M converges to the right answer given K.

Let e satisfy K. Since the presuppositions cover K, let k be least such that e satisfies P_k. Case A: $k = 0$. Then all the P_i are satisfied so for each $i \geq 1$, M_i never rejects. Thus, M always accepts, which is correct because $P_0 = H$ is satisfied. Case B: Suppose $k > 0$. Then for each $i < k$, P_i is false and P_k is true. Hence, for each $i < k$, M_i converges to the wrong answer and, hence, converges to acceptance. M_k converges to the right answer about P_{k-1}, and hence rejects. Thus, M converges correctly to rejection.

$(2) \Rightarrow (1)$: Let M refute H with certainty given K. Let $M = M_1$ and for each $i > 1$, let M_i accept no matter what.

Proof of Proposition 5 $(3) \Rightarrow (2) \Rightarrow (1)$: Immediate, since a certain verifier is a limiting decider which is, in turn, a limiting refuter.

$(1) \Rightarrow (4)$: Let P_0 denote H. Suppose we are given an infinite sequence $((M_1, P_1), \ldots, (M_i, P_i), \ldots)$ of meta-methods, such that for each $i \geq 1$, method M_i is an aspiring limiting refuter given K, and for each $i \geq 0$, M_{i+1} refutes P_i in the limit under P_{i+1}. Also, suppose that for each $i \geq 0$, P_i entails P_{i+1} and that the P_i cover K. We must construct a single M that refutes P_0 in the limit given K.

The constructed method M works as follows. Let $f(0), f(1), \ldots, f(k), \ldots$ be an infinitely repetitive enumeration of the natural numbers. Initialize counter $p := 0$. On the first k data points, suppose that $p := i$. Feed the first k data points to $M_{f(i)+1}$ and see if $M_{f(i)+1}$ accepts. If so, increment $p := i + 1$ and accept, and otherwise leave p set to $p := i$ and reject.

Let e satisfy K. Then by assumption, for some $i \geq 1$, P_i is satisfied by e. Let k be the least such i. Case A: Suppose that $k = 0$, so each P_i including $P_0 = H$ is satisfied. Then each M_{i+1} accepts infinitely often along e, since M_{i+1} verifies P_i in the limit when its presupposition P_{i+1} is satisfied. Hence, the counter p is incremented infinitely often, so M accepts infinitely often, as required. Case B: Suppose that $k > 0$. Then by nesting and choice of k, $H = P_0, \ldots, P_{k-1}$ are not satisfied but P_k is. Hence, M_k converges to rejection along e, say by the time n data points have been read. Since $f(0), \ldots, f(n), \ldots$ is infinitely repetitive, there is an $m \geq n$ such that $f(m) + 1 = k$. Thus, p is never incremented past m, so M converges correctly to rejection.

$(4) \Rightarrow (1)$: Suppose that M refutes H in the limit given K. We need to construct an infinite sequence $((M_1, P_1), \ldots, (M_i, P_i), \ldots)$ of meta-methods, such that for each $i \geq 0$, meta-method M_{i+1} aspires to verify P_i with certainty given K and does so under presupposition P_{i+1}. We must also show that for each $i \geq 0$, P_i entails P_{i+1} and that the P_i cover K.

The construction is as follows. For each $i \geq 0$, let R_i denote the proposition that M rejects from stage i onward. In other words,

$$R_i = \{e \in K: \ \forall m \geq i, \ M(e(0), \ldots, e(m)) \text{ rejects}\}.$$

Then define

$$P_0 = H,$$
$$P_{i+1} = H \vee R_i.$$

For each finite data sequence $(x_0, \ldots, x_k)$, define

$$M_1(x_0, \ldots, x_k) = \begin{cases} \text{accept} & \text{if there is a } j \leq k \text{ such that } M(x_0, \ldots, x_j) \text{ accepts,} \\ \text{reject} & \text{otherwise,} \end{cases}$$

and for each $i > 1$, define:

$$M_{i+1}(x_0, \ldots, x_k) = \begin{cases} \text{accept} & \text{if } k \geq i - 1 \text{ and } M(x_0, \ldots, x_{i-1}) \text{ rejects or there} \\ & \text{is a } j \text{ such that} \\ & \qquad i \leq j \leq k \text{ and } M(x_0, \ldots, x_j) \text{ accepts,} \\ \text{reject} & \text{otherwise.} \end{cases}$$

I now verify that the construction works. It is immediate that P_i entails P_{i+1} and that each M_i aspires to verify with certainty. To see that the P_i cover K, observe that P_0 covers H and since M refutes H in the limit given K, M converges to rejection by some finite stage so $K - H$ is also covered by the P_i. For the corollary, observe that each P_i, for $i > 0$, is the disjunction of H with a proposition R_i that is refutable with certainty.

It remains only to check that for each $i \geq 0$, M_{i+1} verifies P_i with certainty under presupposition P_{i+1}. Since we already know that each M_i aspires to verify given K, it suffices to show that for each $i \geq 0$, M_{i+1} converges to the right answer about P_i on data stream $e \in K$ just in case e satisfies P_{i+1}.

By nesting of the P_i, let n be least such that $P_n, P_{n+1}, \ldots$ are all satisfied.

Case A: $n = 0$. Then M accepts infinitely often along e. Hence, M_1 converges correctly to acceptance when M accepts for the first time. Moreover, for $i > 1$, M_i converges correctly to acceptance when for the first $k \geq i$, $M(e(0), \ldots, e(k)) = $ 'accept', in virtue of the second condition for acceptance in the definition of M_i. But this is just what is required, since each P_i is satisfied by e.

Case B: $n > 0$. Then $M(e(0), \ldots, e(n-1)) = $ 'accept', so M_1 converges incorrectly to acceptance. Suppose $1 < i < n$. The by the second condition for acceptance in the definition of M_i, M_i converges incorrectly to acceptance. Suppose $i = n > 0$. Then neither condition for acceptance in the definition of M_i is satisfied, so M_i converges correctly to rejection. Suppose $i > n$. Then $M(e(0), \ldots, e(i-1)) = $ 'reject', so by the first condition for acceptance in the definition of M_i, M_i converges correctly to acceptance. This is just as it should be, since P_n, P_{n+1}, are all satisfied by e whereas $P_0, \ldots, P_{n-1}$ are not.

Proof of Proposition 6 $(1) \Rightarrow (2)$: Suppose that $((M_1, P_1), \ldots, (M_i, P_i), \ldots)$ is an infinite methodological chain of aspiring limiting deciders given K such that for each $i \geq 0$, M_{i+1} decides P_i in the limit under presuppositon P_{i+1}, P_i entails P_{i+1}, $K = (P_2 \vee \cdots \vee P_n \vee \cdots)$, and M_{i+2} converges as fast as M_{i+1} given K. We need to construct an M that decides H in the limit given K.

M simulates $M_1, \ldots, M_n$ on the first n data points to obtain conjecture sequence $c_1, \ldots, c_n$. Next M sets $k := $ the least k' such that the c_i all agree from $c_{k'}$

onward. M agrees with c_1 if 'reject' occurs an even numbers of times in $c_1, \ldots, c_k$, and disagrees with c_1 otherwise.

Let e satisfy K. Since $K = (P_2 \vee \cdots \vee P_n \vee \cdots)$, we may choose k to be least such that e satisfies P_{k+1}. Thus, there is a stage i after which M_{k+1} is correct about P_k. Since later methods converge as fast as and are as reliable as M_{k+1}, they correctly accept from stage i onward. The finitely many methods $M_1, \ldots, M_k$ eventually all converge (possibly incorrectly), by some later stage i', since they aspire to decide in the limit given K. After the stages k and i' are passed, backward induction (cf. the proof of Proposition 2) shows that M is correct.

$(2) \Rightarrow (3)$: In the other direction, let M decide H in the limit given K. Let $M_2, \ldots, M_i, \ldots$ all accept no matter what, without looking at the data. This trivial meta-methodological sequence satisfies all of the required properties.

Proof of Proposition 8 $(1) \Rightarrow (2)$ is immediate. $(2) \Rightarrow (3)$: Suppose we are given an infinite sequence $((M_1, P_1), \ldots, (M_i, P_i), \ldots)$ of aspiring limiting decision procedures given K, such that for each $i \geq 0$, M_{i+1} decides P_i in the limit under presupposition P_{i+1}. Also, suppose that the sequence is positively reliable and positively covers K. We must construct a single method M that decides H in the limit given K.

The constructed method M works as follows. Let $f(0), f(1), \ldots, f(k), \ldots$ be an infinitely repetitive enumeration of the natural numbers. Initialize counter $p := 0$. After seeing the first k data points, suppose that $p = i$. Feed the first k data points to $M_{f(i)+1}$ and see if $M_{f(i)+1}$ rejects. If so, increment $p := i + 1$, and otherwise leave p set to $p = i$. Then return the current output of M_1 if an even number of methods among $M_2, \ldots, M_{f(p)+1}$ reject and return the result of reversing the current output of M_1 otherwise.

Let e satisfy K. Then by the positive covering condition, for some $i \geq 0$, P_{i+1} is satisfied by e. Thus, M_{i+1} converges to acceptance, say at stage j. Then since f is infinitely repetitive, there is some stage $j' \geq j$ after which p is no longer incremented. Let m be the terminal value of p. Thus, M_{m+1} has converged to acceptance by stage j'. By positive reliability, e satisfies P_{m+1}. Since all of the methods are aspiring limiting decision procedures, there is some possibly later stage by which all methods prior to $M_{f(p)+1}$ have converged. Thereafter, by the backward induction argument of Proposition 2, M conjectures correctly about H.

$(3) \Rightarrow (1)$: In the other direction, suppose we are given an arbitrary method M that decides H in the limit given K. We must construct an infinite, positively reliable sequence $((M_1, P_1), \ldots, (M_i, P_i), \ldots)$ of refuting meta-methods that positively covers K.

Without loss of generality, we can assume that M starts out accepting prior to seeing any data (given a limiting decider M', the result M of forcing M' to accept prior to seeing any data is still a limiting decider). Let O be the proposition that M retracts an odd number of times prior to convergence and let E be the proposition that M retracts an even number of times. Let R_i be the proposition that M retracts at most i times. Let $H = P_0$. Now, for each $i \geq 1$, define

$$P_i = \begin{cases} (R_{i-1} \vee O) & \text{if } i \text{ is odd,} \\ (R_{i-1} \vee E) & \text{if } i \text{ is even,} \end{cases}$$

and define, for each finite data sequence ϵ,

$$M_i(\epsilon) = \begin{cases} \text{reject} & \text{if } M(\epsilon) \text{ uses at least } i \text{ retractions,} \\ \text{accept} & \text{otherwise.} \end{cases}$$

By construction, each M_i is an aspiring refuter given K.

Next, we must establish the positive covering condition. Since M decides H in the limit given K, M uses only finitely many retractions on each data stream e satisfing K. Suppose M uses j retractions on e. So e satisfies R_j and hence e satisfies P_{j+1}. Thus, the P_i cover K.

To establish the positive reliability condition, suppose for arbitrary $i \geq 1$ that M_i never rejects along e. We must show that e satisfies P_{i-1}.

Case A: suppose $i = 1$. Then M never retracts along e. Since M starts out accepting, M always accepts. So since M decides $H = P_0$ in the limit, e satisfies P_0.

Case B: suppose $i > 1$. Then M retracts at most $i - 1$ times so e satisfies R_{i-1}. If i is even, then R_{i-1} entails $(R_{i-2} \vee O)$, which is just P_{i-1}. Similarly, if i is odd, then R_{i-1} entails $(R_{i-2} \vee E)$, which is just P_{i-1}. So in either case, e satisfies P_{i-1}, as required.

It remains only to check that for each $i \geq 1$, M_i refutes P_{i-1} with certainty under presupposition P_i.

When $i = 1$, we must show that M_1 refutes $P_0 = H$ under presupposition $P_1 = (R_0 \vee O)$. Let $e \in K$.

Case A: suppose e satisfies $P_1 = (R_0 \vee O)$.

Case A.1: suppose e satisfies R_0. Then M never retracts and hence converges to acceptance. Since M is correct, $H = P_0$ is satisfied by e. Since M never retracts, M_1 converges correctly to acceptance, as required.

Case A.2: suppose e satisfies O. So M retracts an odd number of times starting with acceptance, and hence M converges to rejection. Since M converges to the right answer, e does not satisfy $H = P_0$. But M_1 converges correctly to rejection after the first retraction is observed, as required.

Case B: suppose e does not satisfy $P_1 = (R_0 \vee O)$. Then M uses some even number of retractions greater than zero, starting with acceptance. Thus, M accepts H in the limit, and since M converges to the right answer, e satisfies $H = P_0$. But since M retracts at least once, M_1 converges incorrectly to rejection, as required.

Now consider the case in which $i > 1$.

Case I: suppose i is odd. Then $P_i = (R_{i-1} \vee O)$.

Case I.A: suppose e satisfies $P_i = (R_{i-1} \vee O)$.

Case I.A.1: suppose e satisfies $P_{i-1} = (R_{i-2} \vee E)$. If e satisfies O, then e satisfies R_{i-2}, so M_i correctly converges to acceptance, as required. If e satisfies E, then e satisfies R_{i-1}, so M_i converges correctly to acceptance, as required.

Case I.A.2: suppose e does not satisfy $P_{i-1} = (R_{i-2} \vee E)$. So e satisfies O but not R_{i-2}. Since i is odd, e does not satisfy R_{i-1} either, else M retracts exactly $i-1$ times, which is an even number of times. Hence, M retracts at least i times, so M_i converges correctly to rejection, as required.

Case I.B: suppose e does not satisfy $P_i = (R_{i-1} \vee O)$. Then e satisfies E and hence satisfies $P_{i-1} = (R_{i-2} \vee E)$. Also, e does not satisfy R_{i-1}, so M retracts at least i times, and hence M_i converges incorrectly to rejection, as required.

Under case II, in which i is even, the same argument works, if one switches O with E everywhere.

NOTES

[1] The basic idea of providing a computational analysis of the problem of finding the truth was independently proposed by Hilary Putnam (1963) and E.M. Gold (1965). For book length presentations, cf. (*LRI*, Osherson *et al.* 1986, Martin and Osherson 1998).

[2] I think Popper would have more plausibly preferred 'if you don't want to end up preserving a false hypothesis for eternity, then you ought to avoid *ad hoc* hypotheses'.

[3] Laudan's position recalls Hans Reichenbach's familiar argument for using the straight rule of induction: if any other method works, the straight rule of induction will eventually lead us to follow that method.

[4] Laudan seems to miss this point. After dismissing 'transcendent' goals like finding the truth as appropriate aims for science, Laudan writes: 'My own proposal ... is that the aim of science is to secure theories with a high problem-solving effectiveness. From this perspective, *science progresses just in case successive theories solve more problems than their predecessors*' (*BPR*, p. 78). In this passage, Laudan plays loosely with modality and tense, both of which are crucial to any discussion of the problem of induction. How many problems a theory actually solved in the past is observable. How many problems it could solve given more ingenuity and time is not. But 'effectiveness' concerns the latter, dispositional concept, not the former, empirical one.

[5] Its focus on diachronic utilities separates learning theoretic analysis from other *a prioristic* approaches to normative naturalism, such as Isaac Levi's methodology of maximizing expected true content (Levi 1983).

[6] For book length expositions, cf. (*LRI*, Osherson *et al.* 1986, Martin and Osherson 1998).

[7] I think of this as the core of truth in Popper's 'deductivism'. The same sentiment is reflected in Reichenbach's 'pragmatic vindication' of the straight rule.

[8] The same is true for any specification of a closed interval of such values.

[9] Thus, a gradual refuter and a gradual verifier cannot always be assembled into a gradual decision procedure (cf. *LRI*, chapter 4). This contrasts with the limiting case.

[10] This method performs at worst one retraction and is data-minimal since whatever it conjectures it possibly converges to, but failing to make a conjecture would fail to be data-minimal and conjecturing any grue(n) prior to the green hypothesis might require two retractions (one from grue(n) to green and another from green to some grue(n')) (Schulte 1998).

[11] Kuhn explicitly rejects uniform viability as a goal, since it would reduce the subject to trivial textbook exercises that could not be published. But as a cognitive, rather than a career goal, it would clearly be more desirable than piecemeal viability. For example, Hubert Dreyfus' (1979) objection to the strong A.I. program is that each bit of human behavior can be duplicated by a computer program, but no single program will ever duplicate all of human behavior due to scaling problems.

[12] Indeed, Miller's (1974) counterexample shows that verisimilitude metrics cannot even be preserved under translation. The moral is that metrical concepts should be avoided in defining progress (Mormann 1988). Learning theoretic success criteria are topological rather than metrical, and hence are not subject to this objection.

[13] It is not easy to pin down contemporary Bayesianism on this issue. Decision theoretic analyses of method, along the lines of Levi (1983), are explicitly instrumental. 'Bayesian confirmation theory', based exclusively on the concept of updating by conditionalization, is instrumental insofar as one takes the diachronic Dutch book argument or the 'almost sure' convergence theorems seriously. But it is increasingly fashionable not to do so. Many advocates of conditioning view limiting convergence as a useless idealization, while others object that diachronic Dutch book arguments are invalid. Without such arguments, conditioning is recommended as an explication of practice.

[14] In practice, of course, the method may be too difficult to analyze using computation theoretic techniques, as in the mundane case of an ordinary word processing program with thousands of features. But the kinds of methodological principles that come up in philosophical discussions are much more amenable to formal analysis than the average word processor program is.

[15] In LRI and (Osherson *et al.* 1986) the presupposition of a method is called its *scope*.

[16] The proof of this generalization is left to the reader. Hypotheses assessable by 'mixed' chains of this sort are Boolean combinations of hypotheses that are verifiable in the limit. The complexities of such hypotheses are characterizable in the finite difference hierarchy over Σ_2^0.

[17] The same result also holds for an infinite meta-methodological sequence of *verifiers* that are *negatively* reliable and whose eternal *rejection* propositions cover K. Simply follow the proof for the refutation case, substituting dual notions where appropriate.

[18] Cf. also (Putnam 1965) and (Hàjek 1978).

REFERENCES

Case, J. and Smith, C. 1983: 'Comparison of Identification Criteria for Machine Inductive Inference', *Theoretical Computer Science* 25, 193–220.

Donovan, A., Laudan, L. and Laudan, R. 1992: *Scrutinizing Science*, Johns Hopkins University Press, Baltimore.

Dreyfus, H. (1979) *What Computers Can't Do*, Harper and Rowe, New York.

Feyerabend, P. (1975) *Against Method*, Verso, London.

Gold, E. M. 1965: 'Language Identification in the Limit', *Information and Control* 10, 447–474.

Hàjek, P. 1978: 'Experimental Logics and Π_3^0 Theories', *Journal of Symbolic Logic* 42, 515–522.

James, W. 1948: 'The Will to Believe', in *Essays in Pragmatism*, A. Castell (ed.), Collier Macmillan, New York.

Kelly, K. 1996: *The Logic of Reliable Inquiry*, Oxford, New York.

Kelly, K. and Glymour, C. 1992: Inductive Inference from Theory-Laden Data, *Journal of Philosophical Logic* 21, 391–444.

Kelly, K. and Schulte, O. 1995a: 'Church's Thesis and Hume's Problem.' *Proceedings of the XIIth Joint International Congress for Logic, Methodology and the Philosophy of Science*. Florence 1995.

Kelly, K. and Schulte, O. 1995b: 'The Computable Testability of Theories Making Uncomputable Predictions', *Erkenntnis* 43, 29–66.

Kelly, K., Schulte, O. and Juhl, C. 1997: 'Learning Theory and the Philosophy of Science', *Philosophy of Science* 64, 245–267.

Kuhn, Thomas 1962: *The Structure of Scientific Revolutions*, University of Chicago Press, Chicago.

Kuratowski, K. 1966: *Topology*, Vol. 1. Academic Press, New York.

Laudan, L. 1984: *Science and Values*, University of California Press, Berkeley.

Laudan, L. 1996: *Beyond Positivism and Relativism*, Westview Press, Boulder CO.

Levi, I. 1983: *The Enterprise of Knowledge*, M.I.T. Press, Cambridge.

Martin, E. and Osherson, D. 1998: *Elements of Scientific Inquiry*, M.I.T. Press, Cambridge.

Mayo, D. 1996: *Error and the Growth of Experimental Knowledge*, University of Chicago Press, Chicago.

Miller, D. 1979 'Popper's Qualitative Theory of Verisimilitude. *British Journal for the Philosophy of Science* 24, 166–177.

Mormann, T. 1988: 'Are All False Theories Equally False? A Remark on David Miller's Problem and Geometric Conventionalism.' *British Journal for the Philosophy of Science* 39, 505–519.

Nola, R. 1999: 'On the Possibility of a Scientific Theory of Scientific Method', *Science and Education* 8, 427–439.

Osherson, D., Stob, M. and Weinstein, S. 1986: *Systems that Learn*. M.I.T. Press, Cambridge.

Popper, K.R. 1968: *The Logic of Scientific Discovery*, Harper, New York.

Putnam, H. 1963: '"Degree of Confirmation" and Inductive Logic', in *The Philosophy of Rudolph Carnap*, A. Schilpp (ed.), Open Court, LaSalle, IL.

Putnam, H. 1965: 'Trial and Error Predicates and a Solution to a Problem of Mostowski', *Journal of Symbolic Logic* 20, 49–57.

Schulte, O. 1998: *Hard Choices in Scientific Inquiry*, Doctoral Thesis, Department of Philosophy,
     Carnegie Mellon University.
Schulte, O. 1999a: 'Means–Ends Epistemology'. *The British Journal for the Philosophy of Science*
     50, 1–31.
Schulte, O. 1999b: 'The Logic of Reliable and Efficient Inquiry'. *The Journal of Philosophical Logic*
     28, 399–438.

HOWARD SANKEY

METHODOLOGICAL PLURALISM, NORMATIVE
NATURALISM AND THE REALIST AIM OF SCIENCE*

1. INTRODUCTION

There are two chief tasks which confront the philosophy of scientific method. The first task is to specify the methodology which serves as the objective ground for scientific theory appraisal and acceptance. The second task is to explain how application of this methodology leads to advance toward the aim(s) of science. In other words, the goal of the theory of method is to provide an integrated explanation of both rational scientific theory choice and scientific progress.[1]

Theorists of scientific method may be broadly divided into two main camps: monists and pluralists.[2] Traditional methodologists tend to fall into the monist camp. They see science as characterised by a single, universally applicable method, invariant throughout the history of science and the various fields of scientific study. The two leading versions of monism are inductivism, which takes scientific theories to be grounded in inductive inference from observed data, and Popperian falsificationism, which treats the method of science as the ruthless attempt to refute conjectural hypotheses which scientists propose to explain observed phenomena.

By contrast, recent methodological pluralists argue, against the idea of a fixed method, in favour of a plurality of methodological rules governing theory evaluation.[3] Such methodological rules may vary from time to time, as well as field to field, within science. New rules may be introduced and old ones discarded. Rules may be modified, as they undergo refinement in the course of scientific practice. They may be applied in different ways in different fields of science, and different scientists may interpret the same rules in different ways. Moreover, as there is always a plurality of rules, different scientists may choose to emphasise different rules in the evaluation of alternative theories. On the resulting pluralist conception of methodology, science is not characterised by a single invariant method, but by a set of evaluative rules to which scientists appeal in the context of theory appraisal.[4]

As for the aim of science, a number of alternative approaches may be distinguished here as well. According to scientific realism, the aim of science is to arrive at true, explanatory theories of both observable and unobservable aspects of the world, and the best explanation of the success of science is that considerable

211

Robert Nola and Howard Sankey (eds.), After Popper, Kuhn and Feyerabend, 211–229.
© 2000 *Kluwer Academic Publishers. Printed in Great Britain.*

headway has been made toward that aim. For the empiricist, by contrast, the aim of science is restricted to producing predictively accurate theories which are empirically well-supported by the observed phenomena. Conventionalist philosophers of science, who regard theories as classificatory schemes which impose order on experience, take the main aim of science to be to produce an economical ordering of experience. Philosophers of a pragmatist bent emphasise prediction and control of the environment, in the service of successful achievement of practical goals.[5]

In this paper, I will focus on the relationship between methodological pluralism and scientific realism. In particular, I will consider the question of whether sustained application of a plurality of methodological rules conduces to realisation of the scientific realist aim of truth. This question, which raises issues of both an epistemological and a metaphysical nature, is a special instance of the more general demand for an integrated account of rational theory choice and scientific progress. It is, in my view, *the* most urgent question facing the scientific realist who seeks to derive insights about scientific methodology from the pluralist approach found in the work of T.S. Kuhn and P.K. Feyerabend.

The paper is organised as follows. In section 2, I discuss the threat of relativism which is raised by methodological pluralism. In section 3, I show that Laudan's normative naturalist metamethodology removes the threat of relativism. In section 4, I propose that normative naturalism be incorporated within the framework of scientific realism. Section 5 presents objections due to Laudan against the realist aim of truth, which threaten the incorporation of normative naturalism within a realist framework. In sections 6 and 7, I defend the realist aim of truth against these objections. Finally, I argue in section 8 that use of a plurality of methodological rules promotes the realist aim of science.[6]

2. PLURALISM AND RELATIVISM

The main impetus for a pluralist conception of method derives from the historical philosophy of science notably championed by Kuhn and Feyerabend. By contrast with earlier monist orthodoxy, advocates of the historical approach argued that science should be conceived as a developmental process, which takes place in a variety of historical circumstances using a variety of methods, rather than the implementation of an invariant, universal method. Kuhn, who initially argued that standards of theory appraisal vary with scientific paradigm, later came to argue that science is governed by a set of cognitive values (e.g., accuracy, breadth, simplicity, coherence, fertility) which guide theory choice. Feyerabend, for his part, argued not only that all methodological rules are routinely violated in the course of scientific practice, but that there are often good grounds for the violation of such rules.

Some writers suppose that the historical approach of Kuhn and Feyerabend entails wholesale rejection of scientific method. However, I prefer to draw a more positive moral. What is to be rejected, if one adopts the historical approach, is not method as such, but a monistic *theory* of method. Ample scope remains to develop a more adequate account of method within the framework of the historical approach. In particular, what emerges from the historical approach is a pluralist conception of method, on which the principles of method are not unique and invariant, but multiple and subject to variation in the history of science.

I have elsewhere attempted to sketch the main outlines of a pluralist theory of method (1997a, chapter 7). For present purposes, it suffices to characterise the pluralist account by means of the following five theses, which represent central themes of the historical school:

- *Multiple rules*: scientists utilise a variety of methodological rules in the evaluation of theories and in rational choice between alternative theories.[7]
- *Methodological variation*: the methodological rules utilised by scientists undergo change and revision in the advance of science.[8]
- *Conflict of rules*: there may be conflict between different methodological rules in application to particular theories.[9]
- *Defeasibility*: the methodological rules, taken individually rather than as a whole, are defeasible.[10]
- *Non-algorithmic rationality*: rational choice between theories is not governed by an algorithmic decision procedure which selects a unique theory from among a pool of competing theories.[11]

These five theses constitute the basic elements of a pluralist conception of methodology, according to which scientific theory appraisal is governed by an evolving set of methodological rules. Because the rules may conflict in practice, and are individually defeasible, appeal to the system of rules need not uniquely determine the outcome of theory choice. Accordingly, scientists who place differential weight on various rules may come thereby to decide in favour of opposing theories.

It is precisely the scope that methodological pluralism affords for rationally grounded disagreement between scientists that makes it controversial. For it brings it into tension with methodological monist accounts which restrict rational divergence of opinion to that allowed by compliance with a single method. The opposition between monist and pluralist accounts of method is at the root of much recent concern with epistemological relativism in the philosophy of science. For, on the one hand, it is widely held that a monistic theory of method avoids relativism by grounding theory choice in a shared, invariant method.[12] On the other hand, it is also widely assumed that pluralism entails relativism, since the existence of a plurality of methods would provide scientists with rational justification for the acceptance of opposing theories on the basis of alternative sets of rules.

However, it is a mistake to suppose that rational disagreement due to variation of methodological rules necessarily leads to relativism. For that would be to suppose that mere difference in the rules employed by scientists entails relativism. And that in turn would be to suppose that mere compliance with operative rules suffices for rational justification. Yet the latter assumption is surely mistaken. It overlooks the crucial distinction between rules which provide rational justification and those which do not. Not all methodological rules that may be proposed or employed are capable of providing rational justification. Some provide no justification at all. Given the distinction between rules which provide justification and those which fail to do so, relativism is not entailed by pluralism, since mere satisfaction of a methodological rule does not suffice for rational justification.

Yet, while a plurality of methodological rules may not entail relativism, the challenge of relativism now arises in a novel form. For the distinction between rules which provide justification and those which do not is a distinction that is itself in need of defence. After all, how can one rule be shown to provide greater rational justification than another? The relativist challenge, therefore, is to show how one methodological rule may be epistemically superior to another.

3. NATURALISM AND RELATIVISM

The question of how to assess the epistemic merits of a methodological rule is a metamethodological question about the nature of epistemic normativity. One of the most promising approaches to this issue is a form of epistemic naturalism which grounds normativity in the facts of inquiry.[13] This approach involves two key elements. On the one hand, it treats methodological rules as empirical hypotheses about how to pursue inquiry, which may be evaluated in light of empirical evidence. On the other hand, such rules are conceived as instruments or tools of inquiry, the epistemic function of which is to advance cognitive ends. The two elements are combined by grounding evaluation of methodological rules in empirical evidence about performance of epistemic function.

As a special case of this naturalist approach to epistemic normativity, I turn to the *normative naturalist* metamethodology of Larry Laudan.[14] Laudan is critical of the scientific realist view defended here that the aim of science is advance on truth. In the sections to follow I will explore the possibility of incorporating Laudan's normative naturalism within a scientific realist framework. However, in this section my concern is with the normative naturalist account of epistemic normativity as a response to relativism.

As a naturalist, Laudan treats metamethodology as an empirical discipline continuous with natural science. In order to ground methodology empirically, it must be possible to treat methodological rules as normative claims about the conduct of inquiry which are capable of empirical evaluation. Accordingly, Laudan proposes that methodological rules be construed in instrumental fashion as recommendations of means of realising desired cognitive ends. This enables such rules to be formulated as conditional claims with the following hypothetical imperative form:

If one wishes to attain aim *A*, then one ought to employ method *M*.

As an example of how a methodological rule may be cast in hypothetical form, Laudan offers the following formulation of Popper's rule against *ad hoc* hypotheses:

> [I]f one wants to develop theories which are very risky, then one ought to avoid *ad hoc* hypotheses.
> (Laudan 1996, p. 133)

Such an analysis permits the recommendation of a methodological rule to be based on historical evidence. For it reveals how such rules may be supported by claims of statistical covariance between past use of method and achievement of results. Where use of a method has historically proven to be a reliable means of achieving a given end, the method may be recommended on the basis of past performance as means to that end. In this manner, empirical evidence from the history of science may serve as the normative ground of a methodological rule.[15]

The normative naturalist analysis of the justificatory basis of methodological rules enables the distinction to be sustained between rules which provide epistemic support and ones that do not. For if use of one rule reliably conduces to a given aim and use of another fails to, then it provides greater epistemic support than the other. But if one rule may have greater epistemic merit than another, the challenge of relativism may be met. For where there may be variation in the epistemic credentials of rules, rational justification does not reduce to mere compliance with operative methodological rules. Hence, one theory may enjoy a higher degree of support than another, despite a plurality of methodological rules.

4. SCIENTIFIC REALISM AND NORMATIVE NATURALISM

Laudan is a well-known critic of the realist view that truth is the aim of science. Accordingly, Laudan develops normative naturalism within the context of an axiology that allows a multiplicity of scientific aims, rather than being limited to the realist aim of truth. However, in contrast with Laudan, I seek to combine methodological pluralism with scientific realism precisely by incorporating normative naturalism into a realist framework. In so doing, I wish to preserve the normative naturalist response to epistemic relativism while providing an integrated account of both the methodology of science and its progress.

The core of the normative naturalist analysis of methodological rules is that rules may be construed as hypothetical imperatives linking epistemic means and ends. This enables such rules to be treated instrumentally as cognitive tools, which may be utilised to advance the aims of science. Such an instrumental analysis of methodological rules leaves the nature of the epistemic or cognitive aims unspecified. As a critic of realism, Laudan rejects the realist aim of truth, for reasons to be considered in the next section. However, Laudan does not offer any one, unique alternative to truth as the correct analysis of the constitutive aims of science. Rather, he argues that scientists' cognitive aims vary historically as part of the continual process of adjustment and correction of theories, methods and aims which characterises scientific inquiry.[16]

Because the instrumental analysis of rules is neutral with respect to the nature and number of aims that scientists may pursue, I hold it to be possible to set the analysis within a realist framework. In particular, if we treat truth as the paramount aim of science, we may then suppose that the cognitive aim that is to be fulfilled by a proposed rule is advance on the truth about the world.[17] On such a realist construal of normative naturalism, a methodological rule conveys epistemic warrant to the extent that fulfilment of the rule conduces to the aim of truth. As such, normative naturalism emerges as a species of reliabilist epistemology once it is placed within the context of scientific realism. For it is reliability in

leading to the truth which is then the basis of the epistemic warrant of methodological rules.[18]

Where the realist sees truth as the aim of science, Laudan allows that a multiplicity of aims may be pursued by scientists. However, in speaking of truth as the
aim of science, the realist need not deny that scientists pursue multiple aims.
Instead, the realist need only conceive truth as the paramount aim that constitutes the ultimate goal of science. The various other cognitive aims which may
be pursued by scientists may be understood as subordinate aims which subserve
the overriding realist aim of truth. This permits the realist to preserve an additional aspect of Laudan's analysis of the epistemic warrant of methodological
rules. Where Laudan holds that the warrant of a rule consists in reliable pro-
motion of cognitive ends, the realist need not insist that the specified aim of the
rule be truth. Rather, provided that the specified aim subserves the overriding
goal of truth, a rule which immediately conduces to a lower level aim may still
convey epistemic warrant.[19]

On the assumption that employment of methodological rules conduces to
truth, or to aims that subserve truth, the present proposal offers an integrated
account of both the methodology and progress of science. However, as I now turn
to Laudan's objections to realism, we are about to see that this assumption is in
need of defence.

5. LAUDAN AND THE AIM OF TRUTH

Laudan has argued against scientific realism on a number of occasions. Perhaps
most notable is his attack on convergent epistemological realism, in which he
attempts to sever the explanatory connections drawn by realists between reference, truth and the success of science.[20] Here, however, I focus on two specific
objections raised by Laudan against the realist aim of truth. These objections
pose a serious threat to my proposal to set the normative naturalist account of
epistemic warrant within the context of a realist account of the aim of science.

Laudan's objections turn crucially on what he takes to be the transcendental
nature of truth. He assumes that we can tell neither that a theory is true nor that
progress toward truth has occurred. Given this initial assumption, Laudan
develops two separate arguments that truth cannot serve as a suitable aim for
science. He argues, first, that it is not rational to pursue a goal which cannot
recognisably be attained or even approached. Second, he rejects transcendental
aims such as truth as unsuited to a naturalistic treatment of the methodology of
science. Before presenting these two objections, I will examine Laudan's view of
the transcendence of truth.

For Laudan, a transcendental aim or property is one to which we have no
epistemic access. He describes truth as a 'transcendental property', and contrasts
it with an 'immanent' goal such as 'problem-solving effectiveness', which '(unlike
truth) is not intrinsically transcendent and hence closed to epistemic access'
(Laudan 1996, p. 78). The distinction between immanent and transcendent states
corresponds more or less to that between what can be empirically shown to be the
case and what cannot. Laudan's grounds for taking truth as transcendental
appear to be twofold. On the one hand, he contrasts transcendental aims with the

'detectable or observable properties' (*ibid.*, 1996, p. 261, fn. 19) that provide evidence of methodological means/ends relationships, implying thereby that a transcendental state is one that cannot be directly observed to obtain. On the other hand, he claims that 'knowledge of a theory's truth is radically transcendent', since 'the most we can hope to "know" about [a theory...] is that [it is] false' and 'we are never in a position to be reasonably confident that a theory is true' (*ibid.*, 1996, pp. 194, 195).[21] The epistemically transcendent therefore emerges as that which transcends the empirical either by being unobservable or by being based on an ampliative inference that extends beyond the observed data. Accordingly, that is what I shall mean when I speak in what follows of the transcendence of truth or theoretical truth.

Laudan accords truthlikeness a status similar to truth. Since the truth of a theory transcends our capacity for knowledge, we can be in no position to judge how closely an actual theory approximates the truth (*ibid.*, 1996, p. 78). The problem is aggravated by lack of a clear conception of approximate truth. On the Popperian account of verisimilitude, for example, a theory may have high verisimilitude and yet display little or no empirical success (Laudan 1984, p. 118). More generally, Laudan claims that there is no known means to measure or estimate how close a theory is to the truth. Consequently, truthlikeness transcends our capacity to know it every bit as much as does truth.

Given the transcendence of truth and truthlikeness, Laudan objects to the role accorded to such notions within realist accounts of scientific progress. He develops his first objection in the context of a discussion of the rational evaluation of cognitive goals in his (1984, pp. 50–55). According to Laudan, a crucial consideration in evaluating a goal is whether it may be realised. He takes it as a requirement of rationality that there be grounds to suppose it possible to achieve the goals one pursues (1984, p. 51). Goals which are unable to be achieved may be rejected as 'utopian'. Laudan distinguishes three ways in which goals may be utopian: goals that can be shown to be unrealisable are 'demonstrably utopian'; ones that are overly vague or imprecise are 'semantically utopian'; and goals which cannot be shown to obtain are 'epistemically utopian'. Laudan's objection to truth as a cognitive goal is that it is epistemically utopian.

As a prime instance of an epistemically utopian goal, Laudan takes the 'goal of building up a body of true theories' (1984, p. 53). He allows that such a goal may not be demonstrably utopian, and that the concept of truth admits of clear analysis. However, he asks us to consider the case in which one 'has no idea whatever how to determine whether any theory actually has the property of being true' (1984, p. 51). (Of course, as we have just seen, Laudan takes this to be our actual epistemic situation, given the transcendence of truth.) In such a case, where value is placed on an unrecognisable property, Laudan says that 'such a value could evidently not be operationalized' (1984, p. 53), meaning by the latter that no procedure is known which would lead to its attainment (cf. 1984, p. 51). He then concludes that:

if we cannot ascertain when a proposed goal state has been achieved and when it has not, then we cannot possibly embark on a rationally grounded set of actions to achieve or promote that goal. In the absence of a criterion for detecting when a goal has been realized, or is coming closer to realization, the goal cannot be rationally propounded even if the goal itself is both clearly defined and otherwise highly desirable. (1984, p. 53)

Given that Laudan takes truth and truthlikeness to be transcendental, I suggest he is to be understood here as proposing the following argument against the realist aim of truth: (a) it is not rational to pursue an aim which may neither be recognised to obtain nor to be close to obtaining; (b) the goal of true theories may neither be recognised to obtain nor to be close to obtaining; therefore (c) it is not rational to pursue the goal of true theories.[22]

While Laudan's first objection concerns rational pursuit of truth, his second objection derives from his naturalistic view of method. In particular, Laudan argues that transcendental goals such as truth are shown to be illegitimate by the normative naturalist analysis of methodological rules. As we saw in section 3, the normative naturalist construes methodological rules in instrumental fashion as hypothetical imperatives which relate cognitive means and ends. Such an analysis enables methodological rules to be evaluated empirically with regard to their effectiveness in promoting specified aims. According to Laudan, the instrumental conception of method places a premium on the realisability of aims. Aims which cannot be achieved (i.e., utopian aims) are unsustainable, given the goal-directed nature of methodology.

More specifically, Laudan claims that the instrumental conception of method leads to rigorous constraints on the legitimate aims of science:

> any proposed aims for science [must] be such that we have good reasons to believe them to be realisable; for absent that realisability there will be no means to their realization and thus no prescriptive epistemology that they can sustain ... (Laudan 1996, pp. 157–158)

Such constraints have direct bearing on the realist aim of truth:

> one of the corollaries of the instrumental analysis is that those ends that lack appropriate means for their realization become highly suspect. Traditional epistemologists who ... hanker after true or highly probable theories as the aim of science find themselves more than a little hard pressed to identify methods that conduce to those ends. Accordingly, normative naturalism suggests that unabashedly realist aims for scientific inquiry are less than optimal. (ibid., 1996, p. 179)

Thus, the demand of realisability entails the rejection of realist aims as unacceptable for science. The reason, as with the previous objection, turns on the transcendental nature of truth:

> if one has adopted a transcendental aim, or one which otherwise has the character that one can never tell when the aim has been realized and when it has not, then we would no longer be able to say that [a] methodological rule asserts connections between detectable or observable properties. I believe that such aims are entirely inappropriate for science, since there can never be evidence that such aims are being realized, and thus we can never be warrantedly in a position to certify that science is making progress with respect to them. (ibid., 1996, p. 261, fn. 19)

In short, because methodological rules derive their epistemic support from underlying empirical means/end connections, there may be no evidence capable of showing that a rule promotes a transcendental aim, since no empirical evidence may show that a transcendental aim has been reached or is close to being reached.

Based on the lack of possible evidence for advance on truth, Laudan concludes that the realist aim of truth fails to be a legitimate goal for science. While it is not entirely clear how the various strands of Laudan's thoughts on this topic fit together, I propose the following reconstruction of his argument: (a) the methods of science are instruments for the realisation of the aims of science; (b) given this, a legitimate aim of science must be such that it may be realised *and* there may be evidence of its realisation; (c) because truth is transcendental there may be no evidence that the end of truth is realised; hence (d) truth is not a legitimate aim of science.

In sum, Laudan rejects the realist aim of truth on the grounds that it is neither rational to pursue the truth nor is the truth a legitimate aim of science. Both of these objections turn on the basic assumption that truth is transcendental. Let us now see if these objections may be met.

6. IS TRUTH TRANSCENDENT?

The two objections canvassed in the preceding section stem from the common premise that theoretical truth is transcendent. In this section I will challenge this premise by arguing that it is possible to have theoretical knowledge. In the next section, I will address the negative consequences which Laudan derives from the premise about the rationality and legitimacy of pursuit of truth.

As we have seen, Laudan regards theoretical truth as transcendent in the sense that such truth transcends our capacity to know it. However, it is by no means evident that theoretical truth is unknowable, as Laudan claims it to be. That this is so may be readily shown on the basis of the standard analysis of knowledge as justified true belief. On such an analysis, a knowing subject S knows a theoretical proposition P *iff* three conditions are fulfilled:

1. S believes that P is true,
2. S's belief that P is true is rationally justified,
3. P is true.

Given such an analysis of knowledge, there is no apparent reason in principle why a theoretical proposition may not be known to be true. For in order to know that P is true, it suffices that there be good grounds for the belief that P and that P in fact be true.

More specifically, let us suppose that a scientist believes a theoretical proposition P (e.g., 'Electrons have negative charge') to be true. On the assumption that it is possible for a theoretical proposition to correctly report an actually existing state of affairs (e.g., that electrons in fact have negative charge), then it is possible for P to be true. Provided, moreover, that P satisfies appropriate methodological standards, there may be good rational grounds for the belief that P is true. Given both these assumptions, and the standard analysis of knowledge, it follows that P may be known to be true, for one may rationally believe P and P may be true. Hence, theoretical knowledge is possible.

Against this, it might be objected that one may have a justified true belief that P and yet be unable to tell that P is true. The objection arises because P is a theoretical proposition whose truth is not directly evident. For, while P may well be true, there is no direct means of knowing that this is so. At most, one may have access to the evidence which justifies the belief that *P*. But there is no access to the truth of *P* that is independent of the evidence for *P*. Thus, even if *P* is true, and justifiably believed to be so, one may fail to be in a position to know that it is true. Given this, the fact that the conditions specified for knowledge may be fulfilled in the case of a theoretical proposition does not show that theoretical knowledge is possible.[23]

This objection rests on a confusion between conditions for the possession of knowledge and criteria for the recognition of knowledge. The justified true belief analysis of knowledge provides a set of conditions, satisfaction of which qualifies a subject as having knowledge. It does not provide criteria which enable a subject to recognise that those conditions obtain, and is thereby in possession of knowledge. Thus, it is possible for one to know that *P* without being able to recognise that one knows that *P* or that *P* is true. In short, one may have theoretical knowledge even in the absence of direct epistemic access to the truth of the theoretical proposition that is known.[24]

Such absence of direct access leads to a further potential objection to theoretical knowledge. For if there are no criteria which enable recognition of theoretical truth, then such truth may not be shown with certainty to obtain. One might then object that theoretical knowledge is not certain knowledge, and so not strictly knowledge at all. Such an objection is suggested by Laudan's previously quoted discussion of the 'epistemically utopian' character of truth, where he says that the value of truth cannot be 'operationalized' and that there is no 'criterion for detecting when a goal [e.g., truth] has been realized' (1984, p. 53). However, I am loath to attribute this objection to Laudan, since he is on record as supporting fallibilism (e.g., 1984, pp. 51, 52; 1996, p. 213), and indeed dismisses 'apodictic certainty' as a transcendental property on par with truth (1996, p. 78).[25] In any event, it is a commonplace of the philosophy of science that scientific theories are constantly subject to revision with the advance of science, so that any adequate conception of scientific knowledge must allow that one may have knowledge without certainty.

There remains an additional basis on which to object to the possibility of theoretical knowledge. Laudan might object to the present use of the justified true belief analysis of knowledge on the basis that there may be no grounds which could rationally justify a scientist in believing that a theoretical proposition is true.[26] In other words, he might deny that the grounds which provide rational support for a theoretical proposition provide support for the truth of the proposition. At first blush, this may seem an implausible objection, since, as has been noted by a number of authors, rational grounds for belief that *P* are *ipso facto* rational grounds for the belief that *P* is true.[27] For if one has grounds for the belief that *P*, then, by semantic ascent, one has grounds for the belief that *P* is true. Hence, one cannot sever rational belief from rational belief in truth in the manner that the objection requires.

There is, however, a consistent line of argument available to Laudan here. On the instrumental analysis of rules, the warrant of a methodological rule relates to the end served by the rule. Hence, since there may be no evidence that a rule conduces to theoretical truth, satisfaction of a rule may provide no warrant for belief in such truth. Rather, satisfaction of a rule provides warrant only with respect to the end served by the rule. Thus, when the aim served by a rule is that of predictive reliability, for example, satisfaction of the rule by a theory licenses belief that the theory is predictively reliable, not that it is true. Given that justification always relates to the end served by a rule, it is therefore consistent for Laudan to hold that there may be rational grounds for a theory that are not grounds for believing that the theory is true.

However, while it may be consistently denied that a warrant need be a warrant for truth, the resulting position is unsustainable for several reasons. For one thing, it leads to an implausible restriction on the epistemic states of scientists. For if there may be no warrant for belief in theoretical truth, no scientist who accepts any theory as true may do so rationally, no matter how weighty the evidence or how well-established the theory. For another thing, it rests on an unduly narrow empiricist epistemology.[28] For if there may be no warrant for belief in the truth of any proposition that transcends empirical evidence, then all inferential or indirect knowledge is precluded due to lack of rationally justified belief. Finally, denial that methodological criteria provide warrant for truth removes the rationale for scientists' use of a plurality of such criteria in the evaluation of theories. Scientists who accept a theory which satisfies multiple criteria (e.g., predictive accuracy, explanatory breadth, simplicity, coherence) may do so because they interpret such joint satisfaction of criteria as indicating the likely truth of the theory. But in the absence of such a unifying aim served by criteria, scientists are deprived of a rationale for conjoint use of multiple criteria.

I conclude that there is every reason to suppose that theoretical knowledge is possible. Neither our lack of direct or infallible epistemic access to theoretical truth, nor the possibility of a warrant that is not a warrant for truth, entails that we are unable to have theoretical knowledge. It may not be possible to prove beyond a shadow of a doubt that a theoretical proposition is true. But that does not mean that such truth radically transcends our epistemic capacities, as Laudan suggests.

7. THE PURSUIT OF TRUTH

In this section I will consider Laudan's two objections to truth as the aim of science. As we saw in section 5, Laudan argues that truth is epistemically utopian, hence unable to serve as an object of rational pursuit. Nor is truth admissible as an aim of science, since there may be no evidence of its realisation. Since both objections depend on the transcendence of theoretical truth, they are in large part undermined by the possibility of theoretical knowledge for which I argued in the previous section. However, it remains to show this in detail.

If theoretical knowledge may be acquired by methods employed by scientists, it would seem natural to suppose that acquisition of such knowledge is a legitimate goal for science. Before further scrutinising this assumption, however, I will briefly consider the consequences of denying that theoretical knowledge is possible. One might think that if theoretical truth or knowledge were wholly unattainable, there could be no rationale for their pursuit. For it is futile to attempt the impossible.

However, as Rescher notes against Laudan, there are circumstances in which it is rational to pursue an unattainable ideal (Rescher 1982, p. 227). Moral perfection may be beyond our reach, for example, but striving for such perfection may make one a better person. Similarly, truth may function in the manner of a 'regulative ideal' for science. For, while it may be impossible for science to achieve perfection, the idea of a perfectly true theory may serve to maintain the self-corrective, evolutionary character of the scientific enterprise. In addition, the pursuit of an unattainable ideal may yield indirect benefits which are themselves otherwise unattainable. For example, it is arguably the case that the ideal of a comprehensive, true theory of the world exerts pressure on science to develop systematic theories with real explanatory breadth. Indeed, such lower level values as explanatory breadth would seem to have little independent rationale in the absence of a demand for a comprehensive, true theory.

The possibility of a regulative role and indirect benefits secures for truth a legitimate place in science even if it is unattainable by scientific means. However, if, as argued in the previous section, theoretical knowledge is possible, then truth is in fact an attainable end that lies within the reach of science. This would seem to vindicate theoretical truth as a legitimate goal of rational scientific inquiry. For, on the one hand, if truth is a realisable aim of science, it is possible for an agent to rationally pursue truth as a goal. On the other hand, the attainability of truth means that it satisfies the requirement of the instrumental conception of method that only achievable aims be allowed into science.

But Laudan's principal objection is not that theoretical truth is inappropriate as an aim because it cannot be attained. His main point is that we would be unable to recognise truth even if we were to attain it. Given this, it is not rational for an agent to pursue truth, since there are no criteria which would enable one to recognise attainment of the aim or that it is close to attainment. Similarly, it is because there may be no evidence indicating that a method yields truth that truth is excluded as an admissible aim of science.

Laudan's emphasis on the absence of criteria for the recognition of truth may suggest that he endorses the requirement, rejected in the previous section, that one must be able to recognise that one satisfies the conditions of knowledge in order to possess knowledge. But, in fact, Laudan's claim is not that ability to recognise truth is a requirement of knowledge. Rather, his claim is that it must be possible for one to recognise the fulfilment of an aim in order to rationally pursue that aim. Thus, his objection to the rational pursuit of truth is not that we are unable to possess theoretical knowledge because we cannot recognise truth. It is that we are unable to recognise whether an action furthers an aim, where the aim happens to be truth. Laudan therefore takes ability to recognise achievement of an aim as a requirement for the rational pursuit of that aim, not as a requirement for knowledge.

But, while Laudan may only require recognition criteria for rational pursuit rather than knowledge, similar considerations apply in either case. For Laudan's denial that there are criteria for the recognition of truth is only plausible on the assumption that such criteria must provide an infallible indication of truth. It may readily be conceded that there are no infallible criteria of truth. But it by no means follows that there are no fallible criteria for the recognition of truth. While satisfaction of methodological criteria cannot decisively prove a theory to be true, it may provide good grounds for believing a theory to be true or close to truth. There may well be no criteria which enable a rational agent to know with certainty that they are advancing on truth or have attained it. Nevertheless, such an agent may justifiably believe that a theory which better satisfies the criteria than a rival theory is likelier to be true, or closer to truth, than the alternative theory. Given this, it is entirely possible for an agent to rationally pursue the goal of truth, since satisfaction of methodological criteria may provide a fallible indication of advance on that aim.

Similar remarks apply to Laudan's objection that truth is an inadmissible aim for science, since there may be no evidence that truth is realised by any method. As we saw in section 5, the objection derives from Laudan's instrumental conception of method. What motivates the objection is the thought that if a method functions in the manner of an instrument, then it is to be assessed by how well it brings about the end for which it is proposed. If there is no evidence that it performs its function, then it may not be proposed as a means to that end. The question is whether it is fair to suppose that there may be no evidence that a method leads to truth. It is perhaps true that there may be no direct empirical evidence that use of a method leads to theoretical truth. But there may surely be indirect evidence that a method conduces to such truth. For where the lower level ends served by a method are ends which themselves may be taken to subserve the aim of truth, the success of the method in conducing to such lower level ends may be taken as evidence that the methods conduce to truth. Just as there may be no infallible criteria for the recognition of truth, there may be no infallible evidence that use of a method serves truth. But that is only to say that there is no certain knowledge in theoretical matters.

Finally, a brief remark is in order regarding the basis of the objection. The objection is based entirely on the instrumental conception of method, which entails the demand for realisability. But no independent argument is given for the instrumental conception, other than that it permits the empirical evaluation of methodological rules within a naturalist framework. This is admittedly a powerful point in its favour. But, if the instrumental conception really does entail that truth is an unacceptable aim for science, this may equally well be regarded as a mark against the instrumental conception. In other words, the fact that the instrumental conception excludes truth as an allowable aim may be taken to count against the instrumental conception rather than against the aim of truth. However, since I remain unconvinced that the prospects of finding a place for truth within normative naturalism are as dim as Laudan claims, I see no need at this juncture to put the instrumental conception in question.

8. CONCLUSION

In this paper I have sought to show that a normative naturalist account of epistemic warrant may be combined with a scientific realist conception of the aim of science. On the general picture which emerges, the naturalistic basis of a non-relativist methodological pluralism may be sustained within a scientific

realist framework. As such, the present approach affords a unified account of
the method of science and its progress. However, since methods may cohere
with aims without promoting them, it remains to show that use of a plurality
of methodological criteria advances the realist aim of truth.

Some philosophers deny there is a problem relating method to truth. Internal
realists such as Ellis (1990) and Putnam (1981) define truth as maximal (or ideal)
satisfaction of methodological criteria. For internalists, advance on truth is the
inevitable result of the use of criteria. Truth is not something separate from
method to which its use may or may not give rise. Rather, for internalists, continued application of methodological criteria produces theories which increasingly satisfy the criteria. The result is advance on truth, since truth simply is
maximal satisfaction of the criteria.

As a realist, I hold that the objective world in no way depends on thought.
Therefore I do not equate truth with satisfaction of criteria.[29] The relation between
method and truth is not an internal or conceptual relation, but an external or
synthetic one. The sole question is whether the relation is necessary or contingent.
I have elsewhere defended the view that the epistemic warrant of certain enumerative inductions rests on the essential properties of natural kinds of things.[30]
But while I hold that metaphysical necessity grounds the reliability of certain
basic kinds of inductive inference, I do not see an analogous role for metaphysical
necessity in the case of theory appraisal since the latter involves factors beyond
those involved in basic induction. I take the relation between method and truth to
be a contingent relation between epistemic means and ends, which may be known
in the *a posteriori* manner suggested by Laudan's naturalist metamethodology.

However, as Laudan notes, no direct empirical evidence may show that use of a
methodological rule yields theoretical truth. This raises the question why use of
criteria of theory appraisal should be taken to promote the goal of truth. In the
absence of direct evidence linking method to truth, the grounds for such a link
may be at best abductive ones. More specifically, the realist claim that application
of a plurality of methodological criteria leads to progress toward truth rests on an
inference to the best explanation of scientific success. What best explains why
scientific theories increasingly exhibit the epistemic virtues highlighted by
methodological criteria is that such theories are increasingly close approxima-
tions to the truth.

In arguing this way, I seek to extend the argument of McMullin (1987) that we
are warranted in taking a theory to be 'approximately true' if it exhibits 'a high
degree of explanatory success' (1987, p. 59). McMullin takes the explanatory
success of a theory to be determined by how well it satisfies the various methodological criteria of theory appraisal (1987, p. 54). Where a theory exhibits a
high degree of explanatory success, as indicated by satisfaction of the criteria,
there are good grounds to take the general kinds of entities postulated by the
theory to really exist, as well as what the theory says about such entities to be
broadly correct, though open to further development (1987, pp. 59, 60).

I wish to amplify McMullin's argument in two minor respects. First, I do not
wish to say simply that the high degree of explanatory success of a theory, as
measured by methodological criteria, permits us to infer abductively to the

approximate truth of the theory. I wish, in addition, to say that where a theory possesses an impressive range of theoretical virtues (e.g., accuracy, breadth, simplicity), the best explanation of why the theory possesses such an impressive range of virtues is that it is approximately true. Second, I wish to extend McMullin's argument by explicitly applying it to the advance of science. For where a sequence of theories increasingly satisfies the methodological criteria, the best explanation is that the sequence of theories is advancing on truth. In both these ways, then, the reason for taking continued use of methodological criteria to yield advance on truth is that this best explains why our theories increasingly satisfy such criteria. It is in this sense that what is needed to bridge the gap between method and truth is an abductive argument about how best to explain scientific success. Echoing Lakatos on Popper, one might call this 'a plea for a whiff of abduction'.

Such a whiff of abduction may seem to beg the question against Laudan's critique of the realist's success argument (Laudan 1984, chapter 5). Rebuttal of that critique is, of course, beyond the scope of this paper, but I will briefly indicate why no question is begged by the current proposal. In the first place, Laudan's critique does not impugn all use of the success argument, but only the ambitious attempt to forge a wholesale link between reference, truth and the success of science. Application on a case-by-case basis, restricted for example to entities postulated to fill specific causal roles, may escape Laudan's strictures on the success argument. In the second place, the current abduction does not proceed at the object-level from the widespread success of science to a general realist attitude toward theories, but is a metamethodological inference to an explanation of why a theory manifests a range of methodologically desirable features.

In sum, on the view I propose the realist aim of science is added to normative naturalism by an inference to the best explanation which augments lower level cognitive ends with the aim of truth. It is a fair question, of course, why truth is the best explanation. But consider the alternative. Suppose there is a scientific theory which possesses a variety of methodological virtues to an impressive degree. The theory is accurate, reliable, predicts novel facts, unifies diverse domains, and is simple and coherent. But let us also suppose that the theory is completely false. None of the entities or mechanisms it postulates exist, and it erroneously imposes unity on domains which in fact have nothing in common.

In such a situation, it is sheer luck that the theory has any success at all. This is especially the case with respect to predictive reliability. Either such success is sheer luck, or else a benevolent force makes the theory's predictions come out true despite the theory being false. Of course, there may be worlds which reward luck with predictive reliability. But our world is not a world like that. We are lucky some of the time. But if a theory is predictively reliable, the likeliest explanation is not that our world is one that rewards luck but that we have cottoned on to the way the world really is. For this reason, I claim that satisfaction of methodological criteria provides a sound but fallible indication that a theory is on the road to truth, and may even be there already.

University of Melbourne

NOTES

* This paper was written while I held a Visiting Fellowship at the Center for Philosophy of Science at the University of Pittsburgh. I wish to express my gratitude to the Center for the invitation, as well as for hospitality and support during my visit. For discussion, I am grateful to audiences at the Center for Philosophy of Science, as well as at the University of Hanover, the Catholic University of Louvain and Swarthmore College, where I presented talks based on this material. For comments, discussion and correspondence relating to the ideas contained in this paper, I am indebted to Thomas Bonk, John Clendinnen, David Cockburn, Michel Ghins, Bruce Glymour, Paul Hoyningen-Huene, Hugh Lacey, Larry Laudan, Jim Lennox, Timothy Lyons, Michele Marsonet, Robert Nola, Stathis Psillos and Nick Rescher.

1 This view of the task of the theory of method accords, for example, with the two ingredients of a 'rational model of scientific change' identified by Newton-Smith, viz., specification of the goal of science and of principles of theory comparison (Newton-Smith 1981, p. 4). The demand for an integrated response to both tasks is well-exemplified by Lakatos' plea for a 'whiff of induction' in Popper's treatment of the relation between corroboration and verisimilitude (Lakatos 1978, p. 159).

2 The distinction between monist and pluralist theories of method is somewhat crude, since there are also mixed positions. John Worrall, for instance, holds that there is an invariant core of methodological principles, which remains fixed throughout change of lower level principles (Worrall 1988). The issue of methodological variance masks further complexity, as well. For, in principle, one might argue that at any one time science is governed by a single method, though this method may undergo historical variation. Conversely, one might argue that there is a plurality of methods which are historically invariant. Hence, a full taxonomy of methodological views would include variationist and invariationist versions of both pluralism and monism, in addition to mixed positions.

3 The best-known pluralists are Feyerabend (1975), Kuhn (1977) and Laudan (1984). Elements of a pluralist methodology may be found in the work of such authors as Chalmers (1982), Ellis (1990), Lacey (1997), Lycan (1988), McMullin (1987), Newton-Smith (1981), Quine and Ullian (1970) and Thagard (1978). I defend a pluralist stance in the later chapters of my (1997a).

4 *Terminological note:* Some comment is necessary regarding my use of the term 'methodological rule' and related expressions. A variety of terms (e.g., 'criteria', 'norm', 'principle', 'rule', 'standard', 'value') is found in the methodological literature. While there are slight differences of meaning and usage, there is no substantive difference between such terms of relevance to the issues dealt with in this paper. All such terms denote methodologically relevant factors to which appeal is made in theory appraisal and justification of theory choice. The terms might therefore be used interchangeably. However, to reduce scope for confusion I will tend instead to speak either of criteria or of rules, restricting use of related terms to contexts in which another term seems especially apt. I will understand the relation between criteria and rules to be roughly as follows: a criterion is a methodologically desirable feature of a theory (e.g., accuracy, coherence, simplicity); rules are prescriptions typically (but not necessarily) stated in linguistic form (e.g., 'avoid *ad hoc* hypotheses', 'employ double blind tests'). In general, criteria (e.g., simplicity) may be stated in an analogous form as rules (e.g., 'prefer simple hypotheses'). It is also worth noting that for present purposes no decision need be made as to whether rules or criteria are best construed as necessary and/or sufficient conditions of theory acceptance, or merely as factors of relevance to theory appraisal. Hence, I ignore as irrelevant in the present context the otherwise important distinction between rules which dictate theory choice and values which merely guide such choice (cf. Kuhn 1977, p. 331).

5 The relationship between aims and methods is not straightforward. There is scope for a variety of different accounts of such relationships. For example, in contrast with other theories of method, the conventionalist elevates the aim of overall theoretical simplicity into the paramount methodological principle of science. On the other hand, realists and empiricists may agree on the nature of method but disagree on the aims served by the method.

6 While the niceties of the doctrine of scientific realism are inessential to the discussion that follows, there is sufficient variation among realist authors to warrant an indication of what I take to be involved in the doctrine. I take scientific realism to involve four main tenets: (a) *axiological realism*: the aim of science is truth, and scientific progress consists in advance on that aim; (b) *anti-instrumentalism*: the unobservable entities postulated by scientific theories are conceived as real entities rather than mere predictive devices; (c) *correspondence truth*: truth consists in correspondence between what a statement says about the world and the way the world in fact is; (d) *metaphysical realism*: the world investigated by scientists is an objective reality, the existence and nature of which are independent of human mental activity.

[7] See, e.g., Ellis (1990, pp. 244–259), Kuhn (1977, pp. 321, 322), Lacey (1997, pp. 31–33), Laudan (1984, pp. 33 ff; 1996, p. 18), Lycan (1988, pp. 129, 130), McMullin (1987, pp. 53, 54) and Newton-Smith (1981, pp. 226–232).

[8] See, e.g., Feyerabend (1978, pp. 33–39, 98), Kuhn (1970, pp. 103–110, 148; 1977, pp. 335, 336), Chalmers (1990, p. 20) and Laudan (1984, pp. 39, 40, 57–59, 81, 82; 1996, p. 17).

[9] E.g., simplicity may favour one theory, coherence or breadth another (cf. Kuhn 1977, pp. 323, 324; Thagard 1978, p. 92). For qualification of the view that there may be conflict between rules, see Laudan (1996, pp. 93, 94).

[10] That methodological rules are defeasible is, of course, the main thrust of Feyerabend's opening argument in his (1975). However, the defeasibility of all rules, taken singly, does not entail that all such rules may be concurrently violated. Hence, while any particular rule may be violated in appropriate circumstances, it is rationally unacceptable to transgress the entire system of methodological rules. While perhaps not entirely explicit in Kuhn, the inviolability in general of the set of rules is in the spirit of Kuhn (1977). For related discussion, see Laudan (1996, pp. 101–105.)

[11] Explicit rejection of an algorithm of theory choice occurs in Kuhn (1970, p. 200; 1977, p. 326), and Laudan (1984, pp. 5, 6; 1996, pp. 17–19). Chalmers tacitly denies an algorithm of theory choice in his discussion of Feyerabend's critique of universal methodological rules (1982, p. 135). Brown develops a non-algorithmic conception of rationality in his (1988). Explicit formulations aside, however, rejection of an algorithm of theory choice is virtually the defining thesis of the historical school.

[12] As such, however, monism need not be immune to the challenge of relativism, since the question may always be raised of the justification of the monist's purportedly invariant method, as against another possible method. For relevant discussion, see the exchange between Laudan (1989) and Worrall (1988; 1989), as well as my (1997a, chapter 10).

[13] Epistemic naturalism is not, of course, the only approach to epistemic normativity. Among the main alternatives to naturalism in metamethodology, it is worth noting the conventionalism of Popper (1959), the intuitionism of Lakatos (1978) and early Laudan (1977), and reflective equilibrium models which trace back to Goodman (1955). For further analysis of the range of metamethodological approaches, see Nola (1987, 1999) and Nola and Sankey (this volume).

[14] See Laudan (1996, chapter 7). While Laudan's normative naturalism is well-suited for the present purpose of defeating the relativist, it is but one instance of a widespread form of epistemic naturalism. Similar views of both the nature and evaluation of methodological rules may be found in Rescher (1977) and Stich (1990). The idea that methodological rules are tools of inquiry has deep pragmatist roots, which may be traced back, for example, to Dewey's comparison of methods of inquiry with methods of farming (Dewey 1986, pp. 107, 108). Closely related views occur as well in Giere (1989) and Kornblith (1993).

[15] The role here attributed to cognitive ends by Laudan raises the spectre of a relativism due to variation of ends (cf. Psillos 1997, p. 707). However, Laudan's hypothetical imperative account of rules needs to be understood in the context of his remarks on rational adjudication of cognitive goals in his (1984, pp. 50 ff). Laudan there adumbrates a number of means of evaluating cognitive aims, e.g. by showing an aim to be utopian, or in conflict with practice. It should be allowed, therefore, that Laudan seeks to avoid relativism due to variation of cognitive aims. Whether he succeeds is another matter.

[16] As examples of cognitive aims that have been pursued by scientists, Laudan mentions infallible knowledge, high probability, simplicity, elegance, as well as Newton's attempt to reveal divine agency at work within the physical world (cf. Laudan 1984, 51 ff; 1996, p. 129).

[17] To say that science aims for truth is not to be distinguished from saying that it aims for truth about the world. Nor would I distinguish it from saying that the aim of science is knowledge (cf. Rosenberg 1990), since knowledge implies truth. Nor either would I demur if a realist were to argue that the aim of science is explanation, as Ellis (1985) does, since seeking true explanations is part of seeking the truth. (However, I would demur at Ellis' suggestion that we renounce the correspondence theory of truth in favour of a pragmatist concept thereof.)

[18] More specifically, combining the instrumental analysis of rules with the aim of truth yields a form of *method*, rather than *process*, reliabilism (cf. Goldman, 1986, pp. 93–95). However, I do not wish to endorse a pure reliabilism on which warrant is strictly identified with truth conduciveness. Such an account is subject to counterexamples, such as Lehrer's case of Mr. Truetemp, who reliably forms true beliefs about the temperature due to a device implanted in his brain, but is ignorant of both the reliability of his belief and of their cause (Lehrer 1990, p. 163). My view is roughly that reliability is a crucial part of the warrant of methodological rules, but that *use* of rules must meet additional constraints, such as being deliberately employed by a scientist on the basis of awareness of such rules.

[19] As an example of a methodological rule which immediately advances a lower-order aim, and indirectly advances the aim of truth, consider Popper's rule against *ad hoc* hypotheses. Avoidance of *ad hoc* hypotheses serves to increase the falsifiability of theories, which thereby subserves the aim of truth, since the ruthless testing of falsifiable theories is held by Popper to conduce, fallibly, to truth, or at any rate to greater verisimilitude.

[20] See Laudan (1981), reprinted as chapter 5 of his (1984).

[21] Laudan credits the point that we cannot know a theory to be true to Hume and Popper (1996 p. 194). However, he also notes (personal communication) that his point is intended to be stronger than simply saying that theories cannot be shown to be true. He refers to the latter as 'Humean underdetermination' (1996, p. 31). By contrast, his point about the transcendence of truth appears to be a strong version of what he describes as 'ampliative underdetermination' (1996, p. 43 ff). For while Laudan denies that ampliative rules of inference underdetermine rational theory choice, his claim that theories cannot be reasonably held true seems to imply that such rules underdetermine rational belief *in the truth of theory*. The grounds for this thesis would appear to be either a version of the 'pessimistic meta-induction' (cf. 1977, p. 126) or his related critique of the explanatory connections drawn by realists between scientific success and truth (1984, chapter 5).

[22] It might be objected that Laudan states the argument in conditional form, e.g. 'if we cannot ascertain when a proposed goal state has been achieved'. Hence, it is not to be interpreted as an argument against realism, but merely as an example of a possible epistemically utopian aim. However, since, as we have seen, Laudan holds truth to be transcendental, he is committed to dismissing it as an epistemically utopian aim, which cannot be rationally pursued.

[23] The present objection to the standard analysis differs from Gettier-style objections. Gettier cases show that the standard analysis fails to provide a set of jointly sufficient conditions for knowledge. By contrast, the present objection turns on lack of direct epistemic access to the truth of theoretical propositions. Incidentally, while Gettier cases show that further conditions are needed to obtain sufficient conditions for knowledge, the conditions specified by the standard analysis remain individually necessary and thereby constitute an approximately correct analysis of the concept of knowledge. Given this, it is unproblematic to treat the standard analysis as an adequate working definition of knowledge.

[24] This implies the falsity of the KK-thesis, i.e., the thesis that in order to know one must know that one knows. I take the KK-thesis to be false, since one may know without being aware that one knows, or even knowing what it is to know.

[25] However, it is not completely clear what Laudan takes to follow from fallibilism with respect to the concept of knowledge. He writes at one point that 'the unambiguous implication of fallibilism is that there is no difference between knowledge and opinion: within a fallibilist framework, scientific belief turns out to be just a species of the genus opinion' (1996, p. 213). This might be taken to suggest that knowledge has no greater warrant than any other form of belief. However, since, in the context in question, certainty is the crucial factor which distinguishes opinion from knowledge, knowledge might still be justified true belief and yet belong to the genus opinion.

[26] That this is indeed Laudan's likely objection is suggested by footnote 21 (above).

[27] The point is made specifically with regard to Laudan by Psillos (1997, p. 712). Lycan makes the point in a more general context in response to the claim that one may have evidence for *P* but not evidence for the truth of *P* (Lycan, 1988, p. 137).

[28] The point that Laudan's epistemology is unduly empiricist has been made by a number of authors, including most relevantly (Nola, 1999). It should be noted that Laudan explicitly denies the charge (1996, p. 160). But his denial is difficult to reconcile with his dismissal of theoretical truth as a 'transcendental' aim.

[29] Put simply, my reason is that epistemic theories of truth such as internal realism entail the mind-dependence of reality. For discussion, see Devitt and Sterelny (1987, pp. 195, 196), and Musgrave (1997).

[30] Roughly, the reliability and hence rationale of induction is explained by the fact that members of a natural kind possess their essential properties necessarily. The reason why we are right when we predict that an unobserved member of a kind bears the same essential property as previously observed members is that, being a member of the same natural kind as previous members, the unobserved member necessarily possesses that property. For discussion, see my (1997b), which combines Brian Ellis' recent scientific essentialism with Kornblith's account of the ground of induction.

REFERENCES

Brown, H.I., 1988: *Rationality*, Routledge, London.

Chalmers, A.F., 1982: *What is This Thing Called Science?* (2nd edn.), University of Queensland Press, St. Lucia.

Chalmers, A.F., 1990: *Science and its Fabrication*, Open University Press, Milton Keynes.

Devitt, M. and Sterelny, K., 1987: *Language and Reality*, Blackwell, Oxford.
Dewey, J., 1986: *Logic: The Theory of Inquiry*, in *John Dewey: The Later Works, Vol. 12, 1938*, Southern Illinois University Press, Carbondale.
Ellis, B.D., 1985: 'What Science aims to Do', in Churchland, P. and Hooker, C.A. (eds.), *Images of Science*, University of Chicago Press, Chicago, pp. 48–74.
Ellis, B.D., 1990: *Truth and Objectivity*, Blackwell, Oxford.
Feyerabend, P.K., 1975: *Against Method*, New Left Books, London.
Feyerabend, P.K., 1978: *Science in a Free Society*, New Left Books, London.
Giere, R.N., 1989: 'Scientific Rationality as Instrumental Rationality', *Studies in History and Philosophy of Science* 20, 377–384.
Goldman, A.I., 1986: *Epistemology and Cognition*, Harvard University Press, Cambridge MA.
Goodman, N., 1955: *Fact, Fiction and Forecast*, Harvard University Press, Cambridge MA.
Kornblith, H., 1993: 'Epistemic Normativity', *Synthese* 94, 357–376.
Kuhn, T.S., 1970: *The Structure of Scientific Revolutions* (2nd edn.), University of Chicago Press, Chicago.
Kuhn, T.S., 1977: 'Objectivity, Value Judgment and Theory Choice', in *The Essential Tension*, University of Chicago Press, Chicago, pp. 320–339.
Lacey, H., 1997: 'The Constitutive Values of Science', *Principia* 1, 3–40.
Lakatos, I., 1978: 'Popper on Demarcation and Induction', in Worrall, J. and Currie, G. (eds.), *The Methodology of Scientific Research Programmes*, Cambridge University Press, Cambridge, pp. 139–167.
Laudan, L., 1977: *Progress and Its Problems*, University of California Press, Berkeley.
Laudan, L., 1981: 'A Confutation of Convergent Realism', *Philosophy of Science* 48, 19–48.
Laudan, L., 1984: *Science and Values*, University of California Press, Berkeley.
Laudan, L., 1989: 'If It Ain't Broke, Don't Fix It', *British Journal for the Philosophy of Science* 40, 369–375.
Laudan, L., 1996: *Beyond Positivism and Relativism*, Westview, Boulder.
Lehrer, K., 1990: *Theory of Knowledge*, Routledge, London.
Lycan, W., 1988: 'Epistemic Value', in *Judgement and Justification*, Cambridge University Press, Cambridge, pp. 128–156.
McMullin, E., 1987: 'Explanatory Success and the Truth of Theory', in Rescher, N. (ed.), *Scientific Inquiry in Philosophical Perspective*, University Press of America, Lanham, pp. 51–73.
Musgrave, A.E., 1997: 'The T-Scheme Plus Epistemic Truth Equals Idealism', *Australasian Journal of Philosophy* 75, 490–496.
Newton-Smith, W.H., 1981: *The Rationality of Science*, Routledge Kegan Paul, London.
Nola, R., 1987: 'The Status of Popper's Theory of Scientific Method', *British Journal for the Philosophy of Science* 38, 441–480.
Nola, R., 1999: 'On the Possibility of a Scientific Theory of Scientific Method', *Science and Education* 8, 427–439.
Nola, R. and H. Sankey, this volume: 'A Selective Survey of Theories of Scientific Method', pp. 1–65.
Popper, K.R., 1959: *The Logic of Scientific Discovery*, Hutchinson, London.
Psillos, S., 1997: 'Naturalism Without Truth?', *Studies in History and Philosophy of Science* 28, 699–713.
Putnam, H., 1981: *Reason, Truth and History*, Cambridge University Press, Cambridge.
Quine, W.V.O. and Ullian, J., 1970: *The Web of Belief*, Random House, New York.
Rescher, N., 1977: *Methodological Pragmatism*, Blackwell, Oxford.
Rescher, N., 1982: *Empirical Inquiry*, Rowman and Littlefield, Totowa, New Jersey.
Rosenberg, A., 1990: 'Normative Naturalism and the Role of Philosophy', *Philosophy of Science* 57, 34–43.
Sankey, H., 1997a: *Rationality, Relativism and Incommensurability*, Aldershot, Ashgate.
Sankey, H., 1997b: 'Induction and Natural Kinds', *Principia* 1, 239–254.
Stich, S., 1990: *The Fragmentation of Reason*, MIT, Cambridge, Mass.
Thagard, P., 1978: 'The Best Explanation: Criteria for Theory Choice', *Journal of Philosophy* 75, 76–92.
Worrall, J., 1988: 'The Value of a Fixed Methodology', *British Journal for the Philosophy of Science* 39, 263–275.
Worrall, J., 1989: 'Fix It and Be Damned: A Reply to Laudan', *British Journal for the Philosophy of Science* 40, 376–388.

MALCOLM R. FORSTER

HARD PROBLEMS IN
THE PHILOSOPHY OF SCIENCE:
IDEALISATION AND COMMENSURABILITY

1. INTRODUCTION

Many philosophers underestimate the general disillusionment in the philosophical outlook on science caused, in part, by Kuhn's *Structure of Scientific
Revolutions*. The challenge presented by Hume's problem of induction has always
kept the issue of scientific truth at the forefront of philosophical research. Philosophers expended great energy in defending a broad spectrum of replies to
Hume's scepticism, ranging from the view that theories are merely instruments
for the control and prediction of nature, to realist views of science (which hold
that science aims at the truth about the world, and is rational in the pursuit of this
goal). Kuhn (1970, p. 171) insisted that this approach to studying science is
unhelpful, and many outsiders have followed his lead:

> Does it really help to imagine that there is some one full, objective, true account of nature and that
> the proper measure of scientific achievement is the extent to which it brings us closer to that ultimate
> goal? If we can learn to substitute evolution-from-what-we-know for evolution-toward-what-we-
> wish-to-know, a number of vexing problems may vanish in the process. Somewhere in this maze, for
> example, must lie the problem of induction.

For Kuhn (1970), and his followers, the rationality of science has nothing to do
with truth: 'As in political revolutions, so in paradigm choice – there is no
standard higher than the assent of the relevant community' (Kuhn 1970, p. 94).

   To be a card-carrying philosopher of science it is almost obligatory to reject
Kuhn's point of view. It is natural that any intellectual community attends mostly
to the internal issue that divides the community, rather than defending their
shared beliefs. Kuhn was right about intellectual communities in this regard. But
should they behave in this way? Perhaps, philosophers of science should unite
against the common enemy. If so, then the strongest and most convincing
rebuttal of Kuhn's position must be based on the weakest and most secure premises, which are the ones on which most philosophers of science agree. Almost all
philosophers of science agree that scientific theories are (successful and rational)
instruments for prediction. The strategy of the present essay is to do as much as
possible with as little as possible.

231

Robert Nola and Howard Sankey (eds.), After Popper, Kuhn and Feyerabend, 231–250.
© 2000 *Kluwer Academic Publishers. Printed in Great Britain.*

Many philosophers of science believe that the many replies to Kuhn are already completely adequate for this purpose. I do not share that conviction. To support my viewpoint, I present a problem to philosophers of science – the problem of idealisation (section 5) – which appears to support Kuhn's view that rationality with respect to truth is a bankrupt notion. The problem is not merely that idealisations are used everywhere in science. The problem is that such falsehoods can actually *increase* the predictive accuracy of the resulting equations. There is necessarily a need, in some cases at least, to trade off truth in one respect for truth in another respect. It is this trade-off that threatens the rationality of truth as a univocal goal of science.

The traditional 'solution' to the problem of idealisation is well represented in Musgrave's writings, especially his 1981. There, Musgrave argues that idealisations are rational from a realist point of view if either they are lead to no losses in predictive accuracy ('negligibility' assumptions), or they are only presented for heuristic reasons ('heuristic' assumptions) to make it easier to learn the full theory. The solution assumes that if the idealisation were removed, then the resulting equations would bring us closer to the truth, or at least no further from the truth. That is, the solution assumes that it is possible to simultaneously optimise ever aspect of truth at the some time, so that truth is a univocal goal of science. It is exactly this assumption that has been shown to be false in recent research on idealisations in science (Forster and Sober 1994, Forster 1999, 2000, forthcoming).

To formulate and to solve the problem of idealisation, one needs to make a clear distinction between three levels of theorising (section 2) – *theories*, like Newton's theory of motion, at the most general level, *models* applied to concrete systems in the middle, and *predictive hypotheses* at the lowest level, which result from fitting models to data. The essential point of this tripartite distinction is that predictive accuracy is a property of predictive hypotheses at the very bottom of the hierarchy, and there is trade-off against the truth at the next level up – the level of models.

The same distinction is also useful for the explication of Kuhn's views about science (section 3). In particular, normal science concerns the development of the middle layer of theory – at the level of models. Revolutionary science involves a change of theory at the top. If theory change is rationally motivated by the success or failure of normal science, and normal science consists in the development of models, and models are not evaluated according to their truth, then there is a *prima facie* problem here for realists. Kuhn may not have explicitly pointed to the problem of idealisation, but it supports his view of science, at least on the surface.

Therefore, philosophers of science who want to defend a standard of rationality higher than the assent of the relevant community must address the problem of idealisation. To do this, they need a finer-grained definition of the goal of science than truth *simpliciter* (section 4). Traditionally, philosophers of science have made a distinction between epistemic and pragmatic goals of science. Epistemic goals include all goals that depend on what is true of the world. This includes not only the truth of theories, but also the predictive accuracy of predictive hypotheses. Truth and predictive accuracy operate at different levels of theorising, so depend on each other in complicated ways. The three levels of theorising are essential parts of the problem.

Once these distinctions are in place, a space of possible philosophical positions is opened up, and the core instrumentalist view of science is strengthened in the process. It not only survives the problem of idealisation, but it explains the use of idealisation in a way that makes *essential* reference to epistemic values. It is not that idealisation is epistemically harmless, as Musgrave believes. It has positive epistemic value, which cannot be explained except by reference to predictive accuracy.

Predictive accuracy, like truth of theories, is something that hypotheses do not wear on their sleeves. But unlike the truth of theories, it can be directly tested by seeing whether the predictions come out to be true, or approximately true. This requires that a hypothesis constructed from one set of data is tested against a different set of data. This is quite different from testing hypotheses against the combined set of data. The difference might be described as *diachronic* testing as opposed to *synchronic* testing. The suggestion is that models and theories should be evaluated according to their survival of diachronic tests.

Once the problem of idealisation is resolved, one needs to determine whether the truth of competing theories can be rationally evaluated. The problem that Kuhn presents in this regard is the problem of incommensurability, which, in part, denies the comparability of the theoretical content of rival theories. I have no argument against incommensurability in this sense. Rather, I present it as a non-problem. If theories are to be rationally compared according to their truth, or verisimilitude, then the judgement should supervene on the degree of predictive accuracy that can be obtained within each theory. In section 6, I explain what this means, and describe the difficulties that crop up in making such judgements. Kuhn's incommensurability is not on that list.

In one sense, the solutions presented here are small achievements relative to the wide diversity of methodological issues in science. Modest though they may be, they go beyond the assent of the relevant scientific community in an essential way. They go beyond the assent of any scientific community because our understanding of predictive accuracy is relatively recent – scientists have been unaware of positive *epistemic* benefits of idealisation. It is not therefore a part of any story about the psychological goals of scientists, or of a community of scientists, or their beliefs. Nevertheless, the payoff is real, and its explanation argues for the rationality of science in the objective sense recommended by many philosophers of science, and rejected by Kuhn.

To take the argument further – towards establishing the rationality of science with respect to the full realist goal of truth – is an unsolved problem. However, to respond to the common Kuhnian enemy, this first step is the essential one. The modest problem-solving exercise described in this essay is sufficient to establish that the rationality of science should not be conflated with the rationality of scientists.

2. THEORIES, MODELS AND PREDICTIVE HYPOTHESES

Perhaps the easiest way of introducing the distinction between theories, models and predictive hypotheses is to consider how observational predictions are derived from theories. For this purpose, I will suppose that predictions are *logically deduced* from theories. Such an assumption will not be true in statistical theories, which have only probabilistic consequences. However, the idealised picture is sufficient for the task at hand.

Suppose that E is an observational statement about the position of a planet relative to the fixed stars at some particular time, or the frequency of light emitted by burning a certain substance, or the rate at which a species will colonise a new volcanic island. Let T stand for the fundamental theoretical principles involved in making such a prediction, like Newton's laws of motion, the laws of quantum mechanics, or the principles of population ecology. Everyone agrees that it is impossible to logically deduce E from T because there are missing premises. I will divide these additional assumptions into two kinds. First, there are the *background empirical data* – statements of past observation that are used in the theory to fix initial conditions and estimate parameter values. Let me refer to this background data by the letter D ('D' for data). However, there are other assumptions needed, which are not directly determined by past experience. I will refer to these as *auxiliary assumptions*, denoted by the letter A. On this analysis, a prediction E is deduced from a theory T via the logical entailment $T \& A \& D \Rightarrow E$.

Auxiliary assumptions are typically more theoretical than those included in the background data. They most commonly include simplifying assumptions about the *absence* of interfering factors, like the absence of confounding causal factors in causal modelling, the absence of other forces like air resistance in Newtonian mechanics, the purity of a chemical substance in chemistry, or the absence of genetic mutations in population ecology. Auxiliary assumptions also include the assumptions made by applied mathematicians when they omit high order terms of a Taylor expansion, or when they drop terms on the basis of an order of magnitude analysis. They are often *known* to be false, in which case we refer to them as *idealisations*.

It is important to distinguish between theories and models. Unfortunately, the term 'model' has several unrelated uses in the philosophy of science. Here are three senses in which the term will *not* be used in this essay. (1) A 'model' as in a model aeroplane. Such models do appear in science, such as in the 'model of DNA' Watson and Crick used to 'model' the helical structure of the DNA molecule. But it is not the sense of 'model' used here. (2) 'Model' in the sense used by mathematicians in model theory (e.g., Sneed 1971, Stegmüller 1979). This has a rather technical meaning, which corresponds roughly to what logicians call an *interpretation of a language* (an assignment of objects to names, a set of objects to properties, a set of object pairs to relations, and so on). It is not the sense of 'model' used here. (3) I have heard people speak of Darwin's model of evolution, where they are referring to the core postulates of the theory. 'Model' in this instance refers to what we are calling a 'theory', and is not the sense of term used here.

I am more concerned with the way in which scientists speak of models. To capture their usage, it is better to say that a *model* is a theoretical statement (often in the form of an equation) that is specific enough to be applied to a concrete system. Theories do not have this specificity. For example, Newton's *theory* of gravitation says that every body in the solar system attracts every other body in the solar system in a certain way without making any implications about the number or nature of such bodies. Nor does it say whether the system should be treated as isolated, or whether electromagnetic forces play a role. This is the function of auxiliary assumptions. That is, a model M is obtained from the theory T with the aid of a set of auxiliary assumptions A. In symbols, $(T \& A) \Rightarrow M$. Note that the entailment does not work the other way. Models do not 'contain' the theory from which it is derived – in fact, it is not essential that they be derived from theories at all. I have explained the meaning of models by their relationship to theory only because that is when we need to be careful about the distinction.

Consider a famous case in the history of planetary astronomy. In the sixty years before the discovery of Neptune in 1846, there was a series of Newtonian models of planetary motion which assumed that Uranus is the outermost planet in the solar system. When the discrepancies between the predictions of this model and the observed motions of Uranus remained after the interactions of the known planets were taken into account, Le Verrier and Adams adopted a model that assumed the existence of an eighth planet. Let me label this new model as M'. M' postulated the existence of an eighth planet, but made no precise assumptions about its position. But when it was combined with the data, D, about the past positions of Uranus and the other planets, the new model did predict its position, whereupon Neptune was discovered when telescopes were pointed towards the predicted position of the planet. That is, $M' \& D \Rightarrow E$, where E is a statement about the position of the eighth planet.

A model is unable to make precise predictions because it postulates a number of free parameters, like mass values, or initial conditions, whose values are not given by the theory, or the auxiliary assumptions. Let a *predictive hypothesis* be a *version* of the model together with a precise numerical assignment of values to all adjustable parameters. It is a *predictive* hypothesis because once all parameters have precise numerical values, the hypothesis is able to make precise numerical predictions. There are *many* such versions of the model, so the model is really a *family* of predictive hypotheses. Logically speaking, M is an open-ended disjunction that says that one of its members is true. (Scientists often refer to predictive hypotheses as 'models' as well – 'fitted models' might be the appropriate translation in most instances.)

There are two kinds of models – statistical and non-statistical. Philosophers of science mostly think about non-statistical models, which make precise predictions. Most of the predictive hypotheses in such an M are logically inconsistent with the background data D. Naturally, we only want the unrefuted members of M to play a role in prediction. Ideally, only one member of M is consistent with D, in which case $M \& D$ singles out a unique predictive hypothesis. If no members of M are consistent with D, then M is falsified by D. If many members of M are consistent with D, then the predictions will be imprecise.

In the case of models that make only probabilistic assertions, D may be logically consistent with every member of M (e.g., if we assume Gaussian error distributions), although some members will always fit the data D better than others. In that case, a unique member of M is picked out by choosing the best fitting member of M, where 'best' is defined by some statistical measure of fit, as in the method of maximum likelihood or the method of least squares. Therefore, in either case, *the role of background data is to single out a unique predictive hypotheses from a model*. If we label this predictive hypothesis by H, then $(T \& A \& D) \Rightarrow H$, or equivalently $(M \& D) \Rightarrow H$.

A theory may be thought of as a family of models. Different models are derived from a theory using different idealisations, different simplifying assumptions, and different auxiliary hypotheses. Many different models can be derived from a single theory. For instance, if we assume that there are six planets, which are small point masses, then we get one Newtonian model of the solar system. But if we assume that there are 7 planets, or if we model the Earth as bulging at equator, then we get a different Newtonian models of the solar system.

Not all theories are as precisely formulated as Newton's or Einstein's theories of motion. For example, connectionist modelling (Rumelhart *et al.* 1986) of animal or human behaviour is based on the idea that behaviour is caused by information processed by neural networks. There is a collection of basic models, or what Kuhn would call the exemplars of connectionist science, which serve to guide the construction of new models. But there is no well articulated procedure for constructing models from something like Newton's three laws of motion. That is why science at the level of models and predictive hypotheses is perhaps the most important. Models appear in every science, while theories do not. That is why it is essential to include models as a species of scientific hypothesis – to speak only in terms of theory and auxiliary assumptions is to either exclude such sciences from the discussion, or to conflate models and theories.

3. KUHN'S PICTURE OF SCIENCE

Some of the most notable examples of science are those spawned by the great books in science, such as Copernicus' *De Revolutionibus*, Newton's *Principia, and Darwin's Origin*. These books were not the end, but the beginning, of highly productive periods of science. For Kuhn (1970), these periods of science are examples of *normal* science. Normal scientific research is conducted under a *paradigm, or disciplinary matrix*. He lists four elements of a disciplinary matrix; *symbolic generalisations, metaphysical presumptions, values, and exemplars*. I will assume that a model is a kind of symbolic generalisation, and that the goals of research come under the heading of 'values'.

In contrast, *revolutionary* science is the process by which one paradigm is replaced by another. In the history of science, Kuhn sees periods of normal science punctuated by revolutions followed by new periods of normal science. While this broad picture is consistent with the traditional philosophies of science, Kuhn's explanation of how and why the changes come about is quite different.

In the terminology of the previous section, normal science is about the construction of new *models* or the improvement of old ones, whereas paradigm change or revolutionary science is about *theory* change. Kuhn's account of these two different kinds of change is sketched below.

Anomalies are the driving force behind normal science. For Kuhn (1970, p. 52) an *anomaly* is a violation of 'the paradigm-induced expectations that govern normal science.' In terms of the previous section, an anomaly is one of two things: (a) a *discrepancy* between the best worked-out model of a theory at the time and the known phenomena or (b) a *discrepancy* between two models, each of which is accepted as a good representation of different parts of the phenomena. The 'puzzles' of normal science are puzzles about how to change the 'paradigm-induced' expectations so that an anomaly is removed. Scientists solve these puzzles by constructing new models that remove the anomaly without creating too many new ones. Model construction is the engine of normal science, while anomalies provide the fuel. Kuhn does not use the term 'model' in this sense, but I believe that it does fit well with how contemporary scientists describe the activities of normal science.

Kuhn's account of model construction departs from the deductive account described in the previous section, which is more familiar to philosophers. He tends to downplay the role of the formal derivation of models from a background theory, and, instead, suggests that models are constructed by analogy from *exemplars* of the science taught in textbooks. By an 'exemplar' Kuhn (1970, p. 187) refers to 'the concrete problem-solutions that students encounter from the start of their scientific education, whether in laboratories, in examinations, or at the ends of chapters in science texts.' 'All physicists, for example, begin by learning the same exemplars: problems such as the inclined plane, the conical pendulum, and Keplerian orbits; instruments such as the vernier, the calorimeter, and the Wheatstone bridge.' Exemplars provide the scientist with a kind of *tacit* knowledge that cannot be articulated explicitly, but is nevertheless an essential part of the paradigm. The advantage of 'substituting paradigms for rules' is that it 'should make the diversity of scientific fields and specialties easier to understand' (Kuhn 1970, p. 48). Kuhn therefore rejects the deductivist view that models are the logical consequences of theory and auxiliary assumptions. If we are concerned about the psychology of scientists, then Kuhn is right that scientists do not always follow a rigorous pattern of deduction. However, philosophers of science are not concerned with the psychology of models but with their evaluation. The deductive picture may be useful for this purpose, although I will be neutral on this point during the remainder of this essay.

Kuhn recognises that the removal of anomalies by using 'fudge factors', or *ad hoc* gerrymandered changes in auxiliary assumptions, is not an acceptable puzzle solving strategy in science. At the same time, he is denying the existence of rules for the construction of models, so he is not able to say that *ad hoc* models are disallowed because they violate the *rules* for model construction. So, how are they disallowed? One such example that he considers is Ptolemaic astronomy:

> Given a particular discrepancy, astronomers were invariably able to eliminate it by making some particular adjustment in Ptolemy's system of compounded circles. But ... astronomy's complexity was increasing far more rapidly than its accuracy and ... a discrepancy corrected in one place was likely to show up in another. (Kuhn 1970, p. 68)

This passage is ambiguous. On the one hand, it could point to a practical difficulty in fitting a particular Ptolemaic model (defined by a specific number of epicycles assigned to each celestial body). The adjustment of radii and periods of motion to remove one discrepancy might fail to provide good fit with other known data. This is a problem concerning synchronic fit with data. On the other hand, he may be referring to the fact that after a complex model successfully fits all known data, it was likely to fail in its prediction of new data, and therefore the corrected discrepancy would show up in another place. This concerns a diachronic notion of fit. The second diachronic concept of fit is the epistemologically important notion, for it is a well known character of complex models that accommodation is easy and prediction is hard. This is therefore the more charitable reading of Kuhn.

This brings us to Kuhn's description of revolutionary science, in which the concept of a 'crisis' plays a key role. A *crisis* in normal science occurs when puzzle-solving breaks down; either because no solutions are found, or because the discrepancy corrected in one place shows up in another. Although crisis is necessary to end a period of normal science; it is not *sufficient*. A second requirement is that there is a competing paradigm that shows greater promise in puzzle-solving potential. 'The decision to reject one paradigm is always simultaneously the decision to accept another, and the judgement leading to that decision involves the comparison of both paradigms with nature *and* with each other' (Kuhn 1970, p. 77).

At the time of publication, Kuhn introduced the new and controversial idea that scientists do not see anomalies, or even crises, as *testing the paradigm* itself.

> Though they may begin to loose faith and then to consider alternatives, they do not renounce the paradigm that has led them into crisis. They do not, that is, treat anomalies as counterinstances, though in the vocabulary of philosophy of science that is what they are. (Kuhn 1970, p. 77)

For Kuhn, '... science students accept theories on the authority of teacher and text, not because of evidence' (Kuhn 1970, p. 80). All the standard confirmation theories of the time, assumed that scientists are constantly evaluating predictive hypotheses, models, *and theories* against the latest empirical evidence. However, writers like Popper recognised that the mediation of auxiliary assumptions often protected theories from *direct* falsification. The issue was whether there were ever occasions when the auxiliary assumptions were sufficiently well tested independently of the theory so that the arrow of *Modus Tollens* could be directed at the theory some of the time. If Kuhn is right to claim that no such process actually takes place in normal science, then it is a genuine embarrassment for the Popperian point of view. This is still a controversial issue. However, at best it undermines one particular account of how theories are evaluated. It does not preclude the possibility that theories can be objectively evaluated in a different way.

However, at the level of models, Kuhn (1970, p. 80) concedes that 'Normal science does and must continually strive to bring theory and fact into closer agreement, and that activity can easily be seen as testing or as a search for confirmation or falsification.' Scientists may try out a number of solutions to a puzzle, 'rejecting those that fail to yield the desired result' (Kuhn 1970, p. 144). Scientists do, therefore, test their *models*. This is an important difference between theories and models on Kuhn's account. *Models are constantly evaluated in normal science, whereas the theory is only evaluated in times of crisis, and only against a competing theory.*

The lesson is clear: If Kuhn is right, then there is a huge difference between the way scientists evaluate theories and the way they evaluate models. Philosophers of science have paid too little attention to normal science. If one has no clear concept of 'model', then one has no clear conception of normal science. To consider only the conjunction of theory and auxiliary assumptions, $T \& A$, will not do, because a 'model' in the proper sense does not imply such a conjunction. Otherwise it would be impossible to derive *true* models from a false theory, or false auxiliary assumptions. The proper concept therefore allows for the separation of the questions: (A) Is the theory true, and (B) Are the models true? If normal science aims at true models, then it may not matter that the theory is false. It may make sense that the truth of theory is not questioned in normal science, because its falsehood does not preclude the success of normal science.

Nevertheless, there is a need to refine the question. As I shall argue in the following sections, the use of idealisations makes it difficult (though not impossible) to defend the view that normal science aims at true models. It is better to argue that normal science aims at predictively accurate models, and then to ponder how this can lead to truth at a higher level of theorising. To defend the objective rationality of science, I believe that it is important to decompose the problem in this way.

4. HEMPEL'S CRITICISM OF POPPER AND KUHN

Hempel (1979, pp. 50, 51) makes the point that a procedure 'can be called rational or irrational only relative to the goals the procedure is to attain.' He notes that 'Popper, Lakatos, Kuhn, Feyerabend, and others have made diverse pronouncements concerning the rationality or irrationality of science ... without ... giving a reasonably explicit characterisation of their conception of rationality which they have in mind and which they seek to illuminate or to disparage in their methodological investigations.' The point is not that there is one unique sense of scientific rationality, for there are many goals of science (some of which are arguably more essential to science, *qua* science, than others). The point is merely that clarity demands that the goals are made explicit, and that rationality with respect to different goals should be discussed separately, one at a time.

Suppose that we can agree that *one* goal of planetary astronomy, from Ptolemy to Einstein, was to search for the *true* trajectories of the planets in the future and the past. That seems clear enough. But is it entirely clear? I think that the goal implied by this statement is the *predictive accuracy* (Forster and Sober 1994) of a predictive hypothesis, rather than its truth. The trouble with 'truth' as a goal is that there is no automatic criterion of partial success. The only obvious criterion for achieving truth is black and white – you either achieve it, in which case you are 100% successful, or you do not, in which case you are entirely unsuccessful. This is not the way we understand the 'search for true trajectories'. Some false trajectories are better than others, and predictive accuracy defines what counts as better. Some false hypotheses are predictively more accurate than other false hypotheses, even though none is truer than any other. This feature of predictive accuracy is good.

Other features of predictive accuracy appear to be bad.[1] Predictive accuracy is defined by first fitting a model to one data set, D_1, and then considering the accuracy of its predictions in another data set, D_2. The predictive accuracy is the expected fit with respect to D_2, or equivalently, the fit with the true hypothesis within the domain of data defined by D_2. For example, suppose that D_1 is the set of observations of Halley's comet available to a group of scientists at the present time. That set will include observations at an assortment of times over a fixed period of times. Suppose we find the Newtonian hypothesis that best fits this data. There are at least two kinds of predictive questions we may ask: (1) If there were other observations of Halley's comet during the past that we have not seen, how well does our hypothesis predict these data. (2) How well will our hypothesis predict future positions of Halley's comet. There are two distinct kinds of predictive accuracies at issue here – the first involves the interpolation of our observations to the past, while the second involves their extrapolation to the future.

So, predictive accuracy is subject to Hempel's warning – one should be precise about the notion of predictive accuracy appealed to. Moreover, the distinction amongst different kinds of accuracy allows philosophers of science to raise an interesting variety of methodological questions. For example, suppose that there is no method that will do the best job at optimising the accuracy of interpolation and extrapolation at the same time. Then there is no unique answer to questions about objective rationality. The objective answers are conditional in nature: If you are interested in interpolation, then use method 1; and if you are interested in extrapolation, then use method 2.

In Forster (2000) I describe one computer simulation that shows that such trade-offs do exist (see also Busemeyer and Wang 2000). One of the two methods compared involved synchronic fit with data together with a complexity factor. The standard methods of model selection, including the method of maximum likelihood, AIC, and BIC, are of this kind. They performed reasonably well at interpolation, but performed poorly at extrapolation *even in the limit of infinitely large data sets* (sorry, convergence theorems (Earman 1992) do not help here because you cannot converge on the truth if the true predictive hypothesis is not in any of your models). The second method involved a diachronic method of fit, whereby a model fitted to one data set was tested against new data. Unsurprisingly, perhaps, the past predictive success of models provided a better indicator of future predictive success than the standard methods of model evaluation.

Not only does the standard Bayesian philosophy of science (Earman 1992) not *answer* these harder methodological questions, it does not allow for the formulation of the questions so long as it defines the rationality of science in terms of the truth or the probability of truth of hypotheses. Hempel's criticism of Popper, Lakatos, Kuhn, and Feyerabend was that the goals of science are not clearly specified. The problem with Bayesianism, in its standard form, is that it specifies a single goal, and is unable to consider other epistemic achievements (this criticism does not apply to decision-theoretic Bayesianism, but this is not the standard form of Bayesianism).

The point of this section has been to argue that the multi-faceted nature of predictive accuracy is actually one of its biggest advantages. For it provides a fine-grained analysis of the epistemic effectiveness of the many methods actually used in real science.

5. THE PROBLEM OF IDEALISATION

What follows is an example of the explanatory work done by looking at pre-dictive accuracy (only the kind of predictive accuracy associated with interpolation is considered here). The question is: why should idealisations be used in science even when more 'realistic' models are available? Why should complicated models *not always* supersede simpler ones? Also, why should Newtonian science flourish today even though Newton's *theory* is false? In brief, the explanation is that false theories and false models may sometimes help, rather than hinder, the search for truth at the level of predictive hypotheses. To understand when, and why, this should be the case, we need to examine the relationship between predictive hypotheses and models.

Recall that a model M must make use of background data D in order to make predictions. Suppose that there is a unique predictive hypothesis in M, namely H, that best fits that data D. Then it is this best fitting hypothesis, H, that is used to make predictions, and it is therefore the predictive accuracy of H that defines the predictive accuracy of the model M at that particular time. The predictive accuracies of the other members of M are irrelevant.

In particular, it is irrelevant that there are some predictive hypotheses in the model that are more predictively accurate than H, and this is the key point. If we denote the most predictively accurate member of M by H^*, then H may not be close to H^*, in which case the predictive accuracy of the model is below its *potential* predictive accuracy.

Potential predictive accuracy is irrelevant if it is not actualised. Think carefully about this last statement. It implies the possibility that a true model (in which H^* is true) may achieve less accurate predictions than a false model. Let CIRCLE and ELLIPSE be competing models of a planet's trajectory, and suppose that the true trajectory is actually an ellipse. Then ELLIPSE is true, and CIRCLE is false. However, the data may be such that best fitting circle may be closer to the true trajectory than the best fitting ellipse. If this happens, it is because the best fitting ellipse is not close to the best ellipse. There are at least three reasons why this may happen.

One reason is that there are errors of observation in the background data, D. Consider the fact that it is always possible to fit an n-degree polynomial through n data points *exactly*. Thus, the H obtained from M will achieve perfect fit, but is unlikely to have high predictive accuracy. This is similar to the case of Ptolemaic or Copernican astronomy, where a model with a sufficient number of epicycles can fit any finite set of observational data to an arbitrary degree (as proven by Fourier's theorem). In such cases, it is necessary to sacrifice the *potential* predictive accuracy of the model in order the maximise the *actual* predictive accuracy of the model. Note that this makes sense of Kuhn's observation (section 3) that the complexity of Ptolemaic astronomy was increasing far more rapidly than its accuracy in the sense that a discrepancy corrected in one place was likely to show up in another.

The second reason why H may not be close to H^* has nothing to do with observational error. Suppose that M is not true, and that there are no errors of observation. Then M will never fit the data perfectly. H, by definition, is the member of M that fits D the best, and different D will lead to different H, for no other reason than that they are sampled differently. It is impossible for all of these H's to be equal to H^*. Therefore, H may be quite different from H^*.

The third reason is the data D may be unrepresentative of the domain over which the predictive accuracy is defined. In that case, even an infinite number of data may fail to pick out an H that is close to H^*. This is the case of 'extrapolation error' discussed in the previous section.

The magnitude of this effect is relatively small when two models have close to the same degree of complexity, such as CIRCLE and ELLIPSE. However, the effect is significant when comparing models of widely different degrees of complexity. This is philosophically important because it means that a theory should not be blamed for the poor predictive performance when the idealisations are removed. Or to put it another way, theories must be compared by the predictive success of their idealisations. Exactly how this can be done has yet to be worked out in detail.

Philosophers of science might take the ubiquitous use of idealisations in science to mean one of two things: (a) Science should avoid idealisations, because the goal of science is to obtain true theories and models. (b) Science should continue using idealisations, in which case truth is not the goal of scientific theorising. For example, Cartwright (1983) opts for (b) in a book called *How the Laws of Physics Lie*. Our discussion shows that there is a third possibility: (c) Science should continue using idealisations, because they are necessary in order to optimise the predictive accuracy of our models. The explanation is that science seeks *predictive hypotheses* that are as close to the truth as possible.

This shift of focus from truth *simpliciter* to predictive accuracy has subtle but important consequences for any view of testing or confirmation in science. First, there is a problem for any form of Bayesianism that assumes that theories, models, and predictive hypotheses should be evaluated by their probability of truth. If scientists ought to maximise the *probability of truth* of their models, then why should scientists be so indifferent about the falsity of their models? It is no good appealing to posterior probabilities to get around this objection, for the background data, D, will most often confirm that the world is really 'messier' and more complicated than the model assumes. There appears to be a conflict between what scientists actually believe about the probability of their models being true, and the decisions that Bayesians recommend in response to those beliefs. Are we to say that scientists are irrational, or should we resolve the conflict by supposing that their goal is something other than truth?

Philosophers of science frequently talk about hypothesis testing and selection in terms of 'confirmation', 'justification', 'proof', 'warrant', 'credence', 'support', 'verification', and 'corroboration.' All of these terms suggest that hypotheses are evaluated with respect to their truth or falsity. It is time that these terms were replaced by words that do not build in that assumption from the start. For that reason, I prefer to talk about hypothesis *testing, evaluation, appraisal, or assessment*. The main thesis of this section is that the problem of idealisation *cannot* be solved by any philosophical theory of confirmation, given the way that the term 'confirmation' is usually understood.

6. KUHNIAN COMMENSURABILITY

Kuhn (1970) and Lakatos (1970) say a lot about the comparison of programs and paradigms *as a whole*, but say little about the pairwise comparison of *models* in different programmes or paradigms. Why is this? Einstein's solution to the precession of the perihelion of Mercury was surely evaluated against the attempted Newtonian solutions. Planck's model of black body radiation (which introduced the quantum hypothesis for the first time) was surely evaluated against the best classical solutions of the day. These are examples of inter-theory model comparison.

Perhaps the obstacle is Kuhn's famous incommensurability thesis (IT), which says, roughly, that there is a failure of translatability between paradigms – that is, the puzzle solutions of one paradigm cannot be translated and understood in terms of another paradigm. Since I plan to argue that IT is not an obstacle to model comparison, it is appropriate for me to examine IT in more detail.

Kuhn traces his idea back to Butterfield (1962, pp. 1–7), who claims that 'of all the forms of mental activity' in scientific revolutions, 'the most difficult to induce ... is the art of handling the same data as before, placing them in a new system of relations with one another by giving them a different framework.' Kuhn (1970, p. 85) then carries on to remark that 'Others who have noted this aspect of scientific advance have emphasised its similarity to a change in visual gestalt: the marks on paper that were first seen as a bird are now seen as an antelope, or vice versa.'

I am among those who found the gestalt analogy extremely vague, until I mapped it onto the Butterfield quote. Let me use the familiar duck–rabbit visual gestalt as an example. The marks on the paper represent the data, and they are *intrinsically* the same whether or not the drawing is seen as a duck or as a rabbit. There is no incommensurability at that level. However, the different modes of perception imply a difference in the *significance* or the *salience* of features. For example, the kink at the back of the duck's head is 'noise' when it is seen as a duck, while it is an essential feature when it is seen as the mouth of the rabbit. The same is true about the relationships amongst features. The fact that one ear of the rabbit lies above the other is a matter of accident when it is seen as a rabbit, whereas it is essential that one part of the duck's bill is above the other when it is seen as a duck.

We see exactly these kinds of changes across scientific revolutions. Copernicus saw great significance in the fact that the retrograde motions of superior planets occurred when and only when those planets were in opposition to the sun. Ptolemaic astronomers did not, even though they had no trouble agreeing that it was a fact. Darwin saw great significance in the structural similarities (homologies) across species, whereas non-evolutionists did not. It is exactly these kinds of differences that Kuhn (1970, pp. 118, 119) finds in the Aristotelian and Galilean views of pendulum motion:

> To the Aristotelians, who believed that a heavy body is moved by its own nature from a higher position to a state of natural rest at a lower one, the swinging body was simply falling with difficulty. Constrained by the chain, it could achieve rest at its low point only after tortuous motion and a considerable time. Galileo, on the other hand, looking at the swinging body, saw a pendulum, a body that almost succeeded in repeating the same motion over and over again *ad infinitum*. And having seen that much, Galileo observed other properties of the pendulum as well and constructed many of the most significant and original parts of this new dynamics around them. From the properties of the pendulum, for example, Galileo derived his only full and sound arguments for the independence of weight and rate of fall, as well as for the relationship between vertical height and terminal velocity of motions down inclined planes. All of these natural phenomena he saw differently from the way they had been seen before.

The gestalt analogy, and the scientific examples, are consistent with Butterfield's idea that the new mode of perception handles *the same data as before*. In other words, there may be a failure of translation in some cases, but there is *no incommensurability of the background data, D*. If this is right, then IT claims *only* that the solution to a puzzle in one paradigm cannot be translated into a different paradigm.

Nevertheless, there is an unanswered objection here. If we base inter-theory comparison on predictive accuracy, then aren't we assuming the existence of a theory-neutral language of observation? And aren't there strong arguments against the existence of a theory-neutral observation language, some of which Kuhn himself provided? I am inclined to concede that there is no such thing as an observation language that is neutral with respect to *all* theories, but to deny that this is required for inter-theory model comparisons (Sober 1990). All we need is a formulation of the problem in terms that are *neutral with respect to the competing theories*. Aristotelians, and Galileans alike, had no trouble understanding what was meant by the 'number of full swings of the stone in a given period of time' or whether that number changed when the size of the swings decreased. Einsteinians and Newtonians had no disagreement about how the magnitude of the precession of Mercury's perihelion should be measured, or about its observed value. And Planck did not introduce a new way of plotting observed light intensities against wavelengths when he introduced his new quantum model of black body radiation. In each case, the prediction *problems were easily translated* from one paradigm to the next.

Nor was there any reinterpretation of what counted as successful prediction or 'good fit' with the data. All of these examples confirm that the accuracy or inaccuracy of predictions is measured in the common currency of fit, usually defined by standard statistical methods such as the method of least squares.

Kuhn uses IT to argue that scientific knowledge is not *cumulative* despite the fact that the laws or models of earlier theories appear to be derivable as special cases of the later theory. Kepler's laws appear to be special cases in Newton's theory, and Newton's equations appear to be *special* cases of Einstein's equations. However, for Kuhn, this appearance is illusory because 'the physical referents of these Einsteinian concepts are by no means identical with the Newtonian concepts that bear the same name.' 'Newtonian mass is conserved; Einsteinian is convertible with energy. Only at low relative velocities may the two be measured in the same way, and even then they must not be conceived to be the same.' (Kuhn 1970, p. 102) However, this only serves to reinforce the *previous* interpretation of IT; namely that no Newtonian *model* can be translated into Einsteinian mechanics because they invoke a different set of relations and place them in a new framework.

Moreover, from Kuhn (1970, p. 102) it is clear that the derivations do serve some purpose:

> Our argument has, of course, explained why Newton's Laws ever seemed to work. In doing so it has justified, say, an automobile driver in acting as though he lived in a Newtonian universe. An argument of the same type is used to justify teaching earth-centered astronomy to surveyors.

That is to say, the derivation of 'limiting' cases does serve to explain why the older models were so successful in their *predictions*.

Kuhn's claim that the translatability of models across paradigms is impossible *in principle* is still controversial (Musgrave 1979). For the purposes of this article, I will treat this issue as unresolved. I argue only that the acceptance of IT does not rule out the possibility of comparing Newtonian and Einsteinian models *with respect to the goal of predictive accuracy*.

So, exactly how is progress with respect to the truth defined across revolutions? My suggestion is that at a given time, the achievement of one program is greater than another if and only if its best worked-out model is predictively more accurate than the best worked-out model of its competitor. While the definition is vague if the domain of prediction is not explicitly specified, there is usually no problem in resolving this ambiguity in real cases. Planck's formula was accurate over the full range of wavelengths, whereas its predecessors were only accurate for either the low end of the spectrum or the high end of the spectrum, but not both. Everyone agreed that accuracy over the full spectrum of wavelengths was a goal of the research.

Lakatos (1970) suggested that competing research programmes should be compared according to their rate of progressiveness at the time. So, for example, if one is progressive while another is degenerating, then the first receives a better evaluation than the second. I believe that Lakatos' idea is *plainly* wrong. Such an evaluation should compare the *achievements* of one program with the achievements of another. The rate of *improvement* within each program is not relevant to their current state of achievement, though it may be relevant to the question about how the current comparison should be projected into the future. My point is that those two questions should be clearly separated and Lakatos does not.

The best worked-out models of a young research program may not compete well with those of a more established competitor. Its best models have yet to be worked out, so an estimation of the unproven *potential* of the new program is largely an article of faith, similar to religious faith or blind political allegiance. Kuhn (1970, pp. 157, 158) describes this issue in exactly these terms, and he is right, not because of the incommensurability of competing paradigms, and not because models are incommensurable, but because nobody can predict the future course of science.

This talk of science based on faith brought forth various complaints about Kuhn and the irrationality of science, and from Lakatos (e.g., 1977, p. 7) in particular. Of course, Lakatos' charge is unfair because a decision based on uncertainty is not necessarily irrational. However, I believe that Lakatos and Kuhn were talking past each other in any case, just as philosophers and sociologists of science do today. The rationality of individual scientists, or even of scientific communities as a whole, is a different issue from the one that concerned Lakatos. Lakatos, like many other philosophers of science, was more concerned with whether science made sense as a knowledge-seeking enterprise. In other words (irrespective of what scientists *believe* they are doing) does science achieve knowledge in any sense, and what evidence exists for such a view? With respect to this question, the definition of what it means for one model to be more predictively accurate than another is relevant.

7. THE OBJECTIVITY OF SCIENCE

Kuhn's challenge to the philosophy of science was to defend the rationality of science with respect to the goal of truth. Philosophers have responded to this challenge, and Lakatos' methodology of scientific research programs is one such example. I have tried to argue that there are two obstacles in the way of evaluating this research. One problem is a failure to make a clear distinction between theories, models, and predictive hypotheses (section 2). A second problem is that the goal of scientific research is not always explicit (section 4).

As we have seen, Kuhn (1970) was mainly interested in the social psychology of science, while philosophers of science look to science as an objective source of knowledge. Rationality in this objective sense is not about what scientists believe. The question is not settled by taking a survey of scientists asking 'what is the goal of science?' or 'what are the standards of scientific community?' Nor is it concerned with what scientists think that science *ought* to be. It is about the *achievements* or the potential achievements of science, and the causes responsible for those achievements.

Let me expand upon the notion of causal responsibility. Consider any putative goal of science, whether it be the truth of theories, the predictive accuracy of models, or the economic prosperity of the United States. Call the goal 'X'. Now consider two, or more, ways or methods of doing science. Call them A and B. It is now an *objective* question whether *A* is more effective than *B* in achieving *X*. True, it is a vague question until more is said about what 'effective' means. Secondly, the answer may not be univocal; that is, *A* may be more effective in some circumstances, but not in others. Let me refer to such questions as *goal-oriented* questions. These questions have nothing to do with what scientists believe.

Goal-oriented theses are answers to goal-oriented questions: So, '*A* is more effective than *B* in achieving *X*' is a typical goal-oriented thesis by my definition. Such theses are *weakly normative* in the sense that they imply 'ought' statements *when coupled with goal statements*. For example, if one could establish that *A* is more effective than *B* in achieving *X*, *and X* is the goal of science, then it would follow that one ought to adopt *A* as the methodology of science.

There is a huge difference between social psychological questions and goal-oriented questions, although the distinction is not always clear. For example, compare the following normative arguments:

(1) Scientist *Y* believes that method *A* is better than method *B* at achieving *X*. Scientist *Y* wants *X*. Therefore, scientist *Y* *ought to* use method *A*.
(2) Method *A* *is* better than method *B* at achieving *X*. Scientist *Y* wants *X*. Therefore, scientist *Y* *ought to* use method *A*.

There is an important difference between these arguments. The first provides a *subjective* justification for using method *A*, while the second provides a more *objective* justification for the same action. The goal-oriented claim has nothing to do with the beliefs of scientists, and supports a more objective rationality claim. The hard problems in the philosophy of science concern the objectivity of science as a goal-oriented process.

For example, in the problem of verisimilitude (Popper 1963), philosophers of science seek to (a) define the goal of science in terms of closeness to the truth, and (b) argue that science has made progress with respect to this goal (for a survey, see Niiniluoto 1998). It is this kind of objectivity that is often lost in Bayesian decision theory, which currently dominates the philosophy of science in North America. The Bayesian theory is that a decision is rational only if the decision-maker succeeds in maximising *expected utility*. The issue of whether the maximisation of expected utility is causally effective in maximising utility is the objective side of the problem, and it receives next to no attention. By couching the question of rationality entirely in normative psychological terms Bayesians lose sight of the hard problems in the philosophy of science (e.g., Maher 1993, especially section 9.4 on verisimilitude).

It is therefore important to me that an ambiguity in my formulation of the problem of idealisation is well understood. The question was: Why should scientists use models that they know to be false? There are two different ways of answering this question: one in the style of argument (1) and the other in the style of argument (2). Or more exactly, there are two interpretations of the question. I have attempted to answer the question in the style of argument (2) by arguing for a goal-oriented thesis; namely, that idealised models are effective means to the goal of predictive accuracy. In brief, my claim was that idealised models may often promote the accuracy of predictions because of the way that scientists make predictions from models. Models are first fitted to background data, and this introduces errors that may be far smaller for simpler models, even when the simplicity is obtained at the obvious expense of truth at the level of models. This answer refers to how scientists do science and what is achieved by what they do, and not to what they *believe* they are doing.

This solution to the problem of idealisation is still a good one even if it turns out to be psychologically false. For example, scientists might use idealised models because they (truly) believe that they are mathematically more tractable, take less time to apply, and are far less prone to careless computational mistakes. In fact, I would hazard a guess that this is right in many instances. However, there is no conflict between this explanation and mine because they address different aspects of science.

In a similar vein, I have tried to retrieve some of the objectivity of science, which Kuhn threw away in the name of incommensurability (section 6). I argued that while the normal science *solutions* to a Kuhnian puzzle may not be translatable across paradigms, there is frequently no real problem in translating the puzzle itself. Moreover, if the difference between good and bad solutions is defined only in terms of the common currency of predictive accuracy, then Kuhn's incommensurability thesis is no obstacle. This sense of progress is weaker than that sought in the verisimilitude program (Niiniluoto 1998). But since there is no universally accepted definition of verisimilitude, I believe that some objectivity is better than none at all. And nothing I have said rules out the possibility of finding more.

Beginning students in the philosophy of science often enter our subject with a naïve faith in the objectivity of science. Kuhn's *Structure of Scientific Revolutions* challenges their faith, and many of us use it as a classroom text for that reason. However, the truth is always somewhere in between the two extremes. Kuhn himself tries to restore faith in science by appealing to the standards of the scientific community and the less fickle nature of collective decision-making. However, for Lakatos and many other philosophers of science like myself, the objectivity of science is not rescued by the inter-subjective agreement of scientists within a community (and might well be antithetical to it). Rather, the objectivity of science concerns the properties of science as a knowledge-seeking process. Is there progress in science? Is there any sense in which science provides knowledge of the real world using methods that are reliable to some degree in achieving those goals? These are the hard problems in the philosophy of science, and they will never be answered if the philosophy of science is left in the hands of social psychologists.

ACKNOWLEDGEMENTS

I am grateful to Robert Nola for inviting me to participate in *After Popper, Kuhn and Feyerabend* symposium at the 1997 Australasian Association of Philosophy meetings in Auckland, NZ. Thanks also go to Ellery Eells, Elliott Sober and an anonymous referee for comments on an earlier draft of this paper.

University of Wisconsin – Madison

NOTE

[1] The alleged language variance of predictive accuracy (Miller 1975, Devito 1997) is not on this list. For an explanation of why this is so, see Forster (1999).

REFERENCES

Busemeyer, J.R. and Yi-Min Wang, 2000: 'Model comparisons and model selections based on generalization test methodology', *Journal of Mathematical Psychology*, March 2000.

Butterfield, H., 1962: *The Origins of Modern Science*, The Macmillan Company, New York.

Cartwright, N., 1983: *How the Laws of Physics Lie*, Clarendon Press, Oxford.

Cohen, R.S., Feyerabend, P.K. and Wartofsky, M.W. (eds), 1976: *Essays in Memory of Imre Lakatos*, Dordrecht, D. Reidel.

DeVito, S., 1997: 'A Gruesome Problem for the Curve Fitting Solution,' *British Journal for the Philosophy of Science* 48, 391–396.

Earman, J., 1992: *Bayes or Bust? A Critical Examination of Bayesian Confirmation Theory*, The MIT Press, Cambridge MA.

Forster, M.R., 1999: 'Model Selection in Science: The Problem of Language Variance', *British Journal for the Philosophy of Science* 50, 83–102.

Forster, M.R., 2000: 'Key Concepts in Model Selection: Performance and Generalizability', *Journal of Mathematical Psychology*, March 2000.

Forster, M.R., forthcoming: 'The New Science of Simplicity' in H.A. Keuzenkamp, M. McAleer and A. Zellner (eds.), *Simplicity, Inference and Econometric Modelling*, Cambridge University Press.

Forster, M.R. and Elliott Sober, 1994: 'How to Tell when Simpler, More Unified, or Less *Ad Hoc* Theories will Provide More Accurate Predictions', *British Journal for the Philosophy of Science* 45, 1–35.

Hempel, C.G., 1979: 'Scientific Rationality: Analytic vs Pragmatic Perspectives,' in T.H. Geraets (ed.), *Rationality Today*, The University of Ottawa Press, Ottawa.

Kuhn, T., 1970: *The Structure of Scientific Revolutions* (Second Edition), University of Chicago Press, Chicago.

Lakatos, I., 1970: 'Falsificationism and the Methodology of Scientific Research Programs' in I. Lakatos and A. Musgrave (eds.), *Criticism and the Growth of Knowledge*, Cambridge University Press, Cambridge.

Lakatos, I., 1977: *The Methodology of Scientific Research Programmes: Philosophical Papers*, Vol. 1, Cambridge University Press, Cambridge.

Maher, P., 1993: *Betting on Theories*, Cambridge University Press, Cambridge.

Miller, D., 1975: 'The Accuracy of Predictions', *Synthese* 30, 159–191.

Musgrave, A., 1976: 'Method or Madness,' in Cohen, R.S., Feyerabend, P.K. and Wartofsky, M.W. (eds.), *Essays in Memory of Irme Lakatos*, Dordrecht, D. Reidel.

Musgrave, A., 1979: 'How to Avoid Incommensurability' in I. Niiniluoto and R. Tuomela (eds.) *The Logic and Epistemology of Scientific Change*, North-Holland Publishing Co., Amsterdam, pp. 336–346.

Musgrave, A., 1981: 'Unreal Assumptions in Economic Theory: The F-Twist Untwisted', KYKLOS 34, 377–389.

Musgrave, A., 1995: 'Realism and Idealisation: Metaphysical Objections to Scientific Realism', in J. Misiek (ed.), *The Problem of Rationality in Science and its Philosophy*, Kluwer, Dordrecht, pp. 143–166.

Niiniluoto, I., 1998: 'Verisimilitude: The Third Period', *British Journal for the Philosophy of Science* 49, 1–29.

Popper, K., 1959: *The Logic of Scientific Discovery*, Hutchinson, London.

Popper, K., 1963: *Conjectures and Refutations*, Routledge and Kegan Paul, London.

Resnick, R., 1972: *Basic Concepts in Relativity and Early Quantum Theory*, John Wiley & Sons, New York.

Rumelhart, D.E., McClelland, J. *et al.*, 1986: *Parallel Distributed Processing*, Vols. 1 and 2, The MIT Press, Cambridge MA.

Sneed, J.D., 1971: *The Logical Structure of Mathematical Physics*, D. Reidel, Dordrecht.

Sober, E., 1990: 'Contrastive Empiricism', in W. Savage (ed.), *Scientific Theories*, Minnesota Studies in the Philosophy of Science: Vol. 14, University of Minnesota Press, Minneapolis, pp. 392–412.

Stegmüller, W., 1979: *The Structuralist View of Theories*, Springer-Verlag, Berlin.

NOTES ON CONTRIBUTORS

MALCOLM R. FORSTER was partially educated in Applied Mathematics and Philosophy at the University of Otago, New Zealand, and is currently Professor of Philosophy at the University of Wisconsin–Madison, USA. Interests range from the foundations of quantum mechanics, causal modeling, statistics, neural networks and phylogenetic inference, to Newton and the Whewell-Mill debate. For more information, visit the author's homepage at ⟨http://philosophy.wisc.edu/forster⟩.

JOHN F. FOX teaches in the School of Philosophy at La Trobe University, Bundoora. A former Jesuit, he has published papers on, for example truth, relativism, deductivism, the methodology of philosophy, Tarski and Lakatos.

KEVIN T. KELLY received his Ph.D. degree in History and Philosophy of Science at the University of Pittsburgh and is now an Associate Professor in Philosophy at Carnegie Mellon University. Kelly's research centers on the logical structure of reliability and its significance for epistemology and the philosophy of science. He is the author of *The Logic of Reliable Inquiry* (Oxford, 1996) and is a co-author of *Discovering Causal Structure* (Academic Press, 1987). He has also published numerous articles on learning theory, methodology, and the philosophy of science.

LARRY LAUDAN does free-lance philosophy from his base in colonial Mexico. He has written extensively on scientific change and on some problems of realism and relativism. Recent affiliations include the Institute for Advanced Study (Princeton) and the Dibner Institute (MIT). His hobbies include intemperate attacks on political correctness. He is completing a book on the concept of evidence.

ROBERT NOLA teaches at the University of Auckland and has published papers in philosophy of science and on nineteenth century philosophers such as Marx and Nietzsche. His most recent book is a collection of papers on Foucault (in which he also has a contribution). Current work is on incommensurability and on the rationality debate about science.

Robert Nola and Howard Sankey (eds.), After Popper, Kuhn and Feyerabend, 251–252.
© 2000 *Kluwer Academic Publishers. Printed in Great Britain.*

JOHN D. NORTON is a Professor in the Department of History and Philosophy of Science at the University of Pittsburgh. He is interested in confirmation theory and the history and philosophy of physics, with special emphasis on Einstein and general relativity. His most recent publication (with John Earman) is 'Exorcist XIV: The wrath of Maxwell's Demon,' *Studies in History and Philosophy of Modern Physics*, 1998 and 1999.

ANDREW PYLE is lecturer in Philosophy at the University of Bristol. He has reserach interests in two areas: the History and Philosophy of Science, and the History of Seventeeth-Century philosophy. His first book, *Atomism and its Critics* (Thoemmes, Bristol, 1995) is a study of the Atomic Theory of matter from Democritus to Newton. He is currently working on a book on the philosophy of Malebranche.

HOWARD SANKEY is Senior Lecturer in Philosophy of Science in the Department of History and Philosophy of Science at the University of Melbourne. A native of southern California, he completed a Ph.D. at Melbourne after undergraduate studies in philosophy in New Zealand at the University of Otago. He has held research positions at the Center for Philosophy of Science at the University of Pittsburgh, the Center for Philosophy and Ethics of Science at the University of Hanover and the Philosophy Department at Saint David's University College, Lampeter. His main interests lie in the general philosophy of science, with special emphasis on semantic aspects of conceptual change, scientific realism, rationality of scientific theory choice and naturalized accounts of the epistemic warrant of methodological norms. His major publications include two books, *The Incommensurability Thesis* and *Rationality, Relativism and Incommensurability*. He is the editor of a previous volume in the Australasian Studies in History and Philosophy of Science, entitled *Causation and Laws of Nature*.

JOHN WORRALL is Professor of Philosophy of Science at the London School of Economics and Co-Director of the LSE's Centre for Philosophy of Natural and Social Science. He was editor of *The British Journal for Philosophy of Science* from 1974 to 1983. He has published widely on topics in general philosophy of science and in history and philosophy of nineteenth-century physics and is currently completing a book on the reasons for theory-change in science.

INDEX OF NAMES

253

Robert Nola and Howard Sankey (eds.), After Popper, Kuhn and Feyerabend, 253–256.
© 2000 *Kluwer Academic Publishers. Printed in Great Britain.*

AUSTRALIAN STUDIES
IN HISTORY AND PHILOSOPHY OF SCIENCE

1. R. McLaughlin (ed.): *What? Where? When? Why?* Essays on Induction, Space and Time, Explanation. Inspired by the Work of Wesley C. Salmon. 1982
                                                        ISBN 90-277-1337-5
2. D. Oldroyd and I. Langham (eds.): *The Wider Domain of Evolutionary Thought.* 1983   ISBN 90-277-1477-0
3. R.W. Home (ed.): *Science under Scrutiny.* The Place of History and Philosophy of Science. 1983   ISBN 90-277-1602-1
4. J.A. Schuster and R.R. Yeo (eds.): *The Politics and Rhetoric of Scientific Method.* Historical Studies. 1986   ISBN 90-277-2152-1
5. J. Forge (ed.): *Measurement, Realism and Objectivity.* Essays on Measurement in the Social and Physical Science. 1987   ISBN 90-277-2542-X
6. R. Nola (ed.): *Relativism and Realism in Science.* 1988   ISBN 90-277-2647-7
7. P. Slezak and W.R. Albury (eds.): *Computers, Brains and Minds.* Essays in Cognitive Science. 1989   ISBN 90-277-2759-7
8. H.E. Le Grand (ed.): *Experimental Inquiries.* Historical, Philosophical and Social Studies of Experimentation in Science. 1990   ISBN 0-7923-0790-9
9. R.W. Home and S.G. Kohlstedt (eds.): *International Science and National Scientific Identity.* Australia between Britain and America. 1991   ISBN 0-7923-0938-3
10. S. Gaukroger (ed.): *The Uses of Antiquity.* The Scientific Revolution and the Classical Tradition. 1991   ISBN 0-7923-1130-2
11. P. Griffiths (ed.): *Trees of Life.* Essays in Philosophy of Biology. 1992
                                                        ISBN 0-7923-1709-2
12. P.J. Riggs (ed.): *Natural Kinds, Laws of Nature and Scientific Methodology.* 1996
                                                        ISBN 0-7923-4225-9
13. G. Freeland and A. Corones (eds.): *1543 and All That.* Image and Word, Change and Continuity in the Proto-Scientific Revolution. 1999   ISBN 0-7923-5913-5
14. H. Sankey (ed.): *Causation and Laws of Nature.* 1999   ISBN 0-7923-5914-3
15. R. Nola and H. Sankey (eds.): *After Popper, Kuhn and Feyerabend.* Recent Issues in Theories of Scientific Method. 2000   ISBN 0-7923-6032-X

KLUWER ACADEMIC PUBLISHERS – DORDRECHT / BOSTON / LONDON